THE EXTRAGALACTIC DISTANCE SCALE

Proceedings of a symposium held as part of the
100th Summer Scientific Meeting of the
Astronomical Society of the Pacific in
Victoria, British Columbia, Canada
June 29 to July 1, 1988

A SERIES OF BOOKS ON RECENT DEVELOPMENTS IN ASTRONOMY AND ASTROPHYSICS

Printed by Brigham Young University Print Services, Provo, Utah 84602

First published 1988

Library of Congress Catalog Card Number: 88-82923
ISBN 0-937707-21-X

D. Harold McNamara, Managing Editor of Conference Series

ASTRONOMICAL SOCIETY OF THE PACIFIC
CONFERENCE SERIES

Volume 4

THE EXTRAGALACTIC DISTANCE SCALE

PROCEEDINGS OF THE
ASP 100th ANNIVERSARY SYMPOSIUM

Edited by
Sidney van den Bergh and Christopher J. Pritchet

TABLE OF CONTENTS

III. DISTANCE SCALE CALIBRATORS

IV. GALAXIES BEYOND THE LOCAL GROUP

V. THE VIRGO CLUSTER

VI. DEVIATIONS FROM THE HUBBLE FLOW

INTRODUCTION

> We are not certain how far away they are;
> we are not certain what they are.
> Campbell (1917)

Maarten Schmidt once remarked that the fact that the true nature of quasars was still not well understood 25 years after their discovery proved that this was a difficult subject that was worthy of scientific study. By this criterion the study of the extra-Galactic distance scale is a very worthy subject indeed! The redshift-distance relation for galaxies (Hubble 1929, Hubble and Humason 1931) was discovered only a few years after Hubble (1925) had established the extra-Galactic nature of spiral nebulae. Because galaxies are exceedingly distant calibration of the extra-Galactic distance scale has remained a challenge to the ingenuity of astronomers for almost six decades.

The first attempts to determine the Hubble parameter, which relates the redshifts and distances of galaxies, are now known to have been in error by a factor of between 5 and 10. Major changes in our ideas on the size of the Universe resulted first from Baade's (1952) failure to detect RR Lyrae stars in M31 with the Palomar 200-in telescope, and subsequently from Sandage's (1958) discovery that the brightest "stars" in distant spirals were, in fact, HII regions.

The last major review devoted to problems related to calibration of the distance scale was IAU Colloquium No. 37 (Balkowski and Westerlund 1977). The fact that CCD detectors were not even mentioned in that volume shows how far we have progressed during the last decade!

Infrared photometry of Cepheids (which minimize absorption problems) and CCD observation of RR Lyrae stars in the Magellanic Clouds, M31 and M33, that were reported at this symposium, indicate that the distance scale within the Local Group now appears to be secure at the $\lesssim 10$ percent level. In this volume Jacoby *et al*. report that the luminosity function of the [OIII] $\lambda 5007$ flux of planetary nebulae appears to be universal. This discovery provides a powerful new precision tool for the calibration of the extra-Galactic distance scale. At sites with good seeing this technique could probably be pushed out to distances as great as that of the Virgo cluster.

The present proceedings also show that considerable progress has recently been made in reducing the scatter of the Tully-Fisher relation by utilizing only those galaxies that have CCD surface photometry, and by using the isophotes so obtained to derive accurate galaxy inclinations. This improved accuracy is important because Malmquist bias to distance moduli obtained from the Tully-Fisher relation is proportional to the square of this dispersion.

Finally the last decade has made astronomers realize that deviations from a smooth Hubble flow severely limit the accuracy with which the extra-Galactic distance scale can be calibrated. As Djorgovski emphasizes in this volume it is particularly important to realize that possible differences in the physical characteristics of galaxies in widely separated regions of space might produce systematic distance errors that mimic large-scale motions.

In closing it is our pleasant duty to acknowledge the help of J.E. Hesser and F.D.A. Hartwick on the Scientific Organizing Committee of this symposium. Furthermore we are indebted to the National Research Council of Canada, the Natural Sciences and Engineering Research Council, the University of Victoria, the Canadian Institute for Theoretical Astrophysics, the Dominion Astrophysical Observatory and the Government of British Columbia for financial support of this symposium, which formed part of the centenary celebrations of the Astronomical Society of the Pacific.

SIDNEY VAN DEN BERGH
CHRISTOPHER J. PRITCHET

REFERENCES

Baade, W. 1952, I.A.U. Trans. 8, 395.

Balkowski, C., and Westerlund, B.E. 1977, Décalages vers le Rouge et Expansion de l'Univers = I.A.U. Colloquium No. 37 (Paris, CNRS).

Campbell, W.W. 1917, Science, 45, 413.

Hubble, E.P. 1925, Observatory, 48, 139.

Hubble, E. 1929, Proc. Nat. Acad. Sci., 15, 168

Hubble, E., and Humason, M.L. 1931, Astrophys. J., 74, 43.

Sandage, A. 1958, Astrophys. J., 127, 513.

I

INTRODUCTION

HUBBLE'S DISTANCE INDICATORS

PAUL W. HODGE
Astronomy Department, FM-20, University of Washington, Seattle, Washington 98195

ABSTRACT In attempting to establish the distance scale of the extragalactic universe, Hubble considered a large number of different possible distance indicators. This brief review traces the history of some of these criteria and shows that many of those that Hubble rejected have actually proven to be useful contributors to the development of this difficult subject.

I. HUBBLE'S LIST

In a paper published more than 50 years ago, Hubble (1936) mentioned the possibility of using nine different kinds of objects as distance criteria for galaxies: Cepheids, HS variables, supergiants, novae, globular clusters, open clusters, OB associations, HII regions, brightest stars.

This is a remarkably modern looking list. It will be used here as a means of illustrating some of the progress in the field of extragalactic distance determination. I will take up each item in his list in order, mention something from Hubble's work that represents the status of the criterion as of 50 years ago and also mention a modern study related to the same criterion. This will, I hope, be helpful in putting the current symposium in a historical context, though, of course, the treatment is by no means exhaustive.*

Before discussing these as "Hubble's distance indicators", however, it must be admitted that Hubble actually rejected all but the first and last. He felt that the others were too crude to be useful.

II. CEPHEIDS

Cepheids were used by Hubble to form the foundation of the

* In 1981, one could list 23 extragalactic distance criteria (Hodge 1981), and several more have appeared since.

extragalactic distance scale in the local group. For example, his first paper on this subject dealt with NGC6822, for which his Cepheid period-luminosity diagram led to his determining a distance modulus of 21.65, apparent. A modern result, based on the superior method of infrared photometry (McAlary et al. 1983), gave $(m-M)_0=23.47$. The difference, of course, lay in the error then being made in the calibration of the period-luminosity relation, involving the confusion of Pop I and Pop II Cepheids, which we now know to have quite different absolute magnitudes (Baade 1963).

III. HUBBLE-SANDAGE VARIABLES

Hubble found a few very luminous, irregular variables, now called "Hubble-Sandage variables" or "BLVs". These were among the most luminous objects in local group galaxies. For example, his star V19 in M31 is one of them; it is also known as AF And. We now know that these fascinating objects, though not very useful as standard candles because of their 2-3 magnitude range in mean luminosity, are extremely interesting astrophysically (see, for example, the spectroscopic study of AF And and other HS variables published published by Humphreys, 1975).

IV. SUPERGIANTS

Hubble mentioned supergiants as a possible distance indicator. To illustrate the state of knowledge of these stars at that time, I would like to refer you to his color-magnitude diagram for NGC206, an assemblage of bright stars in M31 (Hubble 1929). It is plotted at right angles to modern convention, but shows a clear concentration of stars on the main sequence, with a few supergiants, which show a wide range in luminosity. By comparison, the reader can examine modern color-magnitude diagrams for similar areas in M31 (e.g., Massey et al. 1986, Odewahn 1987, and Hodge and Lee 1988), based on CCD photometry. From these sources it is clear that the brightness of supergiants depends on the age of the youngest stars in the particular portion of the galaxy being sampled, and is not a good distance indicator, as Hubble had concluded long ago.

V. NOVAE

Novae, on the other hand, were found by Hubble to be useful, but he could detect them only within the local group. A great deal has occurred in the meantime. Novae have been detected at much greater distances, even as far as the Virgo Cluster (van den Bergh, 1988). Also, they are being detected by using narrow band Hα filters by Ciardullo *et al.* (1987), who find that Hα photometry of novae provides a particularly powerful distance criterion.

VI. GLOBULAR CLUSTERS

Hubble gave careful consideration to using globular clusters as distance indicators. For example, his paper on the globulars of M31 (Hubble 1932) discussed their absolute magnitudes and concluded that comparison with Galactic globular clusters indicated that M31's globulars were abnormally faint. It is historically interesting to note that if Hubble had chosen the globulars instead of the Cepheids to measure the distance to M31, he would have obtained a much more accurate answer. The error in the P-L relation, due to miscalibration, did not affect Shapley's absolute magnitudes for Galactic globulars.

Recent work has shown that globular clusters make a useful distance indicator out to the Virgo Cloud and even somewhat beyond (Hanes 1979, Hanes and Whittaker 1987).

VII. OPEN CLUSTERS

Hubble wisely discarded this possibility, which has become even more clearly inappropriate in recent years, when we have seen that different galaxies have very different cluster populations (e.g., van den Bergh 1979). The integrated luminosities of the brightest "open clusters" differ from galaxy to galaxy by as much as four magnitudes.

VIII. OB ASSOCIATIONS

Hubble listed star clouds as possibly useful for distance determinations but did not (knowingly) use them as such. Several recent attempts have been made to refine the use of stellar associations as distance indicators. Their appeal, of course, is that they are often the most luminous objects identifiable in a spiral galaxy and are thus potentially useful out to great distances. Wray and de Vaucouleurs (1980), for instance, found the method to be relatively effective to very large distances, while others (e.g., Efremov 1982, Hodge 1986) find that consistency in their identification is a very difficult problem. There is a strong relationship between the absolute magnitudes and sizes of OB associations and the distances of the host galaxies, the apparent result of the blending together of several associations into star clouds when poorer and poorer resolution is attained.

IX. HII REGIONS

Hubble listed emission nebulae as a possible distance criterion candidate. Sersic (1960) took up the idea enthusiastically, and the method was used extensively by Sandage and Tammann (1974a). However, as shown by Kennicutt (1979) and others, measures of the diameters or integrated luminosities of HII regions are difficult to make consistently, independently of distance, and large uncertainties can result when the criterion is applied much beyond the local group. The relationship between velocity dispersion and integrated Hα luminosity of HII regions is a more promising technique (Melnick 1977).

X. BRIGHTEST STARS

This criterion was Hubble's choice for measuring distances beyond the local group. Although for his time and equipment he did an exemplary job of testing the criterion for reliability, it turned out to be an unfortunate choice. As explained in detail by Sandage (1958), Hubble mistook other kinds of objects (HII regions, OB associations, clusters) for brightest stars in his more distant galaxy sample, thereby grossly underestimating galaxy distances.

Sandage and Tammann (1974b) explored the use of this criterion and tested it for nearby galaxies. Humphreys (1983) and others also have examined the problem recently. The brightest blue stars show a clear dependence on the luminosity of the galaxy, and possibly also on the Hubble type, making them unattractive for distance determinations. The brightest M supergiants, however, show an upper bound to their luminosities, which, except for very poorly populated dwarf galaxies, makes them relatively useful as distance indicators.

The nine "Hubble distance indicators" have shown varied fortunes in the 50 years since Hubble's list was published. Some were rejected by Hubble, and subsequent work has shown the wisdom of this decision. Some of those that he declined to employ have turned out to be useful and are presently important to the distance scale enterprise. It is somewhat ironic that the two that he chose to use almost exclusively turned out to be rather seriously flawed; the Cepheid period-luminosity relation had been incorrectly calibrated and the brightest star criterion suffered from misidentifications in distant galaxies. These together led to Hubble's very considerable underestimate of the size scale of the universe. But it would be inappropriate to emphasize this fact; instead, it would seem to be the right thing to do at this conference to point out that we are still using and perfecting many of the distance criteria that Hubble listed long ago, and that Hubble's universe, though too small by our standards, was far closer to what we consider "correct" than the comparatively tiny universe envisaged by most of his predecessors.

REFERENCES

Baade, W. 1963, Evolution of Stars and Galaxies (Cambridge: Harvard University Press).

Ciardullo, R., Ford, H.C., Neill, J.D., Jacoby, G.H., and Shafter, A.W. 1987, Ap. J., **318**, 520.

Efremov, Y.N. 1982, Astron. Zh., **8**, 663.

Hanes, D.A. 1979, M.N.R.A.S., **188**, 901.

Hanes, D.A., and Whittaker, D.G. 1987, A. J., **94**, 906.

Hodge, P.W. 1981, Ann. Rev. Astr. Ap., **19**, 357.

Hodge, P.W. 1986, in Luminous Stars and Associations in Galaxies, ed. C.W.H. DeLoore et al. (Dordrecht: Reidel).

Hodge, P.W., and Lee, M.G. 1988, Ap. J., **329**, 651.

Hubble, E.P. 1932, Ap. J., **76**, 44.

Hubble, E.P. 1936, Ap. J., **84**, 279.

Humphreys, R.L. 1975, Ap. J., **200**, 426.

Humphreys, R.L. 1983, Ap. J., **269**, 335.

Kennicutt, R.C. 1979, Ap. J., **228**, 394.

Massey, P., Armendroff, T.E., and Conti, P.S. 1986, A. J., **92**, 1303.

McAlary, C., Madore, B., McGonegal, R., McLaren, R.A., and Welch, D.L. 1983, Ap. J., **273**, 539.

Melnick, J. 1977, Ap. J., **213**, 15.

Odewahn, S. 1987, A. J., **93**, 310.

Sandage, A.R. 1958, Ap. J., **127**, 513.

Sandage, A.R., and Tammann, G.A. 1974a, Ap. J., **190**, 525.

Sandage, A.R., and Tammann, G.A. 1974b, Ap. J., **191**, 603.

Sersic, J.L. 1960, Zeit. f. Ap., **50**, 168.

van den Bergh, S. 1979, Ap. J., **230**, 95.

van den Bergh, S. 1988, this conference.

Wray, J., and de Vaucouleurs, G. 1980, A. J., **85**, 1.

II

LOCAL GROUP GALAXIES

THE MAGELLANIC CLOUDS AND THE EXTRAGALACTIC DISTANCE SCALE

M W FEAST
S A Astronomical Observatory, P O Box 9,
Observatory 7935, Cape, South Africa

ABSTRACT. Recent work on the distances of the Magellanic Clouds and their relevance to the extra galactic distance scale are reviewed.

INTRODUCTION

Recent work (e.g. papers by Walker, Freedman and Pritchet in this volume) has shown that using CCD detectors, it is possible to make rather spectacular advances in the accurate photometry of Cepheids and other objects, including RR Lyrae variables, in nearby galaxies. This work is only in its early phases. However, it suggests that if sufficient time on large telescopes can be made available, we will, in a few years, have (at least for Cepheids) data in these nearby galaxies of accuracy approaching that so far obtained only in our own galaxy and in the Magellanic Clouds. In these circumstances, it is important to look carefully at the current status of work on the Magellanic Clouds to see how this helps us in extending the distance scale to other members of the local group and beyond, and also to see what further work on Magellanic Cloud objects would be especially desirable for distance scale problems.

CEPHEIDS

A major source of uncertainty in the Cepheid distances to the Magellanic Clouds remains the calibration of the zero points of the period luminosity and period-luminosity-colour relationships (Feast and Walker 1987, Walker this volume). Taking into account the uncertainty in the Pleiades modulus which is used as the basis of the Cepheid scale (and which is derived from a comparison of the Pleiades main sequence with luminosities and colours of nearby main sequence stars of known parallax) and also allowing for the uncertainties in the main sequence fitting of the Pleiades to the galactic clusters containing Cepheids, it is estimated that the uncertainty in zero point determination from these sources is ~ 0.15 mag (i.e. ~ 7% in the distance). Whilst bearing in mind this uncertainty we should look at other areas of uncertainty, especially those that will be of relevance for other nearby galaxies.

To determine a Cepheid distance we need an estimate of the reddening. This can be obtained from the two colour BVI diagram and this method is likely to be of importance for Cepheids in galaxies beyond the Magellanic Clouds since most current Cepheid work in these galaxies is being carried out in BV and I. However, the intrinsic line in the BVI two colour diagram depends on metal abundance. Thus we need to know the metallicity to derive the reddening. The period-luminosity relation (in V) is rather insensitive to metallicity but the moduli derived from it depend strongly on the adopted reddening, thus introducing a metallicity dependence. The period-luminosity-colour relation in V and B-V is sensitive to metallicity but moderately insensitive to reddening. Figures 1 and 2 show how in these circumstances, it is possible to obtain estimates of both metallicity and modulus for galaxies. The curves were

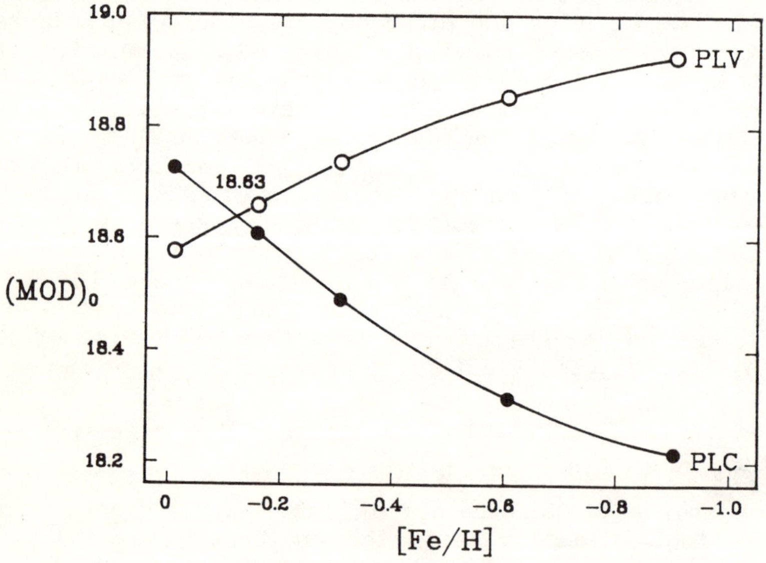

Fig. 1 The LMC true modulus derived from the Cepheid period luminosity relation in V (PLV) and the period-luminosity-colour relation in V and B−V (PLC) as a function of [Fe/H]. This plot takes into account the dependence of the reddening derived from BVI colours on [Fe/H]. Only LMC field Cepheids with individually determined BVI reddenings have been used.

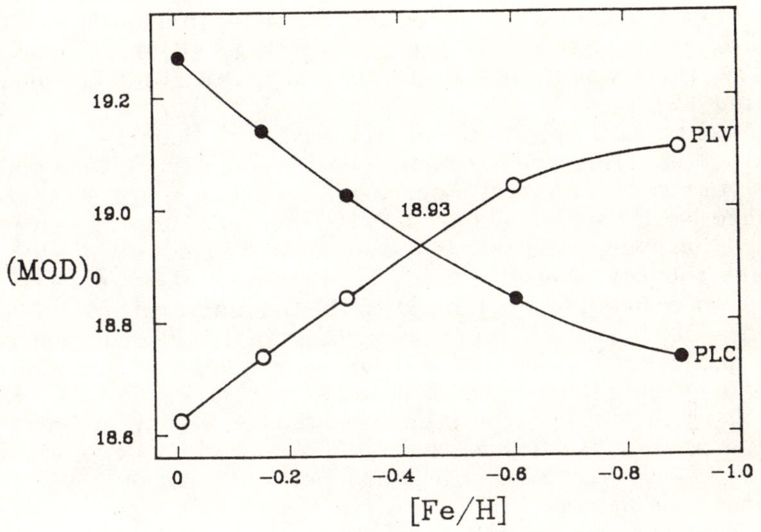

Fig. 2 As Figure 1 but for the SMC.

calculated by Dr J Caldwell as an extention to the analysis in Caldwell and Coulson (1986). They take into account variations of the period-luminosity and period-luminosity-colour relations with metallicity and also the dependence of the BVI reddenings on metallicity. The intersection of the curves gives an estimate of [Fe/H] as well as the true modulus and there is a corresponding set of reddenings for the Cepheids used. It is interesting to note that the mean of the PL(V) and PLC moduli varies rather slowly with metallicity so that it is not necessary to estimate the metallicity with high accuracy to get a good modulus.

Metallicity for Magellanic Cloud Cepheids can also be estimated from BVI reddenings compared with independent reddening estimates and from the period-colour relation. Other Cepheid metallicity estimates come from Walraven photometry, from Washington photometry and from spectral type colour relations. Less direct estimates come from supergiants, HII regions and young clusters in the Clouds. A review of available data (Feast 1988) together with recent work by Russell et al. (1988) on supergiants yields best estimates for [Fe/H] for young objects (including Cepheids) of -0.2 (LMC) and -0.5 (SMC). These are sufficiently close to the values adopted in most recent work on Magellanic Cloud Cepheids (-0.15 and -0.6 respectively) that one need not amend the results derived from the Cepheids. However there is a suggestion that there may be some supergiants in the SMC with distinctly lower [Fe/H] (~ -1.2) (Spite et al.

1986, cf. Feast 1988 and Russell et al. 1988) and it would be important to know if there is a range in [Fe/H] for SMC Cepheids, though the small spread in observed BVI reddenings suggest that most SMC Cepheids have similar abundances (Feast 1985).

A potential problem in the use of Magellanic Cloud Cepheids as distance indicators results from the period distribution of the stars observed. Nearly all the earlier accurate work was on Cepheids with periods greater than 10 days. However, most of the best Galactic calibrators have periods shorter than this. It is obviously desirable to have a good overlap in period with the calibrators. Some work has been done at shorter periods in the Clouds and more is in progress. There is a problem however in that the BVI two-colour intrinsic line makes a relatively small angle with the reddening line for the hotter (shorter period) stars and as a result, unless the observations are of very high accuracy, the reddenings are not well determined. In the past the assumption has often been made that the mean reddening of short period stars was the same as that of the longer period ones. This assumption obviously introduces some uncertainty. An exception is the case of the seven 3-day Cepheids in the LMC cluster NGC 1866 whose reddening can be estimated from that of other cluster members (cf. Walker 1987). Adopting [Fe/H] = -0.15 for the LMC, the results of Caldwell and Coulson (1986) and Walker (1987) can be summarised as in Table I which gives the results for the period-luminosity relation in V and the period-luminosity-colour relation in V and B-V as well as the mean of the two relations (which as noted above varies little with abundance).

Feast and Walker (1987) recently adopted an LMC modulus of 18.47 which is essentially a mean of estimates 2 and 3 of Table I and in effect gives double weight to some of the short period stars. The differences in the mean moduli of the long and short period Cepheids is $\sim$ 0.2 and requires further investigation. It may simply represent an uncertainty in the adopted slopes of the PL and PLC relations. Since the weighted log P of the calibrators is 0.975 the Feast-Walker LMC mean should be relatively insensitive to such uncertainties. Alternatively, the difference may be due to an error in the adopted mean reddening of the short period stars. In that case a larger modulus would be preferred. The result might indicate that the various relationships are nonlinear although this is less likely since, so far as the PLC relation is concerned, the required non-linearity would be of opposite sign (and of larger absolute amount) to that predicted by theory (cf. Caldwell and Coulson 1986).

TABLE I Cepheid Moduli of the LMC

	PL(V)	PLC	Mean
1. Cepheids with individual reddenings (log P 0.9-1.7) (26 stars)	18.66	18.60	18.63
2. Field Cepheids of all periods (log P<1.7) (73 stars)	18.51	18.50	18.50
3. Cepheids with log P<1.0 (including NGC 1866) (43 stars)	18.44	18.41	18.42

A different estimate of the LMC Cepheid modulus can be made from infrared (JHK) photometry. Laney and Stobie (in preparation) have recently completed extensive infrared photometry of 20 Galactic calibrating Cepheids. Interstellar absorption is small at K (2.2μm). However the distances of the clusters and associations containing the calibrating Cepheids are determined from optical photometry where interstellar extinction is often very significant. As a result the infrared Cepheid PL zero point depends rather sensitively on $R = A_V/E_{B-V}$. Laney in fact finds it possible to combine optical values of E_{B-V} with the infrared photometry of the calibrating Cepheids to obtain $R_o = 3.00 \pm 0.07$ (when R_o is the value of R appropriate to early type stars). He has also obtained a zero point for the PL relation at K which when applied to the infrared data on LMC Cepheids (Laney and Stobie 1986, Welch et al. 1987) leads to a modulus of 18.52 ± 0.04 (internal error). The sample of LMC Cepheids with infrared data is roughly similar in its period distribution to that of sample 2 in Table I and the derived modulus is nearly identical.

From a comparison of Cepheids in the same period range and allowing for metallicity effects, one finds the differences in moduli of the centroids of the SMC and LMC to be 0.31 ± 0.03. This result combines optical and infrared data which agree closely (cf. Caldwell and Coulson 1986, Laney and Stobie 1986, Feast and Walker 1987).

CLUSTERS

Beginning with the work of Schommer et al. (1984) it has been recognised that main sequence fitting to Magellanic Cloud clusters is an attractive and important way to determine the moduli of the Clouds. The current situation on main sequence fitting and on distance determinations from other fits to features of the colour-magnitude diagrams (such as the giant clump) is shown in Table II. Where alternative values of the modulus are given these were quoted as possible values by the original workers.

TABLE II Cluster Distances to the LMC

(A) Main Sequence Fits

CLUSTER	METHOD	MODULUS	REF.
NGC 2162/2190	Hyades	18.3 ± 0.2	1
	Theory	18.1	1
H4	Theory	18.1 OR 18.7	2
NGC 2213	Theory	18.2 ± 0.2	3
NGC 1866	Pleiades	18.4/18.5 ± 0.2	4

(B) Colour-magnitude Diagram

NGC 2210 (with $\langle M_V(RR)\rangle=0.6$)		18.42	5
Hodge 11		18.4	6
NGC 1841 (outlier)		18.2	7
E2		18.4-18.5	8+9
NGC 2249/2241		18.4	10
LW79		18.4 ± 0.2	11

Cluster Distances to SMC

(B) Colour-magnitude Diagrams

NGC 411/NGC 121/L113		18.8 better than 19.3	12+13

Refs: (1) Schommer et al. 1984; (2) Mateo & Hodge 1986; (3) Geisler 1987; (4) Walker 1985; (5) Walker 1984; (6) Andersen et al. 1984; (7) Andersen et al. 1987; (8) Gratton & Ortolani 1987; (9) Schommer et al. 1986; (10) Jones 1987; (11) Mateo & Hodge 1987; (12) Da Costa & Mould 1986; (13) Seidel et al. 1986.

Main sequence fitting in LMC clusters tends to give somewhat smaller moduli than the Cepheid work. Perhaps as low as 18.1 if a fit is made to a theoretical main sequence. However, recent models (vandenBerg -- this volume) will increase the modulus derived in this way. Perhaps it is best at present to rely on fits to a Hyades main sequence which give moduli near 18.3. Besides corrections for metallicity effects which introduce some uncertainty, it has been pointed out (e.g. Walker and Laney 1986) that there will be systematic errors both due to the finite accuracy of the photometry and the problem of blending in the crowded fields and this will tend to lead to an underestimate of the modulus. The effect of convective overshooting on the models has also been discussed as a reason for the determination of a low modulus (Chiosi and Pigatto 1986).

Other features in the colour-magnitude diagram tend to give slightly higher LMC moduli (~18.4).

EARLY-TYPE STARS

Table III summarises some determinations of the LMC and SMC moduli using early-type stars. Crampton (1979) used OB spectral types and Hγ equivalent widths in the LMC, and Crampton and Greasley (1982) used OB spectral types in the SMC. Conti et al. (1986) and Garmany et al.(1987) concentrated on O star spectral types in both Clouds as well as fitting a theoretical main sequence for massive stars to the lower envelope of their O star colour-magnitude diagram. This, of course, involved using a spectral type (or colour)-temperature relation and bolometric corrections. Shobbrook and Visvanathan (1987) derived an LMC distance from Hβ photometry of B stars.

TABLE III Early-Type Stars (OB)

(A) LMC Modulus

		Refs
Sp.T. (OB) (25 stars)	18.6 (± 0.1 int)	1
Hγ (OB)	18.7 (± 0.2 int)	1
Sp.T. (O) (75 stars)	18.3 ± 0.4	2
Theory (Main sequence fit)	18.3	2
Hβ	18.4 ± 0.25	3

(B) SMC Modulus

Sp.T. (OB) (19 stars)	19.1 ± 0.25	4
Sp.T. (O) (50 stars)	19.1 (± 0.1 int)	5
Theory (Main sequence fit)	18.9	

Refs: (1) Crampton 1979; (2) Conti et al. 1986; (3) Shobbrook & Visvanathan 1987; (4) Crampton & Greasley 1982; (5) Garmany et al. 1987.

Since these methods are attractive and will no doubt be extended in the future, it is worth considering where the present uncertainties lie. Garmany et al. (1987) note that the difference in distance moduli between the two Clouds, obtained from their O star spectral types, is ~ 0.5 mag. This is somewhat larger than the best estimate from Cepheids (0.30). Further work is required to determine how significant the difference is. Garmany et al. suggest that, if significant, the difference would indicate that a metallicity dependence will have to be included in the O star calibration.

There is a further problem. A large number of workers have contributed over the years to the absolute magnitude calibration of spectral type and luminosity class criteria. The basis of the calibration is primarily the distances of young clusters and associations. It would be useful if all this data could now be re-examined and placed on a consistent scale (referred to the Pleiades for instance) and using a consistent interstellar reddening law. The work of Balona and Shobbrook (1984) showed that there is considerable difference between different calibrations.

In using very bright early type stars as distance indicators one needs to worry about the possibility of some of these objects being compact multiple systems. This was emphasized by Walborn (1977) who found that about one-third of the sample of O and B0 supergiants in the Clouds were in multiple systems. This problem was highlighted by the resolution of R136 in 30 Dor into a number of components (e.g. Weigelt and Baier 1985) and the recent resolution of the O star HD 268743 in the LMC into several components (Heydari-Malayeri et al. 1988). This latter star was known to be unusual in combining broad absorption lines with an (apparrently) high luminosity (Walborn 1977). Nevertheless, it has been clear for a long while (e.g. Feast et al. 1960) that the most luminous early type objects in the Clouds, at a given spectral type, are emission line stars indicating a real upper limit to stellar stability rather than their being unresolved stellar groupings.

The recent results of Shobbrook and Visvanathan (1987) are of special interest since they include relatively faint B stars (to ~ 15 mag) in the LMC and their results are tied to the Pleiades (for which a modulus of 5.57 is adopted here), through the calibration of Balona and Shobbrook (1984). A difficulty with this method is that there is a very strong dependence of M_V on β for luminous B stars so that β measurements of very high accuracy are required. This must be responsible for the fact that even after rejecting some discrepant stars, there is a range of 4 mag in individual moduli. Since the M_V-β relation is non-linear the results are sensitive to the way means are derived and the procedure adopted for rejecting outlying points. A standard error of ~ 0.2 probably fairly represents the uncertainty in the derived modulus (18.41) not taking into account any uncertainty in the M_V-β calibration.

RR LYRAE VARIABLES

Good (CCD based) magnitudes are now available for RR Lyrae variables in three Magellanic Cloud clusters (NGC 2210 (9 variables), NGC 1786 (6 variables) in the LMC and NGC 121 (4 variables) in the SMC) (Walker 1985, Walker and Mack 1988a, b). There still remains considerable uncertainty as to the proper absolute magnitudes to adopt for RR Lyraes. In the present discussion results are given for two possible estimates of the absolute magnitudes. (A) $\langle M_V(RR)\rangle = 0.6$ independent of [Fe/H].

(B) the relation: $\langle M_V(RR)\rangle = 0.92 + 0.2[Fe/H]$.........(1)

This latter relation has some attractions. (1) It was derived by Lee et al. (1987) from synthetic horizontal branch models. (2) A closely similar relation ($\langle M_V(RR)\rangle$ = 0.94 + 0.2 [Fe/H] has recently been derived by Fernley (1988) from Baade-Wesselink data on RR Lyraes. (A relation ~ 0.1 mag fainter $\langle M_V(RR)\rangle = 1.0 + 0.17$ [Fe/H] has also been proposed by Caccairi et al. (1988). (3) Equation 1 yields $\langle M_V(RR)\rangle = 0.72$ at [Fe/H] = -1 which is the mean [Fe/H] of the RR Lyraes in the statistical parallax discussions of Hawley et al. (1986) and Strugnell et al. (1986) which give $\langle M_V(RR)\rangle = 0.68 \pm 0.2$ (cf. Barnes and Hawley 1986, Feast 1987). (4) Table IV shows that in the mean the results of equation 1 agree with the absolute magnitudes of the horizontal branches for four globular clusters with main sequence fits (Richer and Fahlman 1987). (5) The coefficient of [Fe/H] in equation 1 is similar to that derived by Lub (1986) (= 0.24) from an analysis of Walraven photometry of field RR Lyrae stars. Note however that this coefficient is significantly less than that required to explain the Oosterhoff effect (= 0.35) (Sandage 1982).

TABLE IV Absolute Magnitudes of horizointal Branches of Globular Clusters from Main Sequence Fits

(Richer and Fahlman 1987)

Cluster	[M/H]	M (HB)	$M_V(1)$
M15	-2.15	0.44	0.49
M13	-1.40	0.56	0.64
M 5	-1.13	0.81	0.69
M 4	-0.93	0.83	0.73
Mean	-1.40	0.66	0.64

$M_V(1) = \langle M_V(RR)\rangle$ from equation 1

The results are given in Table V. The difference in moduli between the SMC and the LMC agrees well with that derived from Cepheids if equation 1 is used. However given the known large depth of the SMC this does not rule out a solution with constant $\langle M_V(RR)\rangle$.

Laney is carrying out infrared photometry of Type II Cepheids in the LMC and in galactic globular clusters. With a cluster scale based on $\langle M_V(RR)\rangle = 0.6$ he finds a preliminary LMC modulus of 18.42.

TABLE V RR Lyrae Moduli of the Magellanic Clouds

	[Fe/H]	$(MOD)_1$	$(MOD)_2$
LMC			
NGC 2210	-1.9	18.48 (± .06 int)	18.42
NGC 1786	-2.3	18.58 (± .05 int)	18.44
SMC			
NGC 121	-1.4	18.82 (± .07 int)	18.86
	LMC	18.53	18.43
	SMC	18.82	18.86
	SMC-LMC	0.29	0.43

$(MOD)_1$ with $\langle M_V(RR)\rangle$ from equation 1
$(MOD)_2$ with $\langle M_V(RR)\rangle = 0.6$

MIRA AND RED SUPERGIANT VARIABLES

Recent SAAO work (Glass et al. 1987 and in progress) shows that the scatter about the Mira infrared period-luminosity relation is very small (σ = 0.13 mag or less) for M and S type Miras. The calibration of the zero point has been discussed on a number of occasions (e.g. Feast 1987). Most reliance is probably best placed on the data by Menzies and Whitelock (1985) for Miras in globular clusters. The result then depends on the RR Lyrae (or horizontal branch) scale. Miras occur in relatively metal-rich globular clusters for which equation 1 above predicts a mean $\langle M_V(RR)\rangle$ of 0.80. Calibrating the Miras in this way leads to an LMC modulus of 18.28 (± 0.06 internal). This is 0.25 mag smaller than obtained directly from the RR Lyraes in LMC clusters using equation 1 (Table V). On the other hand if $\langle M_V(RR)\rangle = 0.6$ is assumed for all values of [Fe/H] the revised Mira zero point leads to an LMC modulus of 18.48 close to the values obtained directly from the RR Lyraes on this assumption (18.43, Table V). This is a reason for suspecting that there may be rather little dependence of

<M_V(RR)> on [Fe/H].

The C-type Miras in the LMC follow the same PL relation as the M and S type Miras at K (2.2μm) though with somewhat greater scatter (Glass et al.1987). The larger scatter is partly due to the small number of observations per star at present available (so the means are not precisely defined). In addition we know that some carbon Miras, at least, undergo episodes of dust obscuration (e.g. Feast et al. 1984) so that some scatter about the PL relation may be due to this. Evidently changes in the C/O ratio in Mira atmospheres has little effect on the PL relation at K.

One can now combine recent SAAO work (Lloyd Evans et al. 1988) with Mt Stromlo work (Wood et al. 1983) to look at the Mira P-L relation in the SMC. The available data are shown in Figure 3. The line has been drawn with the same slope as the LMC Mira PL relation. Its vertical position has been fixed by a least squares fit to the observations. This gives a difference in the moduli of the two Clouds of 0.30 ± 0.06, a value essentially identical with the Cepheid result (0.31 ± 0.05). Thus any differences in age (or mass)-metallicity relations between the two Clouds is not having any significant effect on the Mira PL zero point.

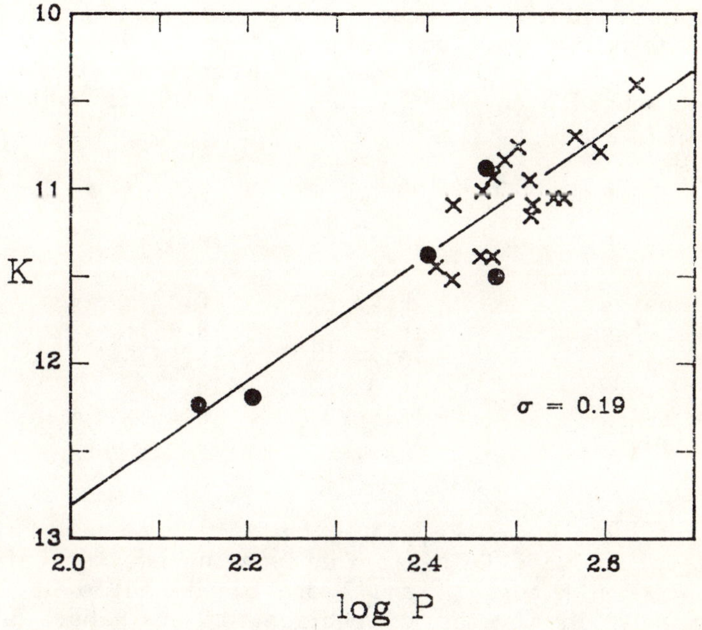

Fig. 3 Period-luminosity relation for SMC Miras Filled circles are M or S type Miras; crosses are C type Miras. The line has the same slope as that derived for LMC Miras.

Whilst Mira variables are especially important distance indicators for old populations, the bright red supergiant variables are likely to be important for distant determinations of some galaxies. Wood et al. (1983) combined SAAO and Mt Stromlo infrared photometry to show that the red supergiant variables were separated in a K-period plot from lower luminosity variables and on this basis it was suggested (Feast 1981) that they would be good distance indicators. With a distance modulus to the LMC of 18.47 they fit the relation:

$M_K = -9.11 - 0.0024P, \quad \sigma = 0.24$ (24 stars).

This could prove a useful relation for nearby galaxies and has already been applied in a preliminary way to M33 by Kinman et al. (1987) (see Mould, this volume).

SUMMARY

Table VI summarises the presently available results on the LMC. Results for the Miras and RR Lyraes are given for both the assumption that $\langle M_V(RR) \rangle = 0.6$ and for the case that equation 1 holds. It will be noticed that the mean of the RR Lyrae and Mira results is rather insensitive to which of the two assumptions is made. The scatter amongst the moduli is now quite small and suggests that the LMC modulus is in the range 18.4 to 18.5 though an uncertainty of ~ 0.15 probably attaches to any mean value. The SMC is 0.31 mag more distant.

TABLE VI LMC Modulus

	$(MOD)_1$		$(MOD)_2$	
Cepheids	18.47		18.47	
RR Lyraes	18.53)	18.40	18.43)	18.46
Miras	18.28)		18.48)	
Clusters	18.3		18.3	
OB Stars	18.4		18.4	

ACKNOWLEDGEMENTS

I am grateful to Dr J Caldwell for carrying out the calculations on which Figures 1 and 2 are based and to Dr C D Laney for data in advance of publication. I have also benefitted from discussions with them and other colleagues at SAAO.

REFERENCES

Andersen, J., Blecha, A. & Walker, M.F. 1984, M.N.R.A.S., 211, 695.

Andersen, J., Blecha, A. & Walker, M.F. 1987, M.N.R.A.S., 229, 1.

Balona, L.A. & Shobbrook, R.R. 1984, M.N.R.A.S., 211, 375.

Barnes, T.G. & Hawley, S.L. 1986, Ap. J., 307, L9.

Cacciari, C., Clementini, G. & Buser, R. 1988, in press.

Caldwell, J.A.R. & Coulson, I.M. 1986, M.N.R.A.S., 218, 223.

Chiosi, C. & Pigatto, L. 1986, Ap. J., 308, 1.

Conti, P.S., Garmany, C.D. & Massey, P. 1986, A. J., 92, 48.

Crampton, D. & Greasley, J. 1982, Pub.A.S.P., 94, 31.

Crampton, D. 1979, Ap. J., 230, 717.

Da Costa, G.S. & Mould, J.R. 1986, Ap. J., 305, 214.

Feast, M.W. & Walker, A.R. 1987, Ann. Rev. Astr. Ap., 25, 345.

Feast, M.W. 1981, in The Most Massive Stars ed. D'Odorico, D. Baade & K. Kjar (ESO) p 217.

Feast, M.W. 1985, in Cepheids: Theory and Observations ed. B.F. Madore (Cambridge) p 157.

Feast, M.W. 1987, in The Galaxy ed. G. Gilmore & B. Casswell (Reidel), p 1.

Feast, M.W. 1988, in Le Monde des Galaxies ed. H.G. Corwin and L. Bottinelli, in press.

Feast, M.W., Thackeray, A.D. & Wesselink, A.J. 1960, M.N.R.A.S., 121, 337.

Feast, M.W., Whitelock, P.A., Catchpole, R.M., Roberts, G. & Overbeek, M.D. 1984, M.N.R.A.S.,211, 331.

Fernley, J.A. 1988, to be published.

Garmany, C.D., Conti, P.S. & Massey, P. 1987, A. J., 93, 1070.

Geisler, D. 1987, A. J. 93, 1081.

Glass, I.S., Catchpole, R.M., Feast, M.W., Whitelock, P.A. & Reid, I.N. 1987, in Late Stages of Stellar Evolution ed. S. Kwok & S.R. Pottasch (Reidel) p 51.

Gratton, R.G. & Ortolani, S. 1987, Astr. Ap. Suppl., 67, 373.

Hawley, S.L., Jeffreys, W.H., Barnes, T.G. & Wan, L. 1986, Ap. J., 302, 626.

Heydari-Malayeri, M., Magain, P. & Remy, M. 1988, submitted to Astr. Ap.

Jones, J.H. 1987, A. J., 94, 345.

Kinman, T.D., Mould, J.R. & Wood, P.R. 1987, A. J., 93, 833.

Laney, C.D. & Stobie, R.S. 1986, M.N.R.A.S., 222, 449.

Lee, W.Y., Demarque, P. & Zinn, R. 1987, in The Second Conference on Faint Blue Stars, ed. A.G. Davis Philip, D.S. Hayes, J.W. Liebert, Davis press, p.137.

Lloyd Evans, T., Glass, I.S. & Catchpole, R.M. 1988, M.N.R.A.S., 231, 773.
Lub, J. 1986. In Stellar Pulsation ed. A.N. Cox, W.M. Sparks & S.G. Starrfield. (Springer-Verlag) p 218.
Mateo, M. & Hodge, P. 1986, Ap. J. Suppl. 60, 893.
Mateo, M. & Hodge, P. 1987, Ap. J., 320, 626.
Menzies, J.W. & Whitelock, P.A. 1985, M.N.R.A.S., 212, 783.
Richer, H.B. & Fahlman, G.G. 1987, Ap. J., 316, 189.
Russell, S.C., Bessell, M.S. & Dopita, M.A. 1988, Preprint.
Sandage, A.R. 1982, Ap. J., 252, 553.
Schommer, R.A., Olszewski, E.W. & Aaronson, M. 1984, Ap. J., 285, L53.
Schommer, R.A., Olszewski, E.W. & Aaronson, M. 1986, A. J., 92, 1334.
Seidel, E., Da Costa, G.S. & Demarque, P. 1987, Ap. J., 313, 192.
Shobbrook, R.R. & Visvanathan, N. 1987, M.N.R.A.S., 225, 947.
Spite, M., Cayrel, R., Francois, P., Richtler, T. & Spite, F. 1986, Astr. Ap., 168, 197.
Strugnell, P., Reid, N. & Murray, C.A. 1986, M.N.R.A.S., 220, 413.
Walborn, N.R. 1977, Ap. J,. 215, 53.
Walker, A.R. & Laney, C.D. 1986, M.N.R.A.S., 224, 61.
Walker, A.R. & Mack, P. 1988a, in press.
Walker, A.R. & Mack, P. 1988b, in press.
Walker, A.R. 1984, Mon. Notes astr. Soc. S. Afr., 43, 89.
Walker, A.R. 1985, M.N.R.A.S., 217, 13p.
Walker, A.R. 1985, M.N.R.A.S., 212, 343.
Walker, A.R. 1987, M.N.R.A.S., 225, 627.
Welch, D.L., McLaren, R.A., Madore, B.F. & McAlary, C.W. 1987, Ap. J., 321, 162.
Wiegelt, G. & Baier, G. 1985, Astr. Ap., 150, L18.
Wood, P.R., Bessell, M.S. & Fox, H.W. 1983, Ap.J., 272, 99.

DISCUSSION

S. VAN DEN BERGH: People who model supernovae seem to insist that they can only make SN1987A if Z(LMC) = $Z_{\odot}/4$. Do the observations of CNO in the LMC permit such a low abundance?

M. FEAST: Work on HII regions indicates that C and N are deficient with respect to other "metals" when a comparison is made with galactic HII regions. So this may go some way to resolving this problem.

D. BRANCH: Apropos SN1987A, what is the maximum back-to-front distance ratio for the LMC?

M. FEAST: The evidence indicates that the young component of the LMC is a flattened structure seen nearly face-on so the thickness along any line of sight is quite small.

B. TULLY: What is your opinion on the proposition that the SMC is quite extended in the line of sight?

M. FEAST: There is good evidence for such an extension (particularly from Cepheid data) although the detailed structure remains a matter of debate. I have tried to refer the SMC results to the same centroid.

J. NEMEC: The Small Magellanic Cloud contains anomalous Cepheids which may be binary systems, and which follow neither the Pop I nor the Pop II Cepheid P-L relations. Are these stars excluded from the regressions used to determine the SMC distance modulus?

M. FEAST: Yes.

J. MOULD: You showed a plot indicating the dependence of the LMC Cepheid P-L relation on metallicity. Is this consistent with recent theoretical predictions by Stothers (Ap. J. **329**, 717)?

M. FEAST: Stothers concludes, as did earlier workers (e.g. Gascoigne, Iben and Tuggle) that the P-L zero point is quite insensitive to metallicity. However, the interstellar absorption has to be estimated to get the true modulus from a P-L relation. If this is derived (as it most frequently is) from BVI reddenings then the result is abundance sensitive and the estimated (true) P-L modulus depends on abundance.

CEPHEID DISTANCES TO NEARBY GALAXIES

WENDY L. FREEDMAN
Mount Wilson and Las Campanas Observatories
Carnegie Institution of Washington
Pasadena, California

ABSTRACT A program is underway to obtain new CCD photometry for Cepheids in nearby galaxies. As a result, reddening-corrected distance moduli can be determined using multicolor data. Adopting the distances for calibrating galaxies available to date from this study, estimates of the Hubble constant using two different secondary methods are presented, and some of the remaining uncertainties discussed.

I. INTRODUCTION

The use of Cepheids as distance indicators has had a long history, and they have continued to fulfil a role as a reliable primary distance since Hubble (1925) first showed conclusively that the Local Group dwarf irregular galaxy, NGC 6822, was in fact an extragalactic system. Primarily due to the early work of Hubble (1925, 1926, 1929), Baade and Swope (1961, 1963, 1965), Sandage and Tammann (1968), and Sandage (1971), for most of the late-type spiral and irregular galaxies in the Local Group (and out to about three times more distant) Cepheids were searched for, found, and their periods determined.

Most determinations of the Hubble constant have relied on Cepheid variables to provide accurate distances for nearby galaxies, and to secure the zero point of the calibration. The current controversy and uncertainty in the Hubble constant at the level of about a factor of two has recently been argued by Aaronson (1986) as primarily due to the choice of distances to the local calibrating galaxies. Sandage (1988) and Kraan-Korteweg *et al.* (1988) have argued instead that the discrepancy lies in a selection bias in the secondary distance indicator samples.

Over the past several years, I have been involved in a program to obtain new Cepheid distances to nearby galaxies with the aim of providing a new calibration for the determination of the Hubble constant. In the course of the program, new CCD photometry is being obtained for all galaxies in which Cepheids have previously been discovered, and where periods have been determined. The advantages of obtaining CCD photometry are many. The use of a linear, panoramic detector allows point-spread-function fitting photometry, and local sky subtraction. Furthermore, many Cepheids can be observed simultaneously, and the study can be undertaken at several wavelengths, from the blue to the red. This latter advantage permits the reddening to be estimated, and provides a correction to the apparent distance modulus. A description of this CCD program, and results to date are presented here. It is argued that distances to the calibrating galaxies are not responsible for the factor of two discrepancy in the Hubble constant. The small number of calibrating galaxies available is a major remaining source of uncertainty.

II. CEPHEID DISTANCES TO GALAXIES

There are now over a dozen galaxies in which Cepheids have been discovered. In the Local Group, these include the SMC, LMC, M31, M33, IC 1613, NGC 6822 and WLM. More distant galaxies include NGC 300, NGC 2403, Sextans A, Sextans B, M81 and M101. Outside of the Magellanic Clouds where extensive *UBVRI* photoelectric photometry is published, little color information for other extragalactic Cepheids has been available. The galaxies in the present program include all galaxies with known Cepheids, excluding the Magellanic Clouds. New *BVRI* CCD photometry is being obtained for Cepheids in these galaxies using the CFHT 3.6m, Palomar 5m, and Kitt Peak 4m telescopes. Much of this work is still in progress, however results for four galaxies in the program, IC 1613, M33, NGC 2403 and M81 are discussed below.

Before moving on to a discussion of new results, some of advantages and the difficulties in the use of Cepheid variables are discussed.

a) Cepheid Variables as Distance Indicators: Advantages and Difficulties

The reasons that Cepheids have been in use for over six decades, and have emerged as the most reliable primary distance indicator are many:

1) Cepheids are intrinsically very bright (*i.e.*, $-7 < M_V < -2$),

2) Cepheids are relatively numerous,

3) their characteristic light curves make them easy to find and subsequently distinguish from other types of variables,

4) Cepheids exhibit a narrow dispersion in the period-luminosity (PL) and period-luminosity-color (PLC) relations, and perhaps most importantly,

5) the physics of their variability is well understood to be a consequence of the pulsation of their atmospheres.

Some of the significant difficulties to be considered in determining Cepheid distance moduli are :

1) Cepheids are Population I objects. As such they may still be closely associated with the gas and dust from which they formed, and in general they will require corrections for extinction internal to the parent galaxy. With multicolor photometry, this difficulty can be addressed, and reddening corrections applied. Alternatively, as discussed extensively by Madore and collaborators (see Madore (1986) for a review), obtaining photometry at near-infrared wavelengths largely circumvents the problem of reddening.

2) The effects of metallicity on the calibration of the PL and PLC relations have not been established empirically. In most determinations of Cepheid distance moduli to date, the possible effects of metallicity have largely been ignored, since no empirical calibration of the effect has been available. Sandage (1972) has argued that the effects of metallicity on the zero point of the PL relation are unlikely to be very large since the slope of the PL relation does not vary from galaxy to galaxy. Theoretical investigations of the problem have been carried out by Iben and Tuggle (1975), Becker, Iben and Tuggle (1977), Iben and Renzini (1984), and most recently by Stothers (1988). All of these studies suggest that the effect of metallicity on the zero point of PL relation is small; however the zero point of the PLC has a significant abundance sensitivity.

In collaboration with B. Madore and the late M. Aaronson, I have been investigating the effect of metallicity on the zero points of the PL and PLC relations. We have been observing Cepheids at varying radial distances in two nearby spiral galaxies, M33 and M31, which have abundance gradients measured from HII region and supernova remnant analyses. The program is underway at the CFHT. Once an empirical calibration is established, then a correction, if necessary, can be applied to the Cepheid distance moduli.

3) The fields in which the Cepheids are found are generally very crowded and photometry can be difficult. Fortunately, with the advent of CCD detectors, many stellar photometry packages have also become widely available which have been specifically written to deal with very crowded fields. The program DAOPHOT (Stetson 1987) has been used in this study.

The crowded nature of these fields also presents a practical limit to obtaining accurate infrared aperture photometry (where an aperture is typically 5 arcsec). Even with the infrared arrays now becoming available, the Cepheids are faint, and the sky is very bright, so that only measurements of very bright extragalactic Cepheids are currently feasible. *I*-band CCD photometry at present offers the best available compromise for simultaneously dealing with the reddening problem, the faint magnitudes, and the image crowding problems.

4) The effects of mass loss. At present these postulated effects remain largely a matter of conjecture.

5) It is assumed for each galaxy, that the Cepheid instability strip is both uniformly filled, *and* uniformly sampled.

6) Cepheids cross the instability strip several times during the course of their evolution, and because of the ensuing mass differences they will obey different PLC relations on each crossing. In practice, this difficulty may not be significant since Cepheids are believed to spend an order of magnitude more time in their second crossing compared to any of the others.

As discussed above, detector and software improvements in the last few years have made it possible to address many of the difficulties listed above, and increase the accuracy with which we can obtain distances to galaxies. Note that difficulties 1) through 4) are not restricted to Cepheids alone, but apply in varying degrees also to all other stellar distance indicators. However, the fact that Cepheids have been so well-studied makes it now possible to analyse and in some cases calibrate the possible systematic effects discussed above. No other bright stellar distance indicator has reached a comparable level of maturity.

b) CCD Photometry of Extragalactic Cepheids

It should be noted that all distance moduli presented in this work are relative moduli with respect to the LMC, which is assumed to be at a distance modulus of 18.5 mag.

IC 1613

IC 1613 is one of the faintest members of the Local Group (providing a low-luminosity tie-in for some secondary distance indicators), it is well-resolved, and appears to have very little either foreground or internal absorption. Historically, the determination the distance to IC 1613 has been of interest because the slope of its PL relation was found to be shallow in comparison to all other measured PL relations, which showed very good agreement (e.g., see Sandage 1971).

A study of the distance to IC 1613 has recently been completed by Freedman (1988), using new CCD photometry of Cepheids obtained at the CFHT. Independent apparent distance moduli obtained at *BVRI* wavelengths, in comparsion with *H*-band photometry from McAlary, Madore and Davis (1984) yield a reddening-corrected distance modulus of 24.3 mag for this galaxy. The slope of the PL relation for the CCD data agrees well with that observed for other Local Group galaxies, lending confidence to the applicability of the Cepheid PL relation as a distance indicator. A plot of the apparent distance modulus as a function of inverse wavelength is displayed in Figure 1.

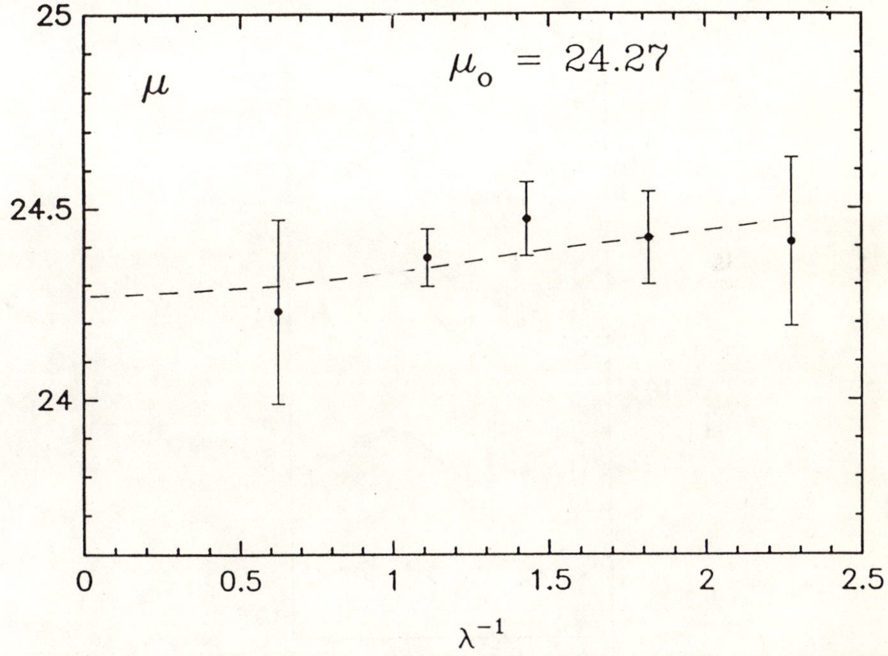

Figure 1. Apparent modulus versus λ^{-1} for IC 1613.
The dashed line is the extinction law from Rieke and Lebofsky (1985).

M33

A recalibration of the original Hubble (1926) photometry, and an independent study of 13 new Cepheids led Sandage (1983) and Sandage and Carlson (1983) to derive an apparent blue distance modulus of 25.3 mag for M33. From near-infrared photometry, Madore *et al.* (1985) concluded that the true distance modulus was actually much closer, with a value of 24.3 mag. Freedman (1985) presented a preliminary modulus based on multicolor CCD data, and Christian and Schommer (1987) have corrected the Hubble photometry using a CCD calibration, and both of these studies also find a smaller modulus in agreement with the infrared study. Mould (1987) finds a modulus of 24.7 mag based on single-epoch *I*-band Cepheid photometry.

In collaboration with C. Wilson and B. F. Madore, five years of observing at *BVRI* wavelengths are now available for several CCD fields in M33. The data can now be phased so that CCD light curves and mean magnitudes are available for several Cepheids. Light curves at four wavelengths for one of Hubble's Cepheids is shown in Figure 2. Although the study is not yet complete, our preliminary result is a true distance modulus of 24.5 mag, with a mean reddening value of $A_V = 0.4$. (Note that this value at present is based on average rather than phased magnitudes for most of the Cepheids).

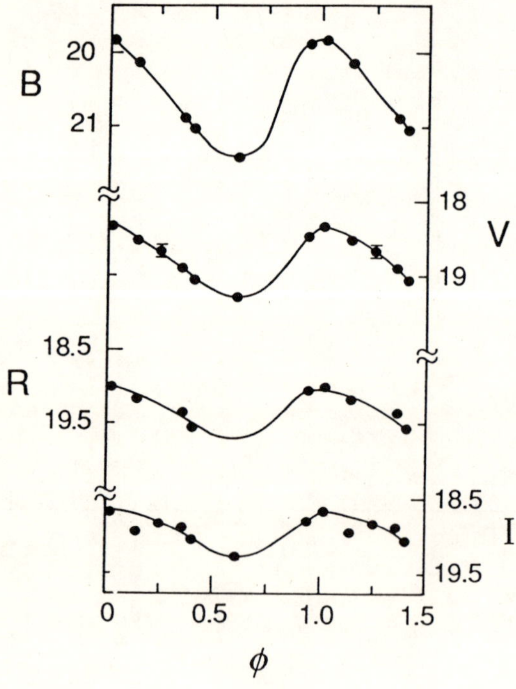

Figure 2. *BVRI* light curves for Cepheid V41 in M33.

NGC 2403 and M81

Tammann and Sandage (1968) discovered 17 Cepheids in NGC 2403, and determined a distance modulus of 27.56 mag (~3 Mpc) for this galaxy. The discovery of Cepheids in M81 proved to be much harder (see Sandage 1984); M81 and NGC 2403 until recently have usually been assumed to be members of the same group. However, Sandage (1984) suggested that M81 is in fact much more distant (μ_B = 28.8 mag, ~6 Mpc), (which significantly affects the calibration of red supergiants as distance indicators). New CCD data for Cepheids in these galaxies is discussed by Freedman and Madore elsewhere in this volume; the conclusion however is that M81 and NGC 2403 appear to be at comparable distances of about 3 Mpc.

III. THE SEARCH FOR CEPHEIDS IN OTHER GALAXIES

The number of galaxies with Cepheid distance moduli available to calibrate the determination of H_0 remains small. To improve this situation, a number of observers are now searching for Cepheids in more distant galaxies. The South Polar Group of galaxies in Sculptor is an ideal place for such a search and contains three galaxies which are suitable Tully-Fisher calibrators: NGC 247, NGC 7793, and NGC 253. Preliminary results of this CTIO 4m survey are described by Freedman *et al.* elsewhere in this volume.

At Palomar, a large consortium of observers has been monitoring M101 and NGC 4214 for variables over the past 3 observing seasons. NGC 4214 is one of the two calibrating galaxies adopted by Sandage and Tammann (1983) for Type I supernovae. (The other galaxy is IC 4182.) Unfortunately, bad weather has prevented much progress on this program. In the meantime, the distinction between various kinds Type I supernovae has now become clear, and the utility of finding Cepheids in NGC 4214 (whose supernova is classified by Branch (1986) as a Type I pec) is diminished. IC 4182, which is a true Type Ia supernova remains an important galaxy to obtain an independent Cepheid modulus for. This year we also included M81 in our search program.

IV. IMPLICATIONS FOR THE HUBBLE CONSTANT

After leaving the primary Cepheid calibrators, one is confronted immediately with the choice of a secondary indicator in the path to the determination of the Hubble constant. As the numerous pages in this volume can attest to, the choice of a secondary distance indicator is not clear cut, and strong opinions, both negative and positive about every indicator have been expressed. As with the choice of a primary indicator, it is desirable to have a well-observed, well-tested indicator, with a dispersion which is both well-understood, and also small. Furthermore, a physical (rather than purely empirical) basis for the distance indicator is desired. Few of the secondary distance indicators available meet all of these criteria very well.

Potential secondary indicators where Cepheid calibrators are useful include red supergiants, long-period variables, Type I supernovae, the Tully-Fisher relation, D_n-σ relation for bulges, novae, HII regions, and globular clusters. With perhaps the exception of the infrared Tully-Fisher relation (e.g., Aaronson *et al.* 1986), the empirical dispersion in the assumed magnitude for a standard candle, or in the relation used for the distance estimates are not yet well-understood or quantified. Supernovae hold promise as a distance indicator where the physical basis is understood; however ideally to quantify the accuracy of the method it is desirable to have an independent check in the same galaxy. The serendipitous nature of their discovery makes this very difficult in practice. Furthermore, supernovae are most easily observed in early-type galaxies, but most easily calibrated in late-type galaxies.

A further very serious problem is the small number of galaxies with reliable distances available to calibrate any of these methods. For example, for the infrared Tully-Fisher (IRTF) relation, the present number of available calibrators is four: M33, M31, NGC 2403, and M81. For the D_n-σ relation for bulges (Dressler 1987), only M31 and M81 are available. Adopting the Cepheid distance moduli given in this paper, and the infrared Cepheid modulus of Welch *et al.* (1986) for M31, the 11 galaxy clusters of Aaronson *et al.* (1986) yield a value for the Hubble constant of 91 km/sec/Mpc, adopting the same corrections to the cluster velocities as given by them. For the two calibrating galaxies available for the bulge D_n-σ relation, a value of H_0 = 67 km/sec/Mpc is obtained. Given that the dispersion in the IRTF relation is empirically established to be 0.45 - 0.5 mag, about 20 calibrating galaxies are required to obtain distances to 0.1 mag or 5% accuracy. Given that other errors are certainly present, this would appear to be a minimum acceptable number of calibrators needed to significantly improve the accuracy of the calibration. For the D_n-σ calibration, the method has only been applied to two clusters to date, and the dispersion is not nearly as well-determined, or the method as well investigated. And although still a matter of controversy for the IRTF, the questions of morphological type or environmental effects have not been addressed at all for Dressler's relation.

It is interesting also to note that by adopting distances to the calibrating galaxies that are higher than given in the present work (for example, those adopted by Sandage and Tammann (see Tammann 1987)) yields a value of 70 km/sec/Mpc when applied to the same IRTF cluster galaxies as discussed above. In other words, the discrepancy between a value of 50 and 90 is not resolved by a different choice of distances to the calibrating galaxies. The remaining difference lies both in the choice and in the treatment of the secondary distance indicator samples.

V. CONCLUSIONS

With the availability of CCD detectors, and their application to the determination of distances to nearby galaxies based on the use of Cepheid variables, accurate, new distances which are corrected for the effects of obscuration are being obtained for all galaxies with known Cepheids. Several programs are now underway to increase the number of calibrating galaxies for the determination of the Hubble constant. However, the numbers of further potential calibrating galaxies for which Cepheids can be discovered from the ground is small, and Space Telescope will be required to continue the effort.

A major source of error in applying the Cepheid zero point calibration for H_0 is the small number of available calibrating galaxies and the consequently poor $\sqrt{N}$ statistics. The zero point of the Cepheid PL relation itself is also a remaining source of uncertainty. For the secondary distance indicators, the accuracy with which the velocity field has been mapped, and present understanding of the scatter in the various relations contribute to the uncertainty with which the Hubble constant can be determined at the present time.

ACKNOWLEDGEMENTS

This work was partially supported by NSF grant AST 87-13889.

REFERENCES

Aaronson, M. 1986, in *Galaxy Distances and Deviations from Universal Expansion*, eds. B. F. Madore and R. B. Tully, (Dordrecht:Reidel), p. 55.

Aaronson, M., Bothun, G., Mould, J., Huchra, J., Schommer, R. A., and Cornell, M. E. 1986, *Ap. J.*, **302**, 536.

Baade, W., and Swope, H. H. 1961, *A. J.*, **66**, 300.

Baade, W., and Swope, H. H. 1963, *A. J.*, **68**, 435.

Baade, W., and Swope, H. H. 1965, *A. J.*, **70**, 212.

Becker, S. A., Iben, I., and Tuggle, R. S. 1977, *Ap. J.*, **218**, 633.

Branch, D. 1986, *Ap. J. (Lett.)* , **300**, L51.

Christian, C. A., and Schommer, R. A., 1987, *A. J.*, **93**, 557.

Dressler, A. 1987, *Ap. J.* , **317**, 1.

Freedman, W. L. 1985, in *IAU Colloquium No. 82, Cepheids: Theory and Observations*, ed. B.F. Madore, (Cambridge University Press), p. 225.

Freedman, W. L. 1988, *Ap. J.* , **326**, 691.

Hubble, E. 1925, *Ap. J.*, **62**, 409.

Hubble, E. 1926, *Ap. J.*, **63**, 236.

Hubble, E. 1929, *Ap. J.*, **69**, 103.

Iben, I., and Tuggle, R. S. 1975, *Ap. J.*, **197**, 39.

Iben, I., and Renzini, A. 1984, *Phys. Rep.*, **105**, 329.

Kraan-Korteweg, R. C., Cameron, L. M., and Tammann, G. A. 1988, *Ap. J.*, in press.

Madore, B. F. 1986, in *Galaxy Distances and Deviations from Universal Expansion*, eds. B. F. Madore and R. B. Tully, (Dordrecht:Reidel), p. 29.

Madore, B. F., McAlary, C. W., McLaren, R. A, Welch, D. L., Neugebauer, G., and Matthews, K. 1985, *Ap. J.*, **294**, 560.

McAlary, C. W., Madore, B. F., and Davis, L. E. 1984, *Ap. J.*, **276**, 487.

Mould, J. 1987, *P. A. S. P.*, **99**, 1127.

Rieke, G. H., and Lebofsky, M. J. 1985, *Ap. J.*, **288**, 618.

Sandage, A. R. 1971, *Ap. J.*, **166**, 13.

Sandage, A. R. 1972, *Q. J. R. A. Soc.*, **13**, 202.

Sandage, A. R. 1983, *A. J.*, **88**, 1108.

Sandage, A. R. 1984, *A. J.*, **89**, 621.

Sandage, A. R. 1988, preprint.

Sandage, A. R., and Carlson, G. 1983, *Ap. J. Letts.*, **267**, L25.

Sandage, A. R., and Tammann, G. A. 1968, *Ap. J.*, **151**, 531.

Sandage, A. R., and Tammann, G. A. 1983, in *Astrophysical Cosmology, Pontif. Acad. Sci. Scripta Varia*, (Vatican), p. 23.

Stetson, P. 1987, *P. A. S. P.*, **99**, 191.

Stothers, R. B. 1988, *Ap. J.*, **329**, 712.

Tammann, G. A. 1987, in *Observational Cosmology*, eds. A. Hewitt *et al.*, (Dordrecht:Reidel), p. 151.

Tammann, G. A., and Sandage, A. R. 1968, *Ap. J.*, **151**, 825.

Welch, D. L., McAlary, C. W., McLaren, R. A, and Madore, B. F. 1986, *Ap. J.*, **305**, 583.

REVIEW OF THE LOCAL GROUP DISTANCE SCALE

JEREMY MOULD
Palomar Observatory, California Institute of Technology
Pasadena, California 91125

1. INTRODUCTION

If our motivation is to determine the cosmic distance scale, rather than any intrinsic interest which these galaxies hold for us, there are two reasons to want to know the distances of galaxies in the Local Group.

The first is to calibrate distance indicators which cannot be calibrated in the Galaxy. The second is to determine the minimum uncertainty to be attached to distance estimates outside the Local Group. I will not discuss in any detail whether these uncertainties are random or systematic.

The list of potentially reliable distance indicators, which cannot be optimally calibrated using only measurements within the Milky Way includes cepheids, red supergiants, and planetary nebulae. I am uncertain whether to include novae (see Cohen, this volume).

For cepheid variables there are two separate methods of attempting a Galactic calibration. These are discussed in detail by Walker (this volume). Here it is sufficient to point out the qualitative difficulties, first with the cluster P-L relation: extinction in the Galactic Plane, cluster membership, and small sample statistics. The promise and difficulties of Baade-Wesselink determination of cepheid distances are examined in full by Gautschy (1987).

2. THE MAGELLANIC CLOUDS

Alternatively, we can use the LMC as the source of the basic cepheid P-L relation, and determine its distance by means of other reliable indicators. Table 1 compares the cepheid distance to the Magellanic Clouds using Feast and Walker's (1987) Galactic calibration with three other indicators.

As Feast (this volume) has just recalled, improved observations of Magellanic Cloud RR Lyrae Stars have recently been reported by Walker and Mack. In a more recent paper, however, Jones, Carney and Latham (1988) have swung the balance in the absolute magnitude of RR Lyrae stars in favor of $M_V = 0.85$ from $M_V = 0.6$. This result is supported by accurate Baade Wesselink analysis of five stars, and recent statistical parallaxes (but not by globular cluster main sequence fitting). There is no evidence for a metallicity dependence of M_V. In this review we adopt $M_V = 0.8 \pm 0.2$. It is worth mentioning in passing that in Baade Wesselink calculations for RR Lyrae stars one is dealing with violent

Table 1. Comparative Distance Moduli of the Magellanic Clouds

Indicators	LMC	SMC	Reference
Cepheids	18.47 ± 0.15	18.78 ± 0.15	Feast & Walker (1987)
RR Lyrae $M_V = 0.8$	18.24 ± 0.2	18.66 ± 0.2	Walker & Mack (1988a,b)
GB tip $M_{\rm bol} = -3.3$*	18.22 ± 0.2	nondetection	Reid & Mould (1987,1989)
MS	18.2 ± 0.5	...	Schommer, *et al.* (1984), Walker (1987), Mateo & Hodge (1987)

* Equation (1) has been corrected to $M_V(\mathrm{RR}) = 0.8$.

accelerations in giant atmospheres ($dv/dt \sim g \sim 10^3$ cm s^{-2}). In extragalactic cepheids, the accelerations are smaller, but the static atmospheres are less well understood ($dv/dt \sim 10; g \sim 10^2$).

The tip of the first giant branch is a useful indicator, stretching perhaps to 10 Mpc with HST. The important caveat is that the RGB tip should be observed at a wavelength where the bolometric correction is not a strong function of temperature, ideally 1–2μ. The best available calibration at present is

$$M_{\rm bol} = -3.82 - 0.26\ [\mathrm{Fe/H}] \tag{1}$$

(Frogel, Persson, and Cohen 1983), which is tied to M_V (RR Lyrae) in the range (0.6, 0.8). Work by Reid and Mould (1987) and Mould and Kristian (1986) probably needs to be extended to a longer wavelength to ensure consistency with this calibration.

Finally, we come to the technique of main sequence fitting to star clusters in the Magellanic Clouds, which has been the driving force behind the reduction in the LMC distance modulus in the last five years. I want to stress the metallicity sensitivity of this technique by showing an isochrone fit (Green, Demarque and King 1987) to probable members of the Hyades cluster (Hanson 1975). Fits with $m - M = 3.29$ and [Fe/H] = 0.1, 0.2, and 0.3 are shown in Figure 1. In practice it may be preferable to fit to the data an observed main sequence such as the Hyades itself for intermediate age clusters, or better still the Pleiades for young clusters (Walker 1987a). But the metallicity difference between the program and the standard cluster must be carefully taken into account. Of course, this has not been ignored by the investigators. But here I want to stress the uncertainty in the available metallicity estimates for the program clusters. These estimates are not much better than guesses. In the mean these may be correct; nevertheless I have attached a large uncertainty to the LMC distance estimate for this reason. For young clusters, the color of the core helium burning loop is not a reliable metallicity indicator (Renzini and Buzzoni 1985). Direct spectroscopic analysis of K giant cluster members should be pursued. Deeper photometry would also help greatly.

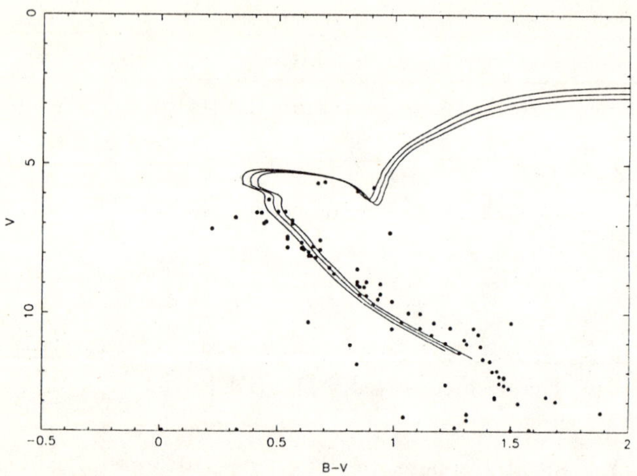

Figure 1. Color-magnitude diagram of the Hyades. Isochrones are shown for 1.5 Gyr and Y = 0.3 (see text).

3. MAGELLANIC AND OTHER DWARF IRREGULARS

Six dwarf galaxies in the Local Group now have cepheid distance moduli. These have been summarized and put on a scale consistent with the LMC modulus in Table 1 by Walker (1987b). Table 2 reproduces these values, indicates the oxygen abundance for the system (Vigroux, *et al.* 1987), and gives the original reference.

Table 2. Cepheid Distance Moduli for Local Group Dwarf Irregulars

Galaxy	$(m-M)_0$	O/H×10^4	Source*
IC 1613	24.1	0.4	McAlary, *et al.* (1984)
	24.3 ± 0.1		Freedman (1988a)
NGC 6822	23.4 ± 0.2	1.7	McAlary, *et al.* (1983)
Sex A and B	25.5 ± 0.1	0.2	Sandage & Carlson (1985b)
Sex A	26.0 ± 0.4		Walker (1987c)
WLM	24.8 ± 0.1		Sandage & Carlson (1985a)
NGC 3109	26.1 ± 0.1		Demers, *et al.* (1985)

* These are the original references. The tabulated values are taken from Walker (1987b).

The galaxies in Table 2 have a special significance in constraining the metallicity dependence of the cepheid P-L relation. Stothers' (1988) recent theoretical study concludes $\partial M_{\rm bol}/\partial Z\ _P = -1.8$ which at the present level of precision can be assumed to be zero, when the cepheid P-L relation is applied. For RR Lyrae stars, however, we already have an empirical constraint on dM_V/dZ. So a comparison of cepheid and RR Lyrae distance moduli for galaxies more metal poor than the LMC can potentially set useful limits on cepheid metallicity dependence. Of course, the RR Lyrae stars must be detected first. This is a feasible proposition for at least three of these galaxies with ground based observations.

In this context it is worth noting that fitting the RGB tip to IC 1613 which has [O/H] = −1.2 yields $(m-M)_0 = 24.2 \pm 0.2$ (Freedman 1988b). One should also note in passing that the distance of NGC 3109 is uncertain with contrasting values of 25.3 ± 0.1 (Elias and Frogel 1985) from the brightest red supergiant and 25.7 from the Tully Fisher relation (Bottinelli, *et al.* 1984). These estimates of distance modulus should be given low weight because of the Scott effect in red supergiants and the difficulty of applying the Tully Fisher relation to dwarf galaxies. Sandage and Carlson, on the other hand, presented at this meeting an estimate from 29 cepheids in NGC 3109 of $(m-M)_B = 26.4$.

These are not all the Magellanic and Irregular galaxies in the Local Group. If we define the Local Group as the set of galaxies which fall within the turnaround radius of the group, then for a reasonable ($\sim 10^{12} M_\odot$) group mass one finds a 2 Mpc radius for membership (Sandage 1986a). Table 3 lists other sometime Local Group candidates, whose modulus estimates should be less than 26.5 for membership by this criterion.

Table 3. Other Possible Dwarf Irregular Members

Galaxy	$(m-M)_0$	O/H$\times 10^4$	Source
IC 10	26.5:	1.5	de Vaucouleurs (1978)
	26.5		Bottinelli, *et al.* (1984)
IC 5152	26	2.0	Sandage (1986a)
	27.4		Bottinelli, *et al.* (1984)
Leo A	26		Sandage (1986b)
Pegasus	27		Sandage (1986b)
	26.1		Hoessel & Mould (1982)
GR 8	25.3	3.8	de Vaucouleurs & Moss (1983)
Phoenix	24.7		Caldwell, Schommer, & Graham (1988), Ortolani and Gratton (1988)
Sag DIG	25.4		Cook & Aaronson (1988)

Note that no attempt has been made to homogenize the distance estimates in Table 3 in respect of reddening models and zeropoints. In addition, there exist further possible dwarf irregular Local Group candidates without distance estimates (beyond their small redshifts). These are DDO 187, 210, and LGS 3 (e.g. Kraan Korteweg and Tammann 1979).

4. M 31 AND M 33

In addition to supplying consistency checks on stellar distance indicators, these galaxies are direct calibrators of the Tully Fisher relation. Table 4 contains the most recent estimates of the distances of M 31 and M 33. For M 31 Walker (1987b) has updated earlier cepheid data to a standard P-L relation and $A_V = 0.33$. This should not be considered inconsistent with $A_B = 0.31$ for halo RR Lyrae stars adopted by Pritchet and van den Bergh (1988), because of the intrinsic reddening of cepheids (e.g., Freedman, this volume).

Table 4. Distance Measurements for M 31 and M 33

Indicator	M 31	Source	M 33	Source
Cepheids B,V	24.2	Baade & Swope (1963)*	24.5 $\pm$0.1	Christian & Schommer (1987)*
Cepheids R,I			24.5 $\pm$0.1	Mould (1987)*
Cepheids H	24.3$\pm$0.1	Welch, *et al.* (1986)*	24.2 $\pm$0.1	Madore, *et al.* (1985)
LPVs K			24.6 $\pm$0.1	Mould (1987)
GB tip	24.4$\pm$0.3	Mould & Kristian (1986)	24.6 $\pm$0.3	MK with $M_V = 0.8$
RR Lyr	24.2$\pm$0.2	Pritchet & van den Bergh (1987,8)	24.45$\pm$0.2	
Novae	24.0$\pm$0.2	Cohen (1985)		

* Using P-L relation and $A_V = 0.33$ (Walker 1987b) for M 31 and $A_V = 0.3$ (Feast 1988) for M 33.

In M 33 $A_V = 0.3$ has been adopted for cepheids in Table 4. The wavelength sequence of $(m - M)$ suggest that a higher value might be appropriate. But the P-L relation for supergiant LPVs in M 33 (Mould 1987) is in opposition to this trend. Application of this form of the LPV distance indicator (Bessell and Wood 1984) without bias requires that supergiant and asymptotic giant branch LPVs should be separable. Observations are continuing at Palomar to secure better measurements of <K> for this purpose. The large number of cepheids found by Kinman, Mould and Wood (1987) in the course of the LPV survey also deserve attention in the infrared.

From Table 4 the mean value of $(m - M)_0$ for M 31 is 24.2. The mean value for M 33 is 24.5. The consistency of these partially separate distance indicators is better than 10% in distance (r.m.s.). There are some obvious gaps in Table 4 to be filled, and there is a need for a better study of the reddening problem.

5. DWARF ELLIPTICALS AND DWARF SPHEROIDALS

The distances of most of the early type dwarf companions of the Milky Way are known from RR Lyrae stars. They are given in Table 5 for $M_V = 0.8$. Two exceptions are the Leo galaxies for which only giant branch fits are available.

Table 5. Distance Moduli of Dwarf Spheroidals

Galaxy	$(m-M)_0$	Source
Fornax	20.4	Buonnano, *et al.* (1985)
U Minor	19.0	Nemec, *et al.* (1988)
Draco	19.2	Stetson(1979)
Sculptor	19.3	Kunkel and Demers (1977)
Leo I	21.8 ± 0.2	Fox and Pritchet (1987)
Leo II	21.65 ± 0.2	Demers and Harris (1983)
Carina	20.0	Saha, *et al.* (1986)

Notes: Where no error bar is given, the uncertainty is 0.1 plus the uncertainty in $M_V(\text{HB}) = 0.8$.

One of the companions to M 31, NGC 147, now has a distance from its RR Lyrae stars 23.85 ± 0.2 (Saha and Hoessel 1987). This agrees with the result from the RGB tip (23.8 ± 0.15, Mould, *et al.* 1983 corrected to $M_V(\text{RR}) = 0.8$), and suggests that this dwarf is on the near side of (and 6° away from) M 31. For the more tightly bound companion NGC 205 Mould, *et al.* (1984 corrected to $M_V(\text{RR}) = 0.8$) found $(m-M)_0 = 24.2 \pm 0.2$, but no RR Lyrae stars have been detected yet. There is very little data on the dwarf spheroidal companions of M 31: And I, II, and III, or the dE galaxy NGC 185.

Finally, it should be noted that a sometime Local Group candidate NGC 404 is placed well outside the group by de Vaucouleurs and Olson (1984), who find $m - M \sim 28$.

Partial support by NSF grants 85-02518 and 87-21705 is gratefully acknowledged. I would like to thank the Aspen Center for Physics for their hospitality during the preparation of this review, my colleagues for preprints, and Helen Knudsen for running a computerized literature search on the Local Group.

REFERENCES

Baade, W., and Swope, H. 1963, *Astron. J.*, **68**, 435.

Bessell, M. and Wood, P. 1984, *Publ. Astron. Soc. Pacific*, **96**, 247.

Bottinelli, L., Gougenheim, L., Paturel, G., and de Vaucouleurs, G. 1984, *Astron. Astrophys. Suppl.*, **56**, 381.

Buonnano, R., Corsi, C., Fusi Pecci, F., Hardy, E., and Zinn, R. 1985, *Astron. Astrophys.*, **152**, 65.

Caldwell, N., Schommer, R. and Graham, J. 1988, this meeting.

Christian, C. and Schommer, R. 1987, *Astron. J.*, **93**, 557.
Cohen, J. 1985, *Astrophys. J.*, **292**, 90.
Cook, K. and Aaronson, A. 1988, this meeting.
de Vaucouleurs, G. 1978, *Astrophys. J.*, **224**, 710.
de Vaucouleurs, G., and Moss, C. 1983, *Astrophys. J.*, **271**, 123.
de Vaucouleurs, G., and Olson, D. 1984, *Astrophys. J. Suppl.*, **56**, 91.
Demers, S., and Harris, W. 1983, *Astron. J.*, **88**, 329.
Demers, S., Kunkel, W., Irwin, M. 1985, *Astron. J.*, **90**, 1967.
Elias, J., and Frogel, J. 1985, *Astrophys. J.*, **289**, 141.
Feast, M. 1988, *Observatory*, in press.
Feast, M., and Walker, A. 1987, *Ann. Rev. Astron. Astrophys.*, **25**, 345.
Fox, M., and Pritchet, C. 1987, *Astron. J.*, **93**, 1381.
Freedman, W. 1988a, *Astrophys. J.*, **326**, 691.
Freedman, W. 1988b, preprint.
Frogel, J., Persson, S. E., and Cohen, J. 1983, *Astrophys. J.*, **275**, 773.
Gautschy 1987, *Vistas in Astronomy* 30, 197.
Green, E., Demarque, P., and King, C. 1987, *The Revised Yale Isochromes and Luminosity Functions* (New Haven: Yale).
Hanson, R. 1975, *Astron. J.*, **80**, 379.
Hoessel, J., and Mould, J. 1982, *Astrophys. J.*, **254**, 38.
Jones, R., Carney, B., and Latham, D. 1988, preprint.
Kinman, T., Mould, J., and Wood, P. 1987, *Astron. J.*, **93**, 833.
Kraan Korteweg, R., and Tammann, G., 1979, *Astron. Nachr.* 300,181.
Kunkel, W., and Demers, S. 1977, *Astrophys. J.*, **214**, 21.
Mateo, M., and Hodge, P. 1987, *Astrophys. J. Suppl.*, **60**, 893.
McAlary, C., Madore, B., Davis, L. 1984, *Astrophys. J.*, **276**, 487.
McAlary, C., Madore, B., McGonegal, R., McLaren, R., and Welch, D. 1983, *Astrophys. J.*, **273**, 539.
Mould, J. 1987, *Publ. Astron. Soc. Pacific*, **99**, 1127.
Mould, J., DaCosta, G., and Kristian, J. 1983, *Astrophys. J.*, **270**, 471.
Mould, J., DaCosta, G., and Kristian, J. 1984, *Astrophys. J.*, **278**, 575.
Mould, J., and Kristian, J. 1986, *Astrophys. J.*, **305**, 591.
Nemec, J., Wehlau, A. and Mendes de Oliveira, C. 1988, *Astron. J.*, **95**, 000.
Ortolani, S., and Gratton, R. 1988, *Publ. Astron. Soc. Pacific*, in press.
Pritchet, C., and van den Bergh, S. 1987, *Astrophys. J.*, **316**, 517.
Pritchet, C., and van den Bergh, S. 1988, preprint.
Reid, I. N., and Mould, J. 1987, *Astrophys. J.*, **323**, 433.
Reid, I. N., 1989, in preparation.
Renzini, A., and Buzzoni, A. 1985, in *"Spectral Evolution of Galaxies"*, C. Chiosi and A. Renzini eds., (Dordrecht: Reidel), p. 195.
Saha, A., and Hoessel, J. 1987, *Astron. J.*, **94**, 1556.
Saha, A., Monet, D., and Seitzer, P. 1986 *Astron. J.*, **92**, 302.
Sandage, A., and Carlson, G. 1985a, *Astron. J.*, **90**, 1464.
Sandage, A., and Carlson, G. 1985b, *Astron. J.*, **90**, 1019.
Sandage, A. 1986a, *Astrophys. J.*, **307**, 1.
Sandage, A. 1986b, *Astron. J.*, **91**, 496.
Schommer, R., Olszewski, E. and Aaronson, M. 1984, *Astrophys. J. (Letters)*, **285**, L53.
Stetson, P. 1979, *Astron. J.*, **84**, 1149.
Stothers, R. 1988, *Astrophys. J.*, **329**, 212.
Vigroux, L., Stasinska, G., and Comte, G. 1987, *Astron. Astrophys.*, **172**, 15.

Walker, A. 1987a, *Monthly Notices Roy. Astron. Soc.*, **225**, 627.
Walker, A. 1987b, *SAAO Circulars*, **11**, 125.
Walker, A. 1987c, *Monthly Notices Roy. Astron. Soc.*, **224**, 935.
Walker, A., and Mack, 1988a, preprint.
Walker, A., and Mack, 1988b, preprint.
Welch, D., McAlary, C., McLaren, R., and Madore, B. 1986, *Astrophys. J.*, **305**, 583.

DISCUSSION

N. CALDWELL: At the same time that Schommer and I were searching for Cepheid variables in the Phoenix galaxy and IC 5152, we also looked for such stars in DDO 210, which you noted as a candidate for such searches. We have yet to find any variables with an amplitude greater than 0.3 magnitude down to an apparent magnitude of about V = 23.

J. NEMEC: Nemec, Wehlau and Oliveria (1988, Astron. J. in press) find that the best studied anomalous Cepheids in dwarf galaxies define two P-L relationships: one for fundamental pulsators, and one for first-overtone pulsators. The anomalous Cepheids in the SMC, Carina and Leo I are seen to approximately follow these relations. It is hoped that these P-L relations can be used to estimate distances to other relatively nearby dwarf galaxies. The method should give distance moduli accurate to ± 0.10 mag.

D. TERNDRUP: Have you calculated the distances to members of the Local Group using any of the proposed $\langle M_V \rangle$, [M/H] relations?

J. MOULD: An interesting test of this idea is to look at $\delta(m-M)_0$ for M31 relative to the LMC. For Cepheids this is 5.8 mag. For RR Lyraes with $\langle M_V \rangle = 0.8$ it is 6.0. With Feast's relation B, which has half the Sandage effect in its coefficient of [M/H], one obtains approximately 5.7 mag, depending on exactly what one adopts for the metallicity of M31's RR Lyraes.

F. FUSI PECCI: It seems to me that the problem of checking the existence of the so-called "Sandage Period Shift Effect" is still open. Moreover, the connections between $\Delta \log P/\Delta$[Fe/H] and ΔM_V(RR)/Δ[Fe/H] or (even worse) ΔM_V(HB)/Δ[Fe/H] may not be so straightforward as it is frequently assumed to be. In fact, it is true as you say, that there is a growing body of observational results (mainly coming from "Baade-Wesselink" measures on field variables) that confirm the non-existence of this effect as indicated by "standard" theoretical models - however, there is also observational evidence suggesting ΔM_V(HB)/Δ[Fe/H] $>$ 0.3. A detailed discussion of this problem is presented in a paper we have submitted to Astron. Astrophys. (Buonanno, Carsi, Fusi Pecci).

F. FUSI PECCI: (continued)

For instance:

1. The results of (CCD) MS-fitting for a wide sample of globular clusters lead to $\Delta M_V(HB)/\Delta[Fe/H] = 0.37 \pm 0.14$ using the 19 clusters in the sample of Buonanno et al., and to 0.33 using the smaller set presented by Richer and Fahlman.

2. A comparison of the data on RR Lyrae stars in M15 and M3 cannot yet be explained even adopting the scenario proposed by Lee et al. (1987). The synthetic procedure they use does not give a good match for M15; the models yield periods that are too short for the hottest RR Lyraes; they have to assume too wide a width for the instability strip; if the period shift is measured at log $T_{eff} = 3.85$ (as is usually done from the models) the derived period shift with respect to the mean locus of M3 decreases to about zero as expected from the HB models but contrary to observations.

PLANETARY NEBULAE AS DISTANCE INDICATORS

GEORGE H. JACOBY and ROBIN CIARDULLO
Kitt Peak National Observatory, National Optical Astronomy Observatories, P.O. Box 26732, Tucson, Arizona 85726–6732

HOLLAND C. FORD
Department of Astronomy, University of Michigan, Ann Arbor, MI 48109-1090

ABSTRACT The [O III] λ5007 fluxes of the brightest planetary nebulae (PN) in the Local Group galaxies are constant to within 20%, suggesting that an upper limit exists for PN luminosities. While this limit may be valuable as a standard candle, we show that the luminosity function for PN is universal and has a sharp bright-end cutoff near $M_{5007} = -4.48$ which provides an even better estimator of distance. The cutoff is a direct result of the small range in masses ($0.61 \pm 0.02 M_{\odot}$) of PN central stars (to the high side of the distribution) in combination with the extremely rapid evolutionary time scales for the more massive stars.

Using the technique of maximum likelihood, the luminosity functions for PN in M81 (Sb), NGC 5128 (E2), the NGC 3379 (E0) group, including NGC 3384 (SB0) and NGC 3377 (E6), are used to derive distances to these and various other galaxies as distant as Virgo. Distances to galaxies in a single group (NGC 3379) having an assortment of Hubble types, and different fields within a single galaxy (NGC 5128) agree to better than 5%, consistent with the formal errors derived using statistical tests.

We propose that PN are excellent distance indicators for galaxies of all Hubble types. Additional advantages include 1) observations are required at only one epoch, 2) they can be observed to distances beyond 15 Mpc with currently available ground based telescopes, 3) the analysis is straightforward in uncrowded fields as a result of continuum suppression by narrow band filters, 4) they reside in dust-free regions of galaxies, and 5) they are relatively insensitive to metallicity variations.

INTRODUCTION

The first identifications of extragalactic planetary nebulae (PN) were reported in the Small Magellanic Cloud by Lindsay (1955) and in M31 by Baade (1955). PN in the Large Magellanic Cloud were later identified by Westerlund and Rodgers (1959). With improved instrumentation, Ford, Jenner, and Epps (1973) identified PN in M32, NGC 205, and NGC 185, marking the start of a productive series of surveys for PN in nearby galaxies (see Ford 1983; Ford *et al.* 1988). Today, the number of PN found in over 30 external galaxies easily exceeds the number of known Galactic PN.

The scientific value of these identifications can be found in abundance and kinematic studies of the older populations of the parent galaxies, and in assessing the rate at which material is returned to the interstellar medium. Additionally, it was suspected by Hodge (1966, 1981) that PN could be used in some circumstances as distance indicators, but it was not until Ford and Jenner (1978) compared the luminosities of PN in M81 with those in M31 that an extragalactic distance was derived from the brightest PN. Shortly thereafter Jacoby and Lesser (1981) derived limits to the distances for a selection of nearby dwarf galaxies, and Lawrie and Graham (1983) derived a distance to NGC 300 using PN.

OBSERVATIONAL FOUNDATION

Ford and Jenner (1979) and Jacoby and Lesser (1981) noted that the brightest PN in Local Group galaxies have comparable luminosities. Improved observational methods using CCD detectors, and recent distance determinations to the Local Group members allowed Ciardullo *et al.* (1988, hereafter Paper II) to re-evaluate the dispersion of the brightest PN in Local Group galaxies. Table I summarizes their findings. We see that the average equivalent M_{5007} ($= -2.5\log(F_{5007}) - 13.74$) for the luminosity of the [O III] λ5007 emission-line is -4.16 with a standard deviation of 0.20 mag. The true dispersion of the PN luminosities must be somewhat smaller due to uncertainties in the distances to the galaxies, uncertainties in the extinction, and with the exception of M31, small number statistics. Furthermore, the effects of metallicity on PN luminosity must not be extreme, since these galaxies span a considerable range. Dependency on the size of the parent population must also be weak, again based on the vast range of luminosity for the galaxies involved.

In order for the brightest PN to have similar luminosities in such a diverse group of galaxies, there must exist a natural cutoff in the planetary nebula luminosity function (PNLF). When we examine the PNLF for several galaxies having a rich PN population, we find that the bright end of the function exhibits a characteristic sharp drop over 0.5 mag. This behavior is evident in the 7 galaxies we have surveyed (see examples in Figure 1), and is consistent with the observed PN populations in the Local Group dwarf galaxies. The universality of the PNLF combined with the rapid bright end cutoff suggests that it provides a good standard candle for distance determination.

TABLE I. Brightest Local Group PN

Galaxy	E(B-V)	Distance (kpc)	M_{5007}
M31	0.11	710	-4.36
LMC	0.03	50	-4.09
SMC	0.20	60	-4.42
M32	0.11	710	-3.94
NGC 205	0.06	760	-3.95
NGC 185	0.17	570	-4.04

THEORETICAL BASIS

Paczynski (1971) modeled the evolutionary history of the central stars of PN and found that the more massive central stars have extremely short lifetimes during their horizontal crossing (luminous period) in the (Log L, Log T_e) plane. For example, a 0.6 $M_\odot$ central star has a high luminosity lifetime of 18,000 years, while for a 1.2 $M_\odot$ central star, this period is only 5 years. Clearly, it will be a rare event to see a bright PN having a massive central star. The rapid evolution of central stars is one key to the bright end cutoff of the PNLF.

To investigate the PNLF behavior in detail, Jacoby (1988, hereafter Paper I) simulated the observation of an ensemble of PN at various stages in their evolutionary history. The model developed by Aller and Keyes (1980), in which the ionization and energy structure of a nebula is carefully detailed, was used to predict the absolute fluxes in the important emission lines. The models were excited by central stars following the luminosity and temperature history predicted by Wood and Faulkner (1986). The resultant values for [O III] λ5007 luminosity as a function of central star mass and age provided a two-dimensional grid from which simulated PN may be selected randomly in age, according to some function in central star mass, G(m). Monte Carlo simulations of observing many PN from the grid generate simulated PNLFs.

If G(m) includes a large fraction of central stars having masses greater than about 0.75 $M_\odot$, then we expect to see PN having [O III] fluxes greater than are observed in the Local Group galaxies, despite their more rapid evolutionary rates. In Paper I, it was assumed that G(m) follows a Gaussian distribution, based on published distributions for white dwarf masses (see for example, Koester, Schultz, and Weidemann 1979). The observed PNLFs can be reproduced only if G(m) describes a narrow function in mass (to the high side of the Gaussian distribution), having $\sigma \approx 0.02 M_\odot$. The most probable mass (the peak of the Gaussian) was found to be 0.61 $M_\odot$, in good agreement with white dwarf estimates.

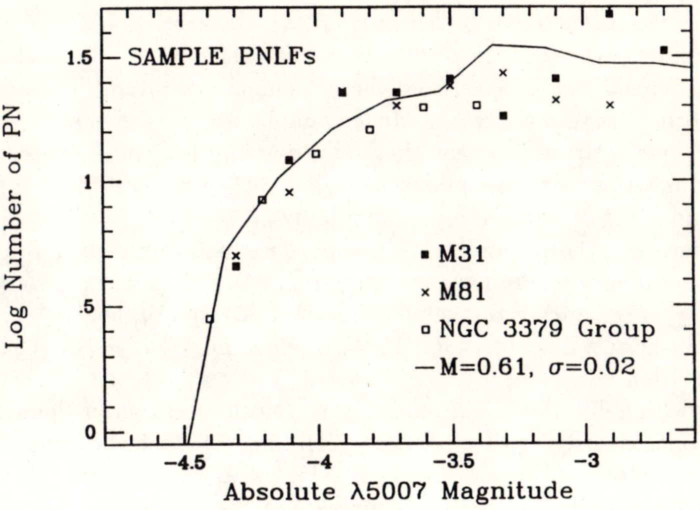

Fig. 1 The PNLFs for M31, M81, and the combination of NGC 3377, NGC 3379, and NGC 3384. The line represents the PNLF derived from a theoretical model from Paper I assuming PN central stars have a mass distribution peak at 0.61 and $\sigma = 0.02 M_\odot$.

Should the assumed central star models underestimate the high luminosity lifetime (Mendez *et al.* 1988) of high mass central stars, then either σ must be reduced, or the entire distribution shifted to lower masses. The opposite is also true; more rapid evolution can be accommodated by allowing a higher percentage of high mass central stars. High mass central stars may also be present in larger numbers without exceeding the maximum observed [O III] luminosities since they derive from high mass progenitors which will dredge up additional coolants (Peimbert and Torres-Peimbert 1983) to compete for central star UV flux, thereby moderating the λ5007 flux. In addition, the nebular model we use does not consider the possibility of non-equilibrium conditions in the ionization structure (e.g. an expanding ionization front due to a high mass central star which is rapidly crossing the (Log L, Log T_e) plane), nor does it include any terms for mechanical dissipation of energy due to expansion of the nebula. Both these effects extract energy which would otherwise heat the nebula and further excite the [O III] line. Thus there exist additional processes which may govern the upper luminosity of PN.

METALLICITY

We see from Table I that metallicity variations cannot have a large effect on the maximum [O III] luminosity. Experiments with the nebular models show that a decrease in the abundances by a factor of 2 (i.e., selecting LMC rather than galactic abundances) results in a 40% reduction in the [O III] flux. The nearly square root dependence derives from the collisional excitation of the emission line of interest. Because a low abundance nebula will be deficient in coolants, the electron temperature will rise, thereby increasing the collisional excitation rate. Although there are fewer ions to radiate, they radiate more often and moderate the effect of reduced ionic abundances.

Furthermore, the central star will suffer less metal line opacity in the far UV as a consequence of the lower abundances, and will radiate more ionizing photons, further enhancing the nebular heating. Although, this is a very small (5%) compensation (Husfeld *et al.* 1984), it does help buffer the [O III] flux from composition variations.

Observationally, there is a tendency to select the brighter, higher abundance PN when identifying them in the [O III] line. This selection effect further moderates abundance variations between galaxies, as the more metal rich population is preferentially selected to participate in the PNLF.

DESCRIPTION OF THE METHOD

Deriving distances to galaxies using PN requires 3 distinct steps: identification, photometry, and statistical analysis of the PNLF.

Identification Technique

The technique for identifications is essentially the same on-band/off-band procedure used by Ford, Jenner, and Epps (1973) in which narrow band images are taken through a filter isolating an important emission line and compared with images taken through a filter suppressing all important emission lines. As a general rule, PN attain their greatest luminosities in the [O III] λ5007 line, and so this line is a natural choice for extragalactic identifications.

If the galaxy of interest has a young population and the presence of HII regions is a possibility, we also obtain an image taken through an Hα filter to provide discrimination between PN and the low excitation compact HII regions. Thus the identification process consists of registering and blinking the on-band frame against the off-band frame, and optionally the Hα frame. Any object appearing in the off-band frame is not considered a PN, and any object which has an Hα intensity greater than about one-third the λ5007 intensity is also rejected.

Several factors must be considered in the filter selection. Because the depth of the survey is a strong function of the on-band filter width, we strive to keep this narrow to suppress the background galaxy light which dominates the noise component. The intrinsic widths of the emission lines in PN are typically less than 100 km-s^{-1}, or about 2 Å, but the velocity dispersion within a galaxy, plus any rotational component implies that filter widths must be greater than about 10 Å. Furthermore, because we do not know, *a priori*, the velocities of individual PN, we must use filters whose transmission characteristics do not change significantly over the velocity range for the objects; otherwise, the subsequent photometric accuracy will be compromised. In practice, this dictates a filter width, FWHM, of not less than 25 Å, assuming it is well centered on the galaxy systemic velocity. The filter characteristics of the off-band continuum filter are much less important; we typically use a filter having a FWHM of about 250 Å centered near 5300 Å.

The temperature in the telescope dome cannot always be controlled, and so the interference filter may shift by several Å (at a rate of roughly +1 Å per +5 C). The filter transmission curve also changes with the focal ratio of the converging telescope beam. This is a small effect for most Cassegrain systems having f/ratios greater than about 7.5, but for prime focus operation where f/ratios may be smaller than 3, the filter changes must be included in the calibration procedures.

Digital Photometry

Having identified the PN candidates in a galaxy, they are first reviewed carefully, one-by-one, to verify that their point-spread-functions (PSFs) are appropriate for stellar objects, and not either radiation events in the detector or slightly extended HII regions. The off-band frame is then scaled to the on-band frame and subtracted from it to remove both large and small scale background variations underlying the candidates. This "sky subtraction" greatly improves the photometry, and can only be done for emission-line objects. A secondary advantage is to remove contributions from stars which are sufficiently bright to image through the narrow band filter, further reducing crowding and variable background effects.

Because the PN candidate images are almost always free from blending or crowding, either PSF fitting or aperture photometry techniques work equally well (Stetson 1987). We generally use the simpler aperture photometry procedures as outlined by Adams *et al.* (1980) but modified to include partial pixel contributions following the algorithms developed by Harvey Butcher in his MPC code.

To place the instrumental magnitudes onto an absolute flux scale, we observe several hot subdwarf standard stars (Oke 1974; Stone 1977). The spectra of these stars are convolved with our filter transmission curves after correction for any telescope beam convergence and temperature effects.

Statistical Analysis

We can quickly and easily derive a distance to an ensemble of PN in a galaxy by comparing the first (or Nth brightest) in the sample to an adopted standard, such as the brightest or the average of the brightest PN in the Local Group. (Actually, since PN are a secondary indicator, this only gives a distance relative to the adopted standard. We use M31 to set the zero point, adopting 24.25 as the true distance modulus from Welch *et al.* 1986). This approach is well known to be riddled with complications and biases.

We prefer to utilize a statistical procedure based on the method of maximum likelihood. Hanes and Whittaker (1987) describe their application of the method to globular clusters; Paper II details our formulation for PN. The method views the PNLF as a description of the probability of finding a PN at any luminosity. A model for the PNLF is convolved with the photometric error function and compared with the observed PNLF. The distance modulus, μ, is then varied until the probability of a match is maximized.

This method is insensitive to occasional PN mis-identifications, and eliminates the uncertainty introduced by binning when dealing with luminosity functions. Perhaps of greatest importance, real confidence limits can be placed on the computed distance moduli.

Because these procedures require that the PNLF be universal, we have performed Kolmogorov-Smirnov tests on the data. In each case, the observed PNLF is found to be consistent with being drawn from the same modeled PNLF at the 99% confidence level.

We include an additional parameter in the maximum-likelihood analysis, the luminosity specific number density of PN, $\alpha = N_{PN}/L_{gal}$. Renzini and Buzzoni (1986) show that the stellar death rate is nearly invariant across stellar populations, and so α should be a constant which is potentially useful to constrain the allowable distance moduli derived from PN. To first order, for a given number of identified PN, a galaxy can be distant if it has a small α, or near if it has a large α. In practice, the depth of the survey along the PNLF and the slope of the PNLF at that point also vary as the galaxy is "moved", and so the relation between μ and α is somewhat complex.

The result of the statistical analysis is a series of confidence level contours in (μ, α) space. Figure 2 shows the confidence level diagram for M31 derived from observations of 104 PN using the Kitt Peak #1 0.9m telescope and RCA CCDs. The most probable distance is constrained so that $\mu = 24.25$, in order to set the PN zero point. The uncertainty in μ at the 3σ level is +0.15, -0.45 mag. At the commonly quoted 1σ level, the uncertainty is only +0.07, -0.18. Note that if we had *a priori* knowledge of the value for α, then these uncertainties would be smaller yet. Because the M31 data is complete to 2.5 mag below the PN maximum, and the distance information is largely derived from the first magnitude of the PNLF, a 4m or 5m class telescope will have a 4-5 mag advantage over the 0.9m observations. Thus, we expect similar accuracies for galaxies as distant at 8 Mpc when observed with existing ground-based large telescopes.

For comparison consider Figure 3 showing the confidence level contours for NGC 205, based on 13 PN. The contour levels now span considerable territory in (μ, α) space, due to the sparse PN population of this dwarf galaxy. Nevertheless, it is clear that the distance to NGC 205 is the same as M31 at the 1σ level, but may be as distant as $\mu = 25$ at the 3σ level. The most probable distance is $\mu = 24.62$, a value in good agreement with 24.40 (Richer, Crabtree, and Pritchet 1984) or 24.30 (Mould, Kristian, and Da Costa 1984).

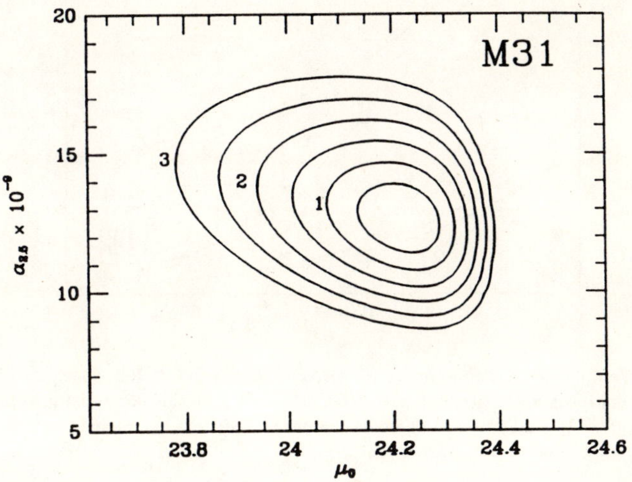

Fig. 2 Confidence level contours in the (μ, α) plane for M31. Contours are shown at 0.5 σ intervals, with levels 1, 2, and 3 as indicated, representing confidence levels of 68%, 95%, and 99.7%. The ordinate, $\alpha_{2.5}$, represents the number of PN contained in the brightest 2.5 mag of the PNLF per galaxy bolometric luminosity.

The asymmetry of the confidence contours is especially noteworthy. It is always easier to pull galaxies closer than to push them away. This is a direct consequence of the upper luminosity limit for PN. Should we try to assign an excessive distance to a galaxy, we reach a hard stop when the brightest few PN begin to exceed this natural intrinsic limit.

OBSERVATIONAL TESTS

We propose 3 verification tests for using PN as extragalactic distance indicators: comparison with other accepted techniques for galaxies in common, comparison of individual galaxies within a single group or cluster, and comparison of different fields within a single galaxy. The first test provides an estimate of the combined external uncertainties involved in both the PN method and the comparison methods. Often the comparison methods are sufficiently inaccurate, or the error estimates so poorly evaluated that the accuracy of the PN method cannot be assessed. Thus far, we have few galaxies in common with other accurate techniques (RR Lyra or Cepheid variables).

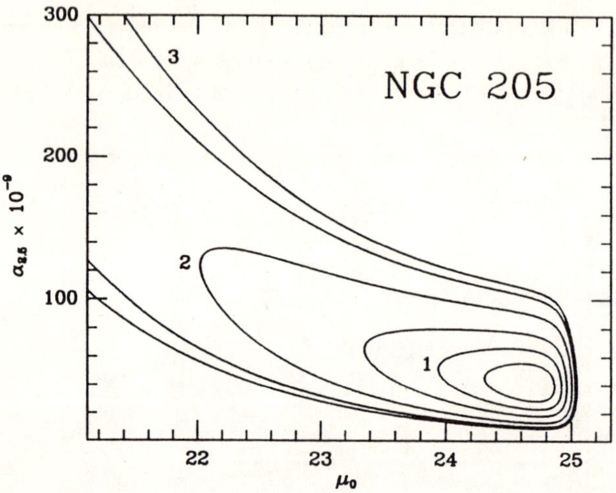

Fig. 3 Confidence level contours in the (μ, α) plane for NGC 205, based on 13 PN. The contours are consistent with NGC 205 being at the same distance as M31, but the PN population of this galaxy does not include a sufficient number of bright PN to derive an accurate distance.

The second test is purely internal; it provides checks on the assumptions concerning metallicity and internal extinction, as well as validates the maximum likelihood procedures and PNLF universality. The third test is also internal; it helps our understanding of the internal calibrations if the separate fields fall at significantly different portions of the rotational velocity curve of the galaxy, thereby requiring accurate filter correction factors.

Comparison for M81

M81 is a good candidate for consideration because it is a mid-type galaxy (Sb) having several standard candles which may be inter-compared. We have identified 185 PN in M81 using the Kitt Peak 4m and the TI2 CCD. Because the PN density follows the light distribution of galaxies (see Paper II) and the scale height is several times that of the young HI and HII populations in our galaxy, we conclude that bright PN will preferentially be unreddened by internal dust. We therefore adopt a foreground extinction correction only; this may bias the derived distance to be slightly too large.

Estimates for the reddening to the nuclear bulge of M81 are typically E(B-V) = 0.10 (Sandage 1976; Peimbert and Torres-Peimbert 1981) to E(B-V) = 0.19 (Peimbert and Torres-Peimbert 1981), although somewhat lower values for the galactic reddening are expected (Sandage 1984). Although some of these estimates may include a contribution from dust internal to M81, we adopt the value of 0.10 from the Balmer decrement of the nuclear emission (Peimbert and Torres-Peimbert 1981) as this region of the galaxy is central to our objects. An equally compelling value is 0.16 derived by Wu *et al.* (1980) from UV spectra of the central 2.5 arcmin, an area comparable to the region over which the PN extend. When combined with a galactic extinction law (Seaton 1979), the extinction at λ5007 is $A_{5007} = 3.56$ E(B-V), or 0.36 to 0.57 mag.

The extinction corrected confidence level diagram for 88 PN in a homogeneous sample of M81 candidates complete to 1.1 mag below the bright PN limit is shown in Figure 4. The most probable distance modulus for M81 is 27.74 with 1σ bounds of +0.05, -0.07 mag, and 3σ bounds of +0.13, -0.20 mag, assuming the extinction, E(B-V), to M81 is 0.10 mag. If we adopt the larger extinction value, E(B-V)=0.16, then $\mu = 27.53$. The uncertainties are astonishingly small, yet quite reasonable considering the small scatter of the PNLF for M81 (Figure 1). The largest uncertainties are in the zero point and the extinction correction.

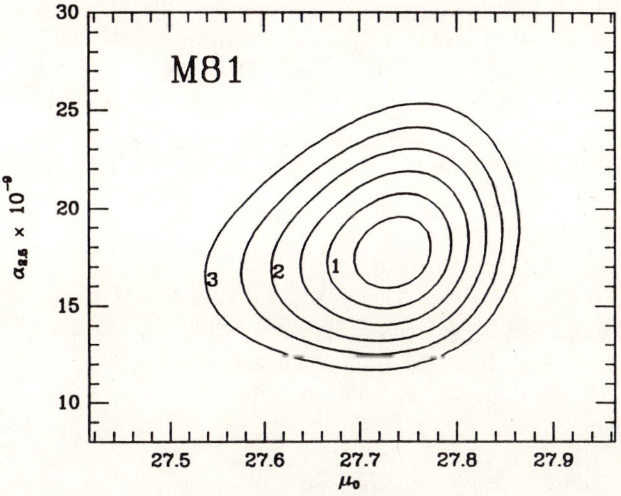

Fig. 4 Confidence level contours in the (μ, α) plane for M81. The 3σ bounds are +0.13, -0.20 mag, a remarkably small uncertainty. The high degree of confidence derives from the large number of PN (88) in the bright portion of the PNLF and the excellent photometric accuracy (0.04 mag) for these PN.

The zero point is derived from M31 and should be accurate to about 0.05 mag. In any case, the difference between μ for M81 and M31 is very well determined to be 3.49 mag, subject to further considerations of the extinction to both galaxies. (In principle, the extinction, whether foreground or internal, can be measured directly from the ratios of the Balmer lines in the bright PN.)

When comparing our results with a recent sampling of other determinations for the distance to M81, we do not include distances derived to companions such as NGC 2403, as this only adds further confusion and uncertainty. Humphreys *et al.* (1986) derive $\mu = 27.55$ based on luminous red supergiants and assuming $A_V = 0.8$ mags. In this and most other distance estimates to M81, internal extinction values are highly uncertain, and clearly vary as a function of position. This is evidenced by the HII region studies by Kaufman *et al.* (1987), and Garnett and Shields (1986) in which reported visual extinctions vary from 0.0 to 2.0 mag.

Bottinelli *et al.* (1984) derive $\mu = 27.60 \pm 0.16$, based on the B-band Tully-Fisher relation. (For comparison, their M31 estimate is 23.93 ± 0.15). The data from Aaronson *et al.* (1983) may be used to derive $\mu = 27.86$ for M81, and $\mu = 24.59$ for M31 based on the IR Tully-Fisher relation. From brightest blue stars, Sandage (1984) derives $\mu_V = 28.75$, which when corrected for E(B-V)=0.10, yields $\mu = 28.44$.

A recent study of Cepheids in M81 by Freedman and Madore (1988) yields $\mu = 27.59 \pm 0.31$, assuming a forwground extinction of $A_I = 0.05$. The quoted uncertainty includes a term for the extinction internal to M81 which has not yet been removed. The authors note that the 2 Cepheids are likely to be selected as a result of low internal reddening (and are therefore apparently brighter and easier to detect than if in regions of high reddening), and that an I band extinction correction is only 36% as large as a B band correction. Therefore, any revision to the distance as a consequence of internal reddening is not likely to be large.

Clearly, it is difficult to draw quantitative conclusions from these comparisons. Yet we can state that the PN based results are consistent with and near the median of distances derived by more established techniques. Furthermore, the PN results include well defined error terms which place the bulk of uncertainty on zero points and extinction estimates, factors which affect most other determinations to a comparable degree.

Galaxies in a Cluster - the NGC 3379 Group

Thus far, we have observed 3 galaxies in the group associated with the well-studied elliptical NGC 3379 (E0). This group is also referred to as the NGC 3368 (M96) group. While NGC 3384 (SB0) forms an optical pair with NGC 3379 on the sky, NGC 3377 (E6) is about 2 degrees away. If these galaxies form a genuinely bound group, we expect the derived distances to agree to within 1 Mpc.

We have identified 54, 95, and 102 PN in NGC 3377, NGC 3379, and NGC 3384. Of these, 41, 77, and 67 PN comprise the complete samples for these galaxies. We assume a foreground reddening, E(B-V), of 0.047 (de Vaucouleurs, de Vaucouleurs, and Corwin 1976). In order to compute values for α, we use the surface photometry from Peletier *et al.* (1988) for NGC 3377, from de Vaucouleurs and Capaccioli (1979) and Peletier *et al.* (1988) for NGC 3379, and Barbon, Capaccioli, and Tarenghi (1975) and Barbon, Benacchio, and Capaccioli (1976) for NGC 3384. Bolometric corrections are based on the photometry by Persson, Frogel, and Aaronson (1979).

TABLE II. Distances to the NGC 3379 Galaxies

Galaxy	μ_0	Distance (Mpc)	σ_μ
NGC 3377	29.93	9.68	+0.12,-0.14
NGC 3379	29.80	9.12	+0.15,-0.15
NGC 3384	29.87	9.42	+0.10,-0.11

Results from the maximum-likelihood analysis are summarized in Table II. We see that these galaxies are about 9.4 Mpc distant and the most probable distances deviate from the mean by less than 0.3 Mpc, well within the definition limits for a cluster. The formal 1σ uncertainties on μ are typically 0.13 mag, or slightly larger than the observed dispersion. Admittedly, 3 galaxies is a small sample, and data for additional objects (NGC 3368 and NGC 3412) in this group are currently being analyzed. Nevertheless, the distances derived from PN are extremely consistent.

Note that the [O III] luminosities for PN at this distance are equivalent to V magnitudes fainter than 25.5, yet the PN are easily identified with 2 hour exposures at the KPNO 4m telescope. The photometric error for a single bright PN is typically about 0.10 mag, but because dozens of PN are used in the PNLF, the derived distances can be much more accurate.

Fields Within a Single Galaxy - NGC 5128

NGC 5128 (Cen A) is a nearby giant elliptical (E2) galaxy having a conspicuous dust lane and HII regions near the center. When identifying PN in NGC 5128 for distance determinations, we avoid the central 8 arcmin to minimize potential errors arising from internal absorption. One expects that a standard candle will yield consistent distance estimates for any two fields in a single galaxy, and that the disagreement is a measure of the internal errors.

We have surveyed 26 fields in the outer halo of NGC 5128 at distances from the nucleus ranging from 8 to 24 arcmin to the NE and SW along the major axis. 62 and 53 PN were identified in each group of fields, respectively; of these, 41 and 28 PN, respectively, form complete homogeneous samples for the maximum-likelihood probability analysis. Surface brightness estimates are taken (and extrapolated) from Dufour *et al.* (1979) to allow calculation of α in deriving the probability contour maps as for Figures 2-4 above.

The formal values for μ to the NE and SW fields (and their associated 1σ uncertainties) are 27.87 (+0.09, -0.16) and 27.98 (+0.10, -0.16), respectively, assuming E(B-V)=0.10. These two values are in excellent agreement considering the small number of PN involved. While the difference in μ is already less than the formal errors, the color maps of NGC 5128 presented by Dufour *et al.* show that the SW side of the galaxy is redder than the NE. Should this be due to a small additional extinction component, then the agreement between the two samples is improved. When combined, the net sample yields $\mu = 27.90$ (+0.07, -0.11). (The values for μ presented here differ slightly from the provisional values given by Ford *et al.* (1988), and will be discussed in detail in a future paper.)

ADVANTAGES OF USING PN

PN offer a number of advantages over more popular distance indicators. For example, PN are:

- Found in all Hubble types.
- Useful over a distance range comparable to Cepheids.
- Measured using simple and accurate photometric methods in which the background can be removed to improve local and global sky flatness.
- Observed at only one epoch.
- Relatively immune to metallicity variations.
- Derived from a population insensitive to internal reddening.
- Future targets for kinematic and composition studies.

There are, of course, some limitations to the use of PN as distance indicators. For late type galaxies (e.g., Sc's) there is potentially serious confusion with compact HII regions, and background noise from large scale diffuse emission cannot be suppressed using the narrow-band filter technique. Also, PN are secondary indicators requiring an extragalactic zero point calibration. We feel, however, that M31 serves this purpose and introduces only a small additional uncertainty.

FUTURE PROSPECTS

With existing ground based telescopes, it is possible under good observing conditions to identify PN and derive extragalactic distances beyond 15 Mpc. With planned 8m and larger telescopes, this can be extended to nearly 25 Mpc. A space-borne telescope should, in principle, be capable of working to similar distances, but the narrow-band filter selection must be chosen to accommodate galaxies at those redshifts, and have characteristics as outlined earlier.

Current work in progress includes PN in M101, NGC 3115, NGC 5102, NGC 5866, and M87 and, we plan to continue probing the Virgo cluster.

ACKNOWLEDGEMENTS

We thank W. L. Freedman for permission to quote results prior to publication, and Don Neill, John Booth, and Xiao Hui for assistance with the data reduction. We are especially grateful to C. D. Keyes for providing the PN model code and assistance in its use, and to R. L. Davies for providing the surface photometry of NGC 3377 and NGC 3379 in advance of publication.

REFERENCES

Aaronson, M., Huchra, J., Mould, J., Tully, R. B., Fidher, J. R., Van Worden, H., Goss, W. M., Chamaraux, P., Mebold, U., Siegman, B., Berriman, G., and Persson, S. E. 1982, Ap. J. Suppl., 50, 241.

Adams, M., Christian, C., Mould, J., Stryker, L., and Tody, D. 1980, Stellar Magnitudes From Digital Pictures, (Tucson: Kitt Peak National Observatory).

Aller, L. H., and Keyes, C. D. 1980, Ap. Sp. Sci., 72, 203.

Baade, A. W. 1955, A. J., 60, 151.

Barbon, R., Benacchio, L., and Capaccioli, M. 1976, Astr. Ap., 51, 25.

Barbon, R., Capaccioli, M., and Tarenghi, M. 1975, Astr. Ap., 38, 315.

Bottinelli, L., Gouguenheim, L., Paturel, G., and de Vaucouleurs, G. 1984, Astr. Ap. Suppl., 56, 381.

Ciardullo, R., Jacoby,, G. H., Ford, H. C., and Neill, J. D. 1988, Ap. J., submitted, (Paper II).

de Vaucouleurs, G., de Vaucouleurs, A., and Corwin, Jr., H. G., in Second Reference Catalogue of Bright Galaxies, (University of Texas: Austin).

de Vaucouleurs, G., and Capaccioli, M. 1979, Ap. J. Suppl., 40, 699.

Dufour, R.J., van den Bergh, S., Harvel, C.A., Martins, D.H., Schiffer III, F.H., Talbot, Jr., R.J., Talent, D.L., and Wells, D.C. 1979, A. J., 84, 284.

Ford, H. C 1983, in IAU Symposium 103, Planetary Nebulae, ed. Y. Terzian (Dordrecht: Reidel), p. 19.

Ford, H. C., and Jenner, D. C. 1978, B.A.A.S., 10, 665.

Ford, H. C., Ciardullo, R., Jacoby, G. H., and Hui, X. 1988, in IAU Symposium 131, Planetary Nebulae, ed. S. Torres-Peimbert (Dordrecht: Reidel), in press.

Ford, H. C., Jenner, D. C., and Epps, H. W. 1973, Ap. J. (Letters), 183, L73.

Freedman, W. L., and Madore, B. F. 1988, priv. comm.

Garnett, D. R., and Shields, G. A. 1987, Ap. J., 317, 82.

Hanes, D. A., and Whittaker, D. G. 1987, A. J., 94, 906.

Hodge, P. W. 1966, Galaxies and Cosmology, (New York: McGraw-Hill), p. 130.
Hodge, P. W. 1981, Ann. Rev. Astr. Ap., 19, 357.
Humphreys, R. M., Aaronson, M., Lebofsky, M., McAlary, C. W., Strom, S. E., and Capps, R. W. 1986, A. J., 91, 808.
Husfeld, D., Kudritzki, R. P., Simon, K. P., and Clegg, R. E. S. 1984, Astr. Ap., 134, 139.
Jacoby, G. H. 1988, Ap. J., submitted, (Paper I).
Jacoby, G. H., and Lesser, M. P. 1981, A. J., 86, 185.
Kaufman, M., Bash, F. N., Kennicutt, R. C., and Hodge, P. W. 1987, Ap. J., 319, 61.
Koester, D., Schultz, H., and Weidemann, V. 1979, Astr. Ap., 76, 262.
Lawrie, D. G., and Graham, J. A. 1983, B.A.A.S., 15, 907.
Lindsay, E. M. 1955, M.N.R.A.S., 115, 248.
Mendez, R. H., Kudritzki, R. P., Herrero, A., Husfeld, D., Groth, H. G. 1988, Astr. Ap., 190, 113.
Mould, J., Kristian, J., and Da Costa, G. S. 1984, Ap. J., 278, 575.
Oke, J. B. 1974, Ap. J. Suppl., 27, 21.
Stone, R. P. S. 1977, Ap. J., 218, 767.
Paczynski, B. 1971, Acta. Astr., 21, 417.
Peimbert, M., and Torres-Peimbert, S. 1981, Ap. J., 245, 845.
Peimbert, M., and Torres-Peimbert, S. 1983, in IAU Symposium 103, Planetary Nebulae, ed. D. R. Flower (Dordrecht: Reidel), p. 233.
Peletier, R. F., Davies, R. L., Illingworth, G. D., Davis, L. E., and Cawson, M. C. M. 1988, in preparation for submission to A. J.
Persson, S. E., Frogel, J. A., and Aaronson, M. 1979, Ap. J. Suppl., 39, 67.
Renzini, A., and Buzzoni, A. 1986, in Spectral Evolution of Galaxies, ed. C. Chiosi and A. Renzini (Dordrecht: Reidel), p. 195.
Richer, H. B., Crabtree, D. R., and Pritchet, C. J. 1984, Ap. J., 237, 138.
Sandage, A. 1984, A. J., 89, 621.
Sandage, A. 1976, A. J., 81, 954.
Seaton, M. J. 1979, M.N.R.A.S., 187, 73p.
Stetson, P. B. 1987, P.A.S.P., 99, 191.
Stone, R. P. S. 1977, Ap. J., 218, 767.
Welch, D. L., McAlary, C. W., McLaren, R. A., and Madore, B. F. 1986, Ap. J., 305 583.
Westerlund, B. E., and Rodgers, A. W. 1959, Observatory, 79, 132.
Wood, P. R., and Faulkner, D. J. 1986, Ap. J., 307, 659.
Wu, C. C., Faber, S. M., Gallagher, J. S., Peck, M., Tinsley, B. M. 1980, Ap. J., 237, 290.

DISCUSSION

M. MATEO: Have you tried to apply your technique to published luminosities of LMC/SMC planetaries?

G. JACOBY: Most published LMC/SMC planetary luminosities are derived from spectroscopic observations rather than photometric, and so the necessary data are not available.

D. OSTERBROCK: I believe this is a very good distance determination method, but can you tell me if there is any danger of confusion of small, dense HII regions with planetary nebulae, particularly in the most distant galaxies where the linear resolution is poor?

G. JACOBY: For late-type galaxies, the danger of confusion with compact HII regions is significant. To discriminate against misidentifications, all PN candidates are checked against Hα images. Since bright planetaries have high excitation and relatively weak Hα, they will not be seen in Hα images, whereas HII regions are strong in Hα.

S. DJORGOVSKI: Elliptical galaxies have x-ray coronae, possibly even cooling flows. Would such an environment modify the lifetimes, and hence luminosity function, of PNe?

G. JACOBY: Because we restrict the observations to the brightest part of the luminosity function, only young dense planetaries are included. They have densities of about 10^4, and so the effects you mention will not have much effect at this point in their lifetimes.

J. MOULD: To what extent is the model you use for the luminosity function of planetaries sensitive to the star formation history of the host galaxy?

G. JACOBY: The results for the Local Group and the NGC 3379 galaxies suggest that this effect cannot be very large.

H. HARRIS: Are the fragmentary data available for Galactic Planetary Nebulae consistent with your luminosity function?

G. JACOBY: The large uncertainties in Galactic planetary nebula distances, and the small number of nearby bright planetaries, make a comparison of the top few magnitudes almost impossible.

M. FEAST: Could you apply your method to the Galactic bulge planetaries?

G. JACOBY: The Galactic bulge planetaries would be very interesting. A reasonable sample of, say, 40-50 bright planetaries would be needed.

J. NEMEC: Have simulations been made to see how successful the technique you use to identify planetary nebulae would be if the Hubble Space Telescope were used?

G. JACOBY: Ground based 4-m telescopes can compete with the estimated performance of the HST in this application. An 8-m class telescope in a good site should do better than HST.

R. HUMPHREYS: You made the comment that there was less of a problem with galactic internal reddening for planetary nebulae because they are not in spiral arms. What about M81, where we know there is high reddening and where dust lanes are observed crossing even in front of its nuclear bulge?

G. JACOBY: Because the planetary nebulae are a bulge population, even for M81, they have a scale height similar to the M81 bulge stars. This is considerably larger than typical scale heights for young Pop I objects such as HI and dust, and so the bright planetaries in M81 will be selected from those in the bulge, well above the dust.

There is some dust seen crossing M81, and to remove effects of local variations such as these, spectra of the planetaries are required to measure the reddening from the Balmer decrement.

RR LYRAE STARS IN NEARBY GALAXIES

C.J. PRITCHET
Department of Physics and Astronomy, University of Victoria, P.O. Box 1700, Victoria, BC V8W 2Y2, CANADA

ABSTRACT RR Lyrae stars play an important role in distance determinations in the Local Group, and provide a distance scale that is independent of that obtained with Cepheids and other Population I objects. In this review we survey the current state of the calibration of the absolute magnitudes of RR Lyrae stars, and discuss recent observations of variables in galaxies beyond the Milky Way (M31, M33, NGC 147, and NGC 185). This review concludes with a brief discussion of the potential of Space Telescope in extending the Population II distance calibration.

1. INTRODUCTION

At the present time there seems to be a consensus that the true global value of H_0 lies somewhere in the range 50 – 100 $km\ s^{-1} Mpc^{-1}$. The reason that this important cosmological parameter has not yet been constrained more precisely is itself the subject of some debate. Some (*e.g.* Aaronson 1986) would hold that the greatest problem in nailing down the extragalactic distance scale lies in *local calibrators*; others (*e.g.* Tammann 1986) would argue that the problem lies in *selection effects* whose magnitude grows with increasing distance. There is probably some truth in *both* of these positions, as the above two papers illustrate. It now seems unlikely that *all* of the factor of two uncertainty in H_0 (corresponding to 1.5 mag in $m - M$) can be attributed to local calibration problems; yet there are some glaring examples of galaxies in the Local Group (*e.g.* LMC, M33) that are of pivotal importance in the calibration of secondary distance indicators, and whose distances have recently been claimed to be in error by more than 0.5 mag!

In this review we will discuss the potential of Population II standard candles such as RR Lyrae stars in checking some of the outstanding calibration problems in the extragalactic distance scale. After reviewing the usefulness of RR Lyrae stars in distance scale problems (§2), the current state of the calibration of RR Lyrae star absolute magnitudes is briefly summarized in §3. Observations of RR Lyrae stars in nearby galaxies (*beyond* the Milky Way) are discussed in §4; future prospects with HST are discussed in §5.

2. IMPORTANCE OF RR LYRAE STARS

There are a number of reasons that RR Lyrae stars are of importance in determining distances to nearby galaxies.

(i) Brightness. RR Lyrae stars are luminous enough to be seen in a number of nearby galaxies. They have of course been detected in Galactic dwarf spheroidals (*e.g.* Saha 1986 and references therein), and in the Magellanic Clouds (*e.g.* Graham 1984; Walker 1985, 1988). Using CCD detectors on ground-based telescopes in sub-arcsecond seeing, it is possible to detect RR Lyrae stars out to about 1 Mpc. As we shall see, the Hubble Space Telescope should be able to detect these objects out to nearly 3 Mpc.

(ii) Frequency. RR Lyrae stars are quite frequent in old ($\geq 10^{10}y$) stellar populations. Based on the statistics of occurrence in Galactic globular clusters, we estimate that there are typically $\sim 10^{2\pm1}$ variables per $M_V = -10$ of luminosity. This number varies tremendously from cluster to cluster because of variations in horizontal branch morphology; however it appears to hold for the halos of both M31 and M33. The implication of this result is that "typical" (≥ 1 $arcmin^2$) CCD fields of galaxies at ≥ 1 Mpc contain ≥ 10 variables (for a surface brightness in the field of $\mu_V = 25$ mag arcsec^{-2}).

(iii) Observability. RR Lyrae stars occur in the halos of galaxies, where *internal* reddening is not a problem (contrary to the situation that obtains for classical Cepheids). The population to which these stars belong possesses a smooth (albeit resolved) structure, so that crowding is not a severe problem.

(iv) Calibration. $< M_V >$ is in principle a well-defined quantity that can be determined from a variety of techniques (see §2). Because many RR Lyrae stars are in the halo, reddening is not a serious problem for the calibration of $< M_V >$. Prospects look excellent for improving the accuracy to which $< M_V >$ can be determined, through various observations with Hipparcos and HST. Especially important is the fact that the calibration of $< M_V >$ is almost completely independent of the calibration of the $P - L$ relation for Cepheids – *i.e.* the Local Group distance scale (and beyond!) can be determined using RR Lyrae stars more or less independently of other techniques. The issue of the calibration of $< M_V >$ is discussed in greater detail in §2.

Summarizing the above, RR Lyrae stars appear to be excellent primary distance indicators within the Local Group, and provide a superb independent check on the calibration of Local Group distances. They may not be as bright as classical Cepheids, but they should be easily visible out to at least 2 Mpc (and possible 3 Mpc) with HST. The biggest problem with the use of RR Lyrae stars as distance indicators at present is the determination of the metallicity sensitivity of $< M_V >$ (see below).

3. CALIBRATION AND COMPLICATIONS

There are two distinct issues in the calibration of RR Lyrae stars for distance determination purposes: *(i)* the determination of $< M_V >$ at fixed metallicity; and *(ii)* the metallicity sensitivity of $< M_V >$. Each of these will be discussed separately below.

3.1 Calibration of $< M_V >$

Until recently, the available evidence suggested a value for $< M_V >$ close to +0.6 (*e.g.* Stothers 1983). More recently the weight of opinion seems to have moved towards $< M_V >= +0.8$, due to several advances.

(i) Statistical parallax solutions for $< M_V >$ (Hawley *et al.* 1986; Strugnell *et al.* 1986) have been obtained for the Wan *et al.* (1981) sample of 159 RR Lyrae stars. Because of the improved proper motions and radial velocities for this relatively large sample of stars, and the adoption of maximum likelihood techniques in their analysis, it is generally accepted that these solutions are the best now available. The Hawley *et al.* result, corrected to an *intensity*-weighted mean magnitude (Pritchet 1986, Barnes and Hawley 1987), and using the Burstein and Heiles (1982) reddening scale, yields $< M_V >= +0.77 \pm 0.14$, a value that we use throughout this paper and which differs negligibly from the Strugnell *et al.* result of +0.75.

(ii) It now seems to be generally recognized that shock waves in the atmospheres of RR Lyrae stars strongly affect their optical colors, especially in the expansion phase. Use of the Baade-Wesselink technique to determine $< M_V >$ therefore requires near-infrared colors, and/or use of a restricted phase interval during which shocks are not a problem. The latest Baade-Wesselink solutions that satisfy these criteria include those referenced in Jones, Carney, and Latham (1988) ($< M_V > = 0.85$), and Cacciari *et al.* (1988) ($< M_V > = 0.83$).

The latest and best results therefore favor a value for $< M_V >$ closer to +0.8 than to the canonical value of +0.6. The reader is referred to Jones *et al.* (1988) for a detailed discussion of other determinations of $< M_V >$.

3.2 Metallicity Dependence of $< M_V >$

Using an empirical period-metallicity relation, Sandage (1982) derived a strong dependence of $< M_{bol} >$ on metallicity, in the sense that metal-poor RR Lyrae stars were predicted to be significantly brighter than metal-rich RR Lyrae stars. The inferred effect was roughly equal in magnitude (but opposite in sign!) to that obtained in previous studies (*e.g.* Carney 1980). Sandage derived a value $\Delta M_{bol}/\Delta[Fe/H] \simeq 0.35$, resulting in a difference between the horizontal branch luminosities of the most metal-rich and most metal-poor globular clusters of ~ 0.5 mag.

The "Sandage effect" has been the subject of considerable controversy since 1982. The Hawley *et al.* (1986) statistical parallax solutions showed no significant difference between the $< M_V >$ values for metal-rich and metal-poor subsamples of stars; but there were not enough stars to exclude such an effect. Jones *et al.* (1988) found a slight dependence of M_{bol} on $[Fe/H]$, but none for M_V. Cacciari *et al.* (1988) has compiled a list of recent "highest quality" Baade-Wesselink determinations of $< M_V >$, and fits a value of $\Delta M_{bol}/\Delta[Fe/H] \simeq 0.22$ (slightly smaller for M_V). This result is in agreement with that derived from CCD color-magnitude diagrams of Galactic globular clusters (Bolte 1988). Theoretical estimates are in the range 0.20–0.24 (*cf.* Sweigart 1987, Lee *et al.* 1988).

Although there appears to be a degree of consensus in the last few values quoted above, this agreement should be tempered by the following considerations. *(i)* The theoretical estimates of the Sandage effect are for *zero-age* horizontal branches only. The wide variation in HB morphology for globular clusters means that stars cross the instability strip at different evolutionary phases in different globulars (Lee *et al.* 1988), thus greatly complicating the determination of $\Delta M_{bol}/\Delta[Fe/H]$. Note that such a complication may not be present for field stars, because such stars are not perfectly coeval, and hence should cluster preferentially near the ZAHB because of lifetime considerations. *(ii)* The theoretical models quoted above do not take into account the variation of [O/Fe] with [Fe/H]; allowing for this effect may lower the expected metallicity dependence of M_{bol} by roughly one third (Vandenberg 1988).

3.3 Future Improvements in Calibration

Many of the calibration issues involving RR Lyrae stars *may* be solved in the near future with the Hubble Space Telescope or with Hipparcos. The absolute magnitudes of *nearby* RR Lyrae stars may be improved by obtaining high precision geometrical parallaxes, as discussed by Monet in this volume. The magnitude of the Sandage effect can be determined directly by observing RR Lyrae stars (or other HB stars) in M31 globulars spanning a range in [Fe/H].

4. CURRENT OBSERVATIONAL SITUATION

In this section we shall discuss what is currently known about RR Lyrae stars in galaxies that are *beyond* the Milky Way and its companions. The reader is referred to papers by Graham (1984), Walker (1985, 1988) and Reid and Strugnell (1986) for information concerning the use of RR Lyrae stars as distance indicators in the Magellanic Clouds.

4.1 M31 (NGC 224)

Baade (1963) has traced the fascinating evolution of our ideas on the distance to the Andromeda nebula (M31). In the late 1940s, the distance of M31 was thought to be $\sim$ 300 kpc; this distance was based primarily on the Cepheid period-luminosity relation, (erroneously) calibrated using Population II Cepheids in Milky Way globular clusters. Although this distance disagreed with that derived from both novae and globular clusters, it was widely expected at the time that exposures with the newly commissioned Palomar 200" telescope would settle this discrepancy by revealing large numbers of RR Lyrae stars at the photographic limit. To quote Baade (1952):

> "The very first exposures on M31 .... showed at once that something was wrong. Tests had shown that we reach .... stars of $m_{pg} = 22.4$ in a 30 min exposure. Hence we should reach in such an exposure the cluster-type variables in M31, at least in their maximum phases. Actually we reach only the brightest stars of Population II in such an exposure."

This revelation led to a re-evaluation of the distance to M31, and the realization that the zero-points of the Population I and II Cepheid P-L relations were very different.

More recently, Pritchet and van den Bergh (1987) have used a CCD in the superb seeing conditions that are found at the Canada-France-Hawaii 3.6 m Telescope to detect, and measure the brightnesses of, RR Lyrae stars in the Andromeda nebula. They obtained approximately 15 hours of CCD data (spread over 6 nights) on a halo field approximately 40' SE of M31 (along the minor axis). This field was chosen to have a high enough halo surface brightness to find many RR Lyrae stars, but not so high as to be severly crowded. (From the observed luminosity function, the surface brightness has now been estimated to be $\sim$ 26.2 mag arcsec^{-2} in B, compared to 27.0 obtained by de Vaucouleurs [1958].)

Candidate variable objects were obtained by blinking the frames. DAOPHOT photometry (Stetson 1987) was performed both on 1 hour subsets of the data, and also on the combined 15 hour mean frame. Periods and light curves were obtained for the variables; regrettably the amount of data available did not allow periods to be determined uniquely. Assuming $< M_V > = +0.77$, $< B - V >= 0.26$, and a Burstein and Heiles (1982) absorption of 0.31 mag in B, Pritchet and van den Bergh obtain $(m - M)_0 = 24.34$. Pritchet and van den Bergh (1988) argue that the stellar content of the M31 halo in this field is analogous to that found in the galactic globular cluster NGC 6171. If this is the case, then the mean color of the RR Lyrae stars should be assumed to be slightly redder, resulting in a true distance modulus of 24.23. *Either* of these values is in fact in quite good agreement with other determinations of the distance to M31 (see Table I).

There are several biases that enter into the calculation of the distance modulus of M31 from RR Lyrae stars. *(i)* The weighting of the light curve in $< B >$ must be the same as that adopted for the calibration of $< M_B >$. The most convenient (for reasons given below) is an *intensity*-weighting of brightness over all phases; a correction must be made to the original work of Hawley *et al.* (1986) to obtain such a weighting (Pritchet 1986, Barnes and Hawley 1987). *(ii)* There can be quite severe scale errors in DAOPHOT photometry near the limit,

TABLE I Some Recent Distance Estimates for M31

Method	Reference	$(m-M)_0$
Novae	van den Bergh (1975)	24.16 ± 0.12
	Cohen 1985	24.04 ± 0.20
GB tip	Mould and Kristian 1986	24.40 ± 0.25
IR Cepheids	Welch *et al.* (1986)	24.26 ± 0.08
RR Lyr stars	Pritchet and van den Bergh (1988)	24.23 ± 0.15

due at least in part to the fact that that those very faint images that *are* found tend to lie on positive noise fluctuations. It therefore makes the most sense to average the data frames (not the magnitudes), and *then* perform DAOPHOT photometry to obtain mean magnitudes, since such magnitudes should be well above the limit. Note that such a procedure naturally produces an intensity-weighted mean magnitude. *(iii)* There are several built in biases in favor of objects that are brighter than average. For example, one is more likely to identify a variable if its phase coverage just happens to produce more times of maximum light. Averaging all the frames together will then produce an overestimate of the brightness. Brighter variables are also easier to detect, so if there is intrinsic dispersion in the distribution of brightnesses of variables, one will tend to find the brightest variables. Numerical simulations show that these effects are probably not significant for our M31 variables, but *may* be important for our M33 field (see below).

4.2 M33

CFHT observations have also been obtained of the halo of M33 in order to search for RR Lyrae stars (Pritchet and van den Bergh, in preparation). Here the results are much more preliminary than those for M31, because only 6 hours of data have so far been obtained. We have blinked the frames for variables, and have identified a number of obvious candidates.

The mean B magnitude (intensity-weighted) for the 7 "best" candidates (those with the most prominent variability when blinking, and with the highest χ^2 of variation) is 25.79 ± 0.15. Using the Johnson and Joner (1987) absorption value of 0.31 mag in B, assuming $< B-V >= 0.26$ (appropriate for metal-poor populations such as the M33 halo – *e.g.* Mould and Kristian 1986), and taking $< M_V > =+0.77$, we obtain a distance modulus $(m-M)_0 = 24.45 \pm 0.2$. The quoted error is from the uncertainty in $< M_V >$, and from the small number of observations that go into the intensity average. This result should be considered *preliminary*, because: *(i)* it does not correct for the "Sandage effect", which could increase the distance modulus by ~ 0.2 mag if theoretical predictions are correct; and *(ii)* light curves cannot be constructed for *any* of the variables because of the paucity of observations, so that we cannot be sure that any of the variables are definitely RR Lyrae stars.

TABLE II Recent Distance Estimates for M33

Method	Reference	$(m-M)_0$
B Cepheids	Sandage and Carlson (1983)	25.23
I Cepheids	Freedman (1985)	24.1 ± 0.20
H Cepheids	Madore *et al.* (1985)	24.3 ± 0.20
B,I Cepheids	Christian and Schommer (1987)	24.2 ± 0.15
R,I Cepheids	Mould (1988)	24.8 ± 0.1
GB tip	Mould and Kristian 1986	24.8 ± 0.3
LPVs	Kinman *et al.* (1987)	24.64 ± 0.1
RR Lyr stars	This paper	$24.45 : \pm 0.2$

Our new (but preliminary!) distance modulus for M33 is compared with other values in Table II. The comparison is greatly confused by the fact that different authors make a variety of assumptions regarding internal reddening (which is *not* a problem here), and the distance to the LMC (which again is irrelevant for RR Lyrae stars). It can be seen that our distance for M33 is intermediate between the results of Kinman *et al.* (1987) and Mould (1988) on Cepheids and long-period variables, and the near-infrared observations of Cepheids by Madore *et al.* (1985). We hope to refine our distance modulus to M33 by obtaining more observations in the upcoming 1988 observing season.

4.3 NGC 147/185

NGC 147 and 185 are two well-known dwarf elliptical companions of M31. Saha and Hoessel (1987) have reported observations of RR Lyrae stars in NGC 147; they found 9 variables, and derived a distance modulus $(m-M)_0 = 23.85$ from these objects. The greatest uncertainty in this study is the assumed absorption in the direction of NGC 147, which lies at very low galactic latitude ($b = -14°$).

Using similar techniques, Saha (private communication) has recently discovered approximately 100 cluster-type variables in the galaxy NGC 185. Again, the preliminary true distance modulus (23.77) turns out to be significantly lower than that of M31, but uncertain reddening may again turn out to be a problem due to the low galactic latitude of this galaxy.

5. DISCUSSION AND FUTURE PROSPECTS

We have seen that observations of RR Lyrae stars in Local Group galaxies are possible out to about 1 Mpc using 4m-class ground-based telescopes with CCD detectors in sub-arcsecond seeing. RR Lyrae stars provide an independent (Population II) check on the distances of Local Group galaxies, and are therefore an extremely important tool in calibrating secondary distance indicators. This is especially important for the Fisher-Tully relation, whose principal calibrators are the two Local Group galaxies M31 and M33. Although there are still some difficulties and uncertainties in the determination of $< M_V >$, there is hope that these difficulties will be overcome using HST and Hipparcos.

To what distance can observations of RR Lyrae stars be pushed with the Hubble Space Telescope? To answer this question, we have simulated HST Planetary Camera CCD frames of a hypothetical galaxy halo at distances out to 3 Mpc. The results show that RR Lyrae stars can be detected out to such distances, and that the expected number of variables per PC frame is more than adequate. Of course real observations will be complicated by cosmic rays, background galaxies, *etc.* Nevertheless we conclude that observations of RR Lyrae stars with HST are feasible out to at least 2.5–3 Mpc. Observations of W Vir stars might be possible to even larger distances!

The implication of these simulations is that the population II distance grid can be extended out to at least the M81 group, and possibly further. Thus it should prove possible to greatly improve the zero-point calibrations of both the Fisher-Tully relation and also the P-L relation for Cepheids.

ACKNOWLEDGEMENTS

This work is part of an ongoing collaboration with Sidney van den Bergh, whose interest and encouragement is gratefully acknowledged. We wish to thank CFHT TAC for granting observing time; without the consistently good seeing of the Mauna Kea site, it would have been impossible to carry out most of the observations that we have reported on. We are also indebted to Pat Waddell and Carol Christian for support of the CFHT CCDs. This work was supported in part by the Natural Sciences and Engineering Research Council of Canada.

REFERENCES

Aaronson, M. 1986, in *Observational Cosmology*, ed. A. Hewitt, G. Burbidge, and L. Fang (Dordrecht: Reidel), p. 187.
Baade, W. 1952, *IAU Trans.*, **8**, 398.
Baade, W. 1963, *Evolution of Stars and Galaxies* (Cambridge: MIT Press).
Barnes, T.G., and Hawley, S.L. 1986, *Ap.J. (Letters)*, **307**, L9.
Bolte, M. 1988, in preparation.

Burstein, D., and Heiles, C. 1984, *Ap.J. Suppl.*, **54**, 33.
Cacciari, C., Clementini, G., and Buser, R. 1988, *Astr. Ap.*, in press.
Carney, B.W. 1980, *Ap.J. Suppl.*, **42**, 481.
Christian, C.A., and Schommer, R.A. 1987, *A.J.*, **93**, 557.
Cohen, J.G. 1985, *Ap.J.*, **292**, 90.
Freedman, W. 1985, Ph.D. Dissertation, University of Toronto.
Graham, J.A. 1984, *Structure and Evolution of the Magellanic Clouds*, ed. S. van den Bergh and K.S. de Boer (Dordrecht: Reidel), p. 207.
Hawley, S.L., Jeffreys, W.H., Barnes, T.G., and Lai, W. 1986, *Ap.J.*, **302**, 626.
Johnson, S.B., and Joner, M.D. 1987, *A.J.*, **94**, 324.
Jones, R.V., Carney, B.W., and Latham, D.W. 1988, *Astrophys. J.*, in press.
Kinman, T.D., Mould, J.R., and Wood, P.R. 1987, *A.J.*, **93**, 833.
Lee, Y.-W., Demarque, P., and Zinn, R. 1988, *IAU Colloquium No. 95, The Second Conference on Faint Blue Stars*, ed. A.G. Davis Philip, D.S. Hayes, and J.W. Liebert (L. Davis Press), p. 137.
Madore, B.F., McAlary, C.W., McLaren, R.A., Welch, D.L., Neugebauer, G., and Matthews, K. 1985, *Ap.J.*, **294**, 560.
Mould, J.R. 1988, preprint.
Mould, J.R., and Kristian, J. 1986, *Ap.J.*, **305**, 591.
Pritchet, C.J. 1986, in *Galaxy Distances and Deviations from Universal Expansion*, ed. B.F. Madore and R.B. Tully (Dordrecht: Reidel), p. 35.
Pritchet, C.J., and van den Bergh, S. 1987, *Ap.J.*, **316**, 517.
Pritchet, C.J., and van den Bergh, S. 1988, *Ap.J.*, in press.
Reid, I.N., and Strugnell, P.R. 1986, *M.N.R.A.S.*, **221**, 887.
Saha, A. 1986, *A.J.*, **92**, 742.
Saha, A., and Hoessel, J.G. 1987, *A.J.*, **94**, 1556.
Sandage, A.R. 1982, *Ap.J.*, **252**, 553.
Sandage, A.R., and Carlson, G. 1983, *Ap.J.*, **267**, L25.
Stetson, P. 1987, *P.A.S.P.*, **99**, 191.
Stothers, R.B. 1983, *Ap.J.*, **274**, 20.
Strugnell, P., Reid, N., and Murray, C.A. 1986, *M.N.R.A.S.*, **220**, 413.
Sweigart, A.V. 1987, *Ap.J. Suppl.*, **65**, 95.
Tammann, G. 1986, in *Observational Cosmology*, ed. A. Hewitt, G. Burbidge, and L. Fang (Dordrecht: Reidel), p. 151.
van den Bergh, S. 1975, in *Stars and Stellar Systems*, Vol. **9**, *Galaxies and the Universe*, ed. A. Sandage, M. Sandage, and J. Kristian (Chicago:University of Chicago Press), p. 509.
Vandenberg, D.A. 1988, *The Harlow Shapley Symposium on Globular Cluster Systems in Galaxies*, ed. J.E. Grindlay and A.G. Davis Philip (Dordrecht: Reidel), p. 107.
Walker, A.R. 1985, *M.N.R.A.S.*, **212**, 343.
Walker, A.R. 1988, in preparation.
Wan, L., Mao, Y-Q., and Ji, D-S. 1981, *Annals of Shanghai Obs.*, No. 2.
Welch, D.L., McAlary, C.W., McLaren, R.A., and Madore, B.F. 1986, *Ap.J.*, **305**, 583.

DISCUSSION

J. GRAHAM: How many "c" type RR Lyraes did you find in M31 relative to "ab" types?

C. PRITCHET: There are no c-type RR Lyrae stars in our sample because their amplitudes are too small to be detected.

J. MOULD: Can you tell us what the basis is for the assumed reddening in your M33 halo field? I think the Burstein and Heiles value is quite a bit lower.

C. PRITCHET: The assumed reddening ($E_{B-V} = 0.077$) was derived from recent work by Johnson and Joner. Their technique was uvbyβ photometry of 151 foreground A and F stars. The Burstein and Heiles (1982) result for this area is $0.03 < E_{B-V} < 0.06$.

CCD PHOTOMETRY OF MAGELLANIC CLOUD RR LYRAE VARIABLES

ALISTAIR R. WALKER
Cerro Tololo Inter-American Observatory, National Optical Astronomy Observatories, Casilla 603, La Serena, Chile.

ABSTRACT: CCD light curves for the RR Lyrae variables in the Magellanic cloud clusters NGC 121, 1786, 2210 and 2257 show that the stars have a mean magnitude of $\langle V\rangle_o = 19.02$ when corrected to the LMC center. If the LMC Cepheid modulus of 18.47 mag is correct then RR Lyraes with [Fe/H] ≈ -2 have $M_V = 0.55$. The difference between the LMC and SMC moduli found from the RR Lyraes is consistent with the the variation of horizontal branch luminosity with metallicity shown by the recent Yale models.

OBSERVATIONS

CCD observations of RR Lyrae variables in several Magellanic Cloud clusters have been made using the 1.0m telescope at the South African Astronomical Observatory and the 0.9m telescope at Cerro Tololo Inter-American Observatory. The RR Lyraes have V magnitudes of ≈ 19 and exposures of ≈ 30 minutes give an accuracy of a few 0.01 mag per observation.

RESULTS

1. The LMC distance is constrained to be 18.42 ± 0.05 mag if metal poor RR Lyraes have the canonical $\langle M_V\rangle = 0.6$. However in order to agree with the LMC Cepheid modulus of 18.47 mag (Feast and Walker 1987), metal poor RR Lyraes with [Fe/H] ≈ -2 (see below) must have $M_V = 0.55$.

2. The agreement of the moduli for NGC 1786, 2210 and 2257 after geometric and absorption corrections are made shows that these clusters lie in the plane of the LMC to better than ±400 pc.

3. The mean magnitude of the NGC 121 RR Lyraes is identical to the mean magnitude of the surrounding field RR Lyraes (Graham 1975) and not 0.2 mag brighter as previously found.

4. Cluster metallicities determined from the RR Lyrae period-amplitude relation are in agreement with [Fe/H] from low dispersion spectroscopy (eg Cowley and Hartwick 1982).

NGC 121 is more metal rich than the LMC clusters by about 0.5 dex.

TABLE I: Magellanic Cloud cluster distances from the RR Lyrae mean magnitudes, and [Fe/H] estimates.

OBJECT	<V>	GEOM. CORR.	$\langle V\rangle_{MC}$	A_V	$\langle V_o\rangle_{MC}$	MODULUS M_V=0.6	[Fe/H]
NGC 121	19.58	?	19.58	0.13	19.46	18.86	-1.6:
NGC 1786	19.27	0.00	19.27	0.23	19.04	18.44	-2.3
NGC 2210	19.12	0.09	19.21	0.19	19.02	18.42	-1.9
NGC 2257	19.02	0.18	19.20	0.19	19.01	18.41	-1.8

5. The difference between the SMC and LMC distance moduli is 0.44 ± 0.09 compared to the Cepheid result (Feast and Walker 1987) of 0.31 ± 0.05 mag. However the most recent Yale models of the horizontal branch (Lee, Demarque and Zinn 1987) give a magnitude - metallicity relation

$$\langle M_V(RR)\rangle = 0.20\ [Fe/H] + 0.92,$$

and upon using this the difference in LMC and SMC moduli becomes 0.31 ± 0.10 mag, in much better agreement. NGC 121 is 4.5 deg. from the SMC center and the difference between the Cepheid and RR Lyrae results could also possibly be due to NGC 121 being more distant than the average for the SMC Cepheids. Unfortunately none of the Cepheids studied by Caldwell and Coulson (1986) in their discussion of the SMC geometry are near NGC 121. NGC 121 does however seem to be at the same distance from us as the surrounding SMC field population.

REFERENCES

Caldwell, J.A.R., and Coulson, I.M. 1986, M.N.R.A.S., 218, 223.

Cowley, A.P., and Hartwick, F.D.A. 1982, Ap. J., 259. 89.

Feast, M.W., and Walker, A.R. 1987, Ann. Rev. Astr. Ap., 25, 345.

Graham, J.A. 1975, Pub. A.S.P., 87, 641.

Lee, Y-K, Demarque, P., and Zinn, R., 1987. IAU Colloquium No. 95, "The Second Conference on Faint Blue Stars", ed. A.G.D. Philip, D.S. Hayes, J. Liebert, (New York: Davis Press).

Walker, A.R. 1985, M.N.R.A.S., 212, 343.

Walker, A.R., and Mack, P. 1988a, A.J., in press.

Walker, A.R., and Mack, P. 1988b, A.J., in press.

AN UPDATE ON THE VELOCITY DISPERSIONS IN THE DRACO AND THE URSA MINOR DWARF GALAXIES

EDWARD W. OLSZEWSKI
Steward Observatory, University of Arizona, Tucson, Arizona 85721, U.S.A.

ABSTRACT In order to believe the claim (Aaronson 1983; see also Aaronson and Olszewski 1988) that the Draco and the Ursa Minor dwarf spheroidal galaxies have velocity dispersions an order of magnitude greater than "expected", we need multi-epoch observations of many, perhaps 20 or more, stars in each system. Observations obtained on 17 observing runs between April, 1982 and June, 1988, are combined to yield 54 observations of 23 stars in Draco, and 37 observations of 18 stars in Ursa Minor. After detecting and rejecting several velocity variables, I derive a velocity dispersion of approximately 10 km s^{-1} for each galaxy. This value supports the earlier results quoted in the above papers.

INTRODUCTION AND OBSERVATIONS

In 1983, Aaronson published high quality velocities of three carbon stars and one K giant in the Draco dwarf spheroidal galaxy, and one carbon star in the Ursa Minor spheroidal.

The velocities of the Draco C stars were -298, -300, and -280 km s^{-1}; trusting his instincts, Marc decided that he was seeing a real velocity dispersion. The later-obtained velocity of the K giant was -286 km s^{-1}, adding weight to Marc's conclusion. He then derived an M/L of ~ 30 for Draco, which is an order of magnitude greater than the M/L of globular clusters.

Let me stress that only six of the 41 stars observed to date are carbon or CH stars. All of these observations were made with the MMT, and its echelle spectrograph and intensified Reticon detector.

Stars in Draco and UMi were chosen from Stetson (1980), and Cudworth, Olszewski, and Schommer (1986), respectively. The few C stars known in each galaxy were also observed.

We have recently supplemented this list with stars outside the area studied for proper motions. We found stars with the correct magnitude and colors to be on the red giant branch of each spheroidal (the Draco candidates were supplied by Stetson, and by McClure). We then took low resolution spectra with John Hill's (Hill and Lesser 1988) MX (multi-fiber) spectrograph on the Steward 90". Stars deemed to be giants were then observed at the MMT. This additional membership list approximately doubled our sample.

The velocities were derived using the SAO software system (see Wyatt 1985). Both the twilight sky and several IAU standard stars were used as templates. In the list of velocities shown at the meeting, the extreme range for the nonvariable stars was 6 km s^{-1}, with an average standard deviation within the 1.5 - 2.0 km s^{-1} estimated measuring error.

CONCLUSIONS

Various ways of inflating a small, true velocity dispersion to this large, measured one have been previously discussed and rejected. These are: 1) binaries; 2) pulsation; 3) rotation of the spheroidals; and 4) tidal destruction of the spheroidals by the Milky Way. The most reasonable conclusion is that Ursa Minor and Draco have a velocity dispersion an order of magnitude higher than that expected for a "normal", globular-cluster-like, old population.

ACKNOWLEDGMENTS

Many of these observations were taken in collaboration with the late Marc Aaronson. This work was partially supported by NSF grant AST 86-11405.

REFERENCES

Aaronson, M. 1983, Astrophys. J. (Letters), 266, L11.

Aaronson, M., and Olszewski, E. W. 1988, in IAU Symposium 130, Evolution of Large Scale Structure in the Universe, in press.

Cudworth, K. M., Olszewski, E. W., and Schommer, R. A. 1987, Astron. J., 91, 1058.

Hill, J. M., and Lesser, M. P. 1988, in Instrumentation for Ground-Based Optical Astronomy: Present and Future, ed. L. B. Robinson (New York: Springer-Verlag), p. 233.

Stetson, P. B. 1980, Astron. J., 85, 387.

Wyatt, W. F. 1985, in IAU Colloq. 88, Stellar Radial Velocities, ed. A. G. D. Phillip and D. W. Latham (Schenectady: L. Davis), p. 123.

DISTANCE MODULUS AND RATE OF NOVAE IN M33

Massimo Della Valle
Department of Astronomy, University of Padova
Vicolo dell'Osservatorio 5, 35122 Padova, Italy

ABSTRACT: A preliminary reduction of seventy blue plates of M33, obtained at the 182 cm telescope of Asiago Observatory has allowed to derive the distance modulus $(m - M)_{M33} = 24.80 \pm .3$ and to give a quantitative estimate of nova rate, $R_{M33} = 4 \pm 2$ novae/yr. Moreover, it is provided an estimate of the galactic nova production and a comparison of the nova rates, normalized to $10^{10} L_{B_{\odot}}$, occurring in the spiral galaxies M31 and M33 and in the giant elliptical galaxies in Virgo.

Recovering a previous work carried out by Rosino and Bianchini (1973a) on 32 plates taken at 122 cm of Asiago Observatory, seventy blue plates of M33, obtained at the 182 cm telescope, between 1975 and 1986, and nine 122 cm-newtonian plates, have been examined to search for extragalactic novae. All material were blinked twice and five novae have been discovered. The analysis of the set of data yields two main results: **a)** using the dependence of the period of visibility of novae on distance, which has been discussed by van den Berg and Pritchet (1986), we have derived a difference between the distance modulus of M33 and M31 of $\Delta(m - M) \simeq 0.3$ mag. Assuming $(m - M)_{M31}$=24.55 (van den Bergh 1988), it follows that $(m - M)_{M33} \simeq 24.8$. This result is consistent with the $(m - M)_{M33}$ =24.9±.3, given by two observed novae (nova 1 and 6) with determined rate of decline ($v_d = 0.16$ mag/day), and calibrated by "maximum magnitude versus rate of decline" relationship (${A^0}_{B_{M33}} = 0.18$ and ${A^0}_{B_{M31}} = 0.32$, Burstein and Heiles 1984). **b)** for the first time, we are able to give a quantitative estimate of the nova production of M33, by means of the *control time* method, normally used for the determination of supernova frequency (Zwicky 1942). The possibility to apply this technique in order to attain the nova rate, has been tested on M31, being available for this galaxy both the " direct" Arp's (1956) determination of nova rate, and the extensive survey carried out in the last thirty years at Asiago Observatory. A preliminary determination of the value of the control time of M33, $t_{c_{M33}} \simeq 550^d$, leads to an observed nova rate of $R_{M33} \simeq 3.5$ novae/yr. This number must be increased by a corrective factor K, for novae outside the observed area, for novae blocked by absorbtion and for novae lost on the central background light. A K value for M33 of ~ 1.2 has been estimated taking in account that Rosino et al.(1988) have found for Andromeda Galaxy $K_{M31} \simeq 1.4$ and considering that for M33 the corrections due to absorbtion and central background light have to be smaller than for M31. Therefore the final value of the nova rate turns out to be $R_{M33} = 4 \pm 2$ novae/yr.

From the knowledge of the nova rate of M33 and of M31 we can draw an indirect estimate of the galactic nova production, whose value is based on the conflicting estimates obtained by Liller and Meyer (1987): $R = 73 \pm 24$ novae/yr and by Sharov (1972): $R = 260$ novae/yr. If the total B-luminosities for M31, M33 and Galaxy are respectively: $L_B \simeq 4.7 \cdot 10^{10} L_{B_\odot}$ (B_T=4.38, A^i=1.03, Sandage and Tamman 1981), $L_B \simeq 0.61 \cdot 10^{10} L_{B_\odot}$ (B_T=6.26, A^i=0.45) and $L_B \simeq 2.3 \cdot 10^{10} L_{B_\odot}$ (van den Bergh 1988), we expect that the galactic nova rate should not be greater than $R \approx 15$ novae/yr *(fig. 1)*. This value disagrees strongly with both the previous determinations.

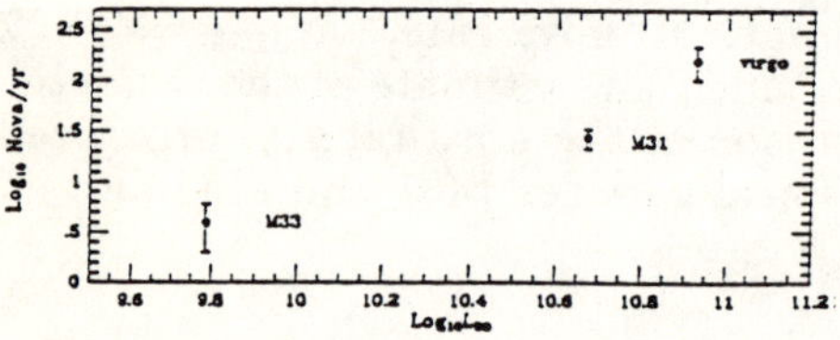

fig. 1 Logarithm of nova rate per year versus total B-Luminosity.

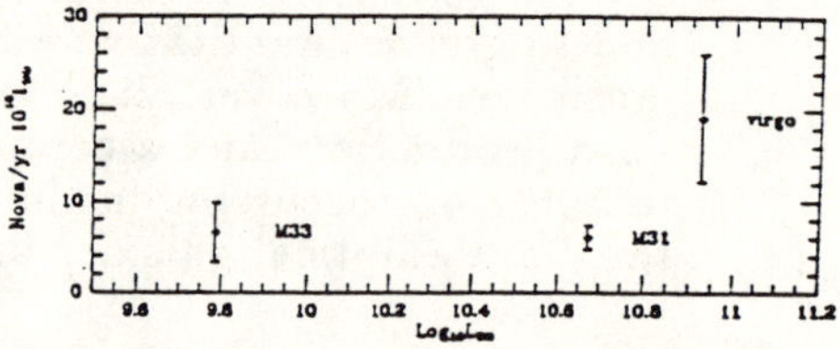

fig. 2 Nova rate per $10^{10} L_{B_\odot}$ per year versus total B-Luminosity.

Finally, the comparison between the normalized (to $10^{10} L_{B_\odot}$) nova rates of M31 and M33 with the one found by Pritchet and van den Bergh (1987) for Virgo galaxies, shows *(fig. 2)* a possible difference in the nova production between the spiral galaxies of the Local Group and the elliptical giants of Virgo cluster.

Acknowledgements

I would like to thank Prof. L.Rosino for helpful discussions.

References

Arp,H.,C. 1956, *A.J.*, **61**, 15
Burstein D., Heiles,C. 1984, *Ap.J. Suppl.*, **54**, 33
Liller,W., Mayer,B. 1987, *Pub.A.S.P.*, **99**, 606
Pritchet,C.J., van den Bergh,S. 1987, *Ap.J.*, **318**, 507
Rosino,L., Bianchini,A. 1973a, *Astron.Astrophys.*, **22**, 461
Rosino,L., D'onofrio, M.,Capaccioli, M.,Della Valle,M. 1988, in prep.
Sandage,A., Tamman,G.A., *A Revised Shapley-Ames Catalog of Bright Galaxies* Carnegie Institution, 1981
Sharov,A.S. 1972, *Soviet Astron.*, **16**, 41
van den Bergh,S. 1988, *Pub.A.S.P.*, **100**, 8
van den Bergh,S., Pritchet,C. 1986, *Pub.A.S.P.*, **98**, 110
Zwicky,F. 1942, *Ap.J.*, **96**, 8

ASYMPTOTIC GIANT BRANCH POPULATIONS IN IC1613 AND THE SAGITTARIUS DWARF

KEM HOLLAND COOK
Mt. Wilson and Las Campanas Observatories, 813 Santa Barbara St., Pasadena, CA 91101, and Steward Observatory, University of Arizona, Tucson, AZ 85721

MARC AARONSON
Steward Observatory, University of Arizona, Tucson, AZ 85721

ABSTRACT The surface distributions of carbon stars in IC1613 and in the Sagittarius Dwarf Irregular have been mapped. These asymptotic giant branch stars trace an intermediate age population which is present in both of these galaxies. The distance to each of these galaxies estimated from the luminosity distribution of the carbon stars is consistent with the distance estimated from the luminosity of the tip of the first giant branch.

INTRODUCTION

We have been using an intermediate-band photometric technique to examine the asymptotic giant branch (AGB) content of Local Group galaxies in an effort to better understand their star formation history, chemical evolution, and distance as well as stellar evolution theory of the AGB (Cook, Aaronson, and Norris 1986). This technique distinguishes M-type from C-type stars and yields metallicity estimates for the M stars (Cook 1987). Since the technique requires a broad-band index, broad-band magnitudes and a color are produced as well an intermediate-band index.

The data presented here were obtained using the CTIO 4-m with an RCA CCD at prime focus. Frames were taken at V, I, 77, and 81. 77 and 81 are intermediate-band filters centered on regions of TiO and CN absorption respectively. Six fields covering the central 96 arcmin^2 of IC1613, one field covering the Sagittarius Dwarf Irregular (Sagdig), and one control field for foreground contamination toward Sagdig were analyzed. The CCD frames were reduced in the usual manner and PSF fitting photometry using DAOPHOT was done for about 6000 stars in the IC1613 fields about 700 stars in the Sagdig field, and about 400 stars in the control field. V−I *vs* I color-magnitude (CM) and 77−81 *vs* V−I color-color (CC) diagrams were formed from these data.

IC 1613

The most prominent feature of IC1613's CM diagram is the presence of the tip of the first giant branch at $I \sim 20.3$ and $V - I \sim 1.6$. The I luminosity function also shows a sharp jump at this magnitude. Following Mould, Kristian, and Da Costa (1983), this feature can be used to estimate the distance as $(m - M)_0 \approx 24.2$.

The CC diagram identifies about 150 carbon stars. There are about the same number of stars in the M star portion of the CC diagram, but ~ 50 are expected to be foreground dwarfs. There *is* a discernible group of low metallicity M stars which are unlikely to belong to The Galaxy. If these are IC1613 M giants, then this population has $[Fe/H] < -1.3$ and the carbon-to-M star ratio (for M3 and later) is $\gtrsim 10$. The luminosity function of the carbon stars is strongly peaked with $< I >= 19.9$. If the tight luminosity distribution of carbon stars is an universal feature, it can be used to estimate distances. There is a 6.0 mag difference between the mean carbon star magnitude in the LMC (Blanco, McCarthy, and Blanco 1980) and IC1613 suggesting $(m - M)_0 \approx 24.3$ for IC1613.

SAGITTARIUS DWARF IRREGULAR

Sagdig was recently discovered (Cesarsky *et al.* 1977), and is interesting because it has a very low intrinsic luminosity— the discovery paper estimated $M_B \approx -9.5$. It also has a large HI mass for its luminosity. Sagdig's low galactic latitude ($l \approx 21°$, $b \approx -16°$) required the use of a control field to correct for foreground contamination.

The CM diagram for Sagdig shows a blue main sequence suggesting recent star formation. The brightest blue stars can be used to estimate the distance (Sandage and Carlson 1985) as $(m - M)_0 \sim 25.3$. Sagdig's luminosity function shows a strong jump at $I = 21.5$ which is probably the tip of the first giant branch suggesting $(m - M)_0 \sim 25.4$.

The CC diagram shows about 25 carbon stars in Sagdig suggesting star formation in the intermediate past. The carbon stars seem to be composed of a fainter, bluer component and a brighter, redder component. The mean, I, magnitude for the C stars is 20.7 which suggests $(m - M)_0 \sim 25.0$ for Sagdig assuming an universal C star luminosity distribution. The spatial distribution of the carbon stars extends well beyond the 'body' of the galaxy which contains all of the blue stars.

SUMMARY

Carbon stars have been found in both IC 1613 and the Sagittarius Dwarf. These stars represent star formation in the intermediate past— 1 to 10 gigayears. Both of these dwarfs have experienced recent star formation as evidenced by their blue stellar content. The carbon star luminosity function was found to be useful in providing a distance estimate to both of these systems. CM diagrams for the fields studied were used to estimate distances by identifying the tip of the red giant branch.

REFERENCES

Blanco, V. M., McCarthy, M. F., and Blanco, B. M. 1980, *Ap. J.,* **242**, 938.

Cesarsky, D. A., Laustsen S., Lequeux, J., Schuster, H-E., and West, R. M. 1977, *Astr. Ap.,* **61**, L31.

Cook, K. H. 1987, Ph.D. Thesis, University of Arizona, Tucson.

Cook, K. H., Aaronson, M., and Norris, J. 1986, *Ap. J.,* **305**, 634.

Mould, J. R., Kristian, J., and Da Costa, G. S. 1983 *Ap. J.,* **270**, 471.

Sandage, A., and Carlson, G. 1985, *Astr. J.,* **90**, 1464.

CEPHEID VARIABLES IN THE PHOENIX GALAXY AND IC 5152

NELSON CALDWELL
F.L. Whipple Observatory
Smithsonian Institution
P.O. Box 97
Amado, Arizona 85645

and

ROBERT SCHOMMER
Department of Physics and Astronomy
Rutgers University
Piscataway, NJ 08854

We have been monitoring several nearby galaxies for variable stars for nearly four years. Our method is to use the Cerro Tololo 4m telescope at Prime Focus with a CCD, imaging the galaxies in the blue passband where the light variation is large in Cepheids. We have twenty epochs for both the Phoenix dwarf irregular galaxy and IC 5152, also an irregular galaxy. To find the variables, we select the 6-10 best seeing frames, and match their point-spread functions and intensity scales. All of those frames are then displayed as a movie on an image display system, and variables are picked out by eye. We find this method to be efficient and less painful than doing photometry on all of the stars on all of the frames. Photometry is then obtained for the candidate variables and selected non-variables from the *original* frames. The collected data for each candidate are analyzed with a period finding program, which usually gives a few possible solutions. Phase diagrams are plotted for those periods, which has always revealed that only one period gives a believable light curve. We intend eventually to use CCD I band measurements to get the luminosities of the Cepheids, for now we use the mean intensity in B.

The Phoenix galaxy was discovered by Schuster and West (1976) who correctly noted that most of the brightest stars were red, and probably giants rather than supergiants. However, their estimates of the star's magnitudes were in error by three magnitudes, and hence they derived a very small distance and surmised that Phoenix was a remote globular cluster. Canterna and Flower (1977) obtained better photometry using CTIO 4m plates, but incorrectly assumed that the brightest red stars in Phoenix are supergiants. They derived a very large distance modulus to the galaxy (m-M=26.3).

It is clear from the resolution of stars on our CCD images that Phoenix is a good deal closer than Canterna and Flower thought. We do not have enough variables with known periods yet to define a P-L relation, but we can still derive a distance from those in hand. The apparent difference of the blue distance moduli bewteen the LMC and Phoenix is 6.2, making the true distance modulus 25.2 for the "long" scale and 24.6 for the "short" scale. Phoenix is positively a new member of the Local Group. The B-V color-magnitude diagram for Phoenix clearly reveals a strong giant branch with a termination at V = 21 (M_V = -3.6). Also present is an extended main sequence, but one that is weaker than any of the other well-studied nearby dwarf irregular galaxies such as IC 1613, NGC 6822, Sextans A, GR 8 or Pegasus. With a total absolute magnitude of M_V = -11, Phoenix is in the luminosity realm of the dwarf spheroidal galaxies. It is interesting to note that the only on-going star formation is confined to a single association, the rest of the galaxy having only an old population of stars. Could Phoenix be a link to the dwarf spheroidals? We must measure the H I content to find out!

IC 5152 is farther away than Phoenix, as is apparent from the resolution on our CCD frames. We calculate that IC 5152 is about 7.5 blue magnitudes farther away than the LMC, putting it at about the same distance as NGC 300. Work on the stellar population is still underway.

Canterna, R., and Flower, P. 1977, Astrophys.J., **212**, L57.
Schuster, H.-E., and West, R., Astron.Astrophys., **49**, 129.

III

DISTANCE SCALE CALIBRATORS

ASTROMETRY AND THE DISTANCE SCALE

DAVID G. MONET
U.S. Naval Observatory Flagstaff Station
P.O. Box 1149, Flagstaff, AZ 86002, U.S.A.

ABSTRACT Currently attainable ground-based astrometric precision is near one milliarcsecond for the measurement of relative parallax. Assuming that the accuracy is similar, direct astrometric measurement of the distance to many stars of astrophysical interest, including those used in distance scale calibration, ought to be possible once technical problems are solved. These issues are discussed and some solutions are presented.

I. INTRODUCTION

Relative optical astrometry by non-photographic techniques appears to offer a significant improvement in accuracy (more than a factor of 2) and limiting magnitude (more than 3 magnitudes) over traditional photographic methods. Two groups have reported accuracies of 1 milliarcsecond or better (Gatewood *et al.* 1988 and Monet *et al.* 1987). Since astrometry requires substantial accuracy, characterization of a detector by a single number is deceiving. One must examine the detector in the context of the problem to which it is applied to understand its capabilities and limitations. Relative astrometry ought to be simple. The position of the object of interest (the program star) is measured with respect to adjacent reference stars or galaxies (the reference frame). The positions of the reference objects are used to compute the transformation of each observation into a common coordinate system. If measurements are made with respect to a stellar reference frame, all objects will have parallax and proper motion and may have other perturbations although the importance of these effects is determined by the accuracy of the measurement.

Because every astrometric detector has a finite dynamic range, a limited field of view, and various noise sources, astrophysically important program stars may pose significant observational problems. Section II presents a discussion of program star and reference object attributes as well as a discussion of the correction of the relative parallax (that measured with respect to the reference frame) to absolute parallax (the reciprocal of the distance). Section III presents a brief comparison of Allegheny Observatory Multichannel Astrometric Photometer (MAP) detector (Gatewood 1987) and the U. S. Naval Observatory Flagstaff

Station (NOFS) charge coupled device (CCD) detector. Further discussion of these and other astrometric detectors was presented by Monet (1988a). Section IV discusses some of the advanced astrometric detectors being developed at NOFS and Section V presents conclusions.

II. CELESTIAL CONSTRAINTS

The choice of a program star is driven by its astrophysical value and the expectation of finding a significant result. Since current astrometric precision is claimed to be near 1 milliarcsecond, Table I lists the number of various stars within a reasonable distance and their typical *V* magnitudes.

TABLE I Candidate Objects North Of The Celestial Equator

Type Of Object	Approximate Number Known Within (pc)	Typical *V* Magnitude
Open Cluster	5 (200)	3-20
Cepheid Variable	8 (500)	3-6
RR Lyrae Variable	1 (500)	8-9
Other Horizontal Branch	≥7 (500)	7-9
Brightest Red Supergiant	0 (500)	–
Brightest Blue Supergiant	1 (500)	1
High Latitude Supergiant	2 (1000)	6-7
Carbon Star	13 (500)	5-21
Planetary Nebula Central Star	10 (500)	11-16
Cataclysmic Variable	29 (350)	12-16
Subdwarf	many (500)	10-20

The use of field stars as reference objects introduces a non-zero value for the correction from relative to absolute parallax. The error of this correction is added in quadrature to the error of the relative parallax to determine the error of the absolute parallax. Consider the following example. According to van Altena's (1974) model for a galactic latitude of 60 degrees, the mean parallax at 10th magnitude is 8 milliarcseconds and that at 16th magnitude is 2 milliarcseconds. If spectrophotometry of the reference stars yields a mean parallax accurate to 10% (30% accuracy for each of 10 stars) and the error of the relative parallax is 0.5 milliarcsecond, the error in the absolute parallax drops from 0.94 to 0.54 milliarcsecond by using the fainter reference frame. This example shows that a suitably faint stellar reference frame is almost as good as an extragalactic reference frame if accurate spectrophotometric parallaxes are measured for each of the stars.

Should an extragalactic reference frame be available, the correction from relative to absolute parallax for Galactic stars would be essentially zero and its error would be the mean measuring error. QSOs would be ideal because of their great distances and because they appear to be (and therefore measure just like) stars. Unfortunately, QSOs are scarce at bright magnitudes. Table II summarizes the results presented by Koo *et al.* (1986) and lists B magnitude and the integral QSO surface density three ways, the number per square degree and the number in the field of view of the NOFS 61-inch astrometric telescope with either the Texas Instruments 800x800 or the Tektronix 2048x2048 CCD.

TABLE II QSO Surface Density

B	Number Brighter Than B (Koo *et al.* 1986)		
Mag	per square degree	N61+TI800	N61+TK2048
18	0.5	0.001	0.02
19	3	0.006	0.13
20	16	0.03	0.7
21	45	0.09	1.9
22	100	0.2	4.3

The use of reference galaxies would obviate the QSO surface density problem. Unfortunately, few quantitative studies of the astrometric properties of galaxies have been done. The Lick survey (Klemola *et al.* 1987) was highly selective in choosing reference galaxies, typically using 80 per 36 square degree field. They found that brightness was a major contributor to the astrometric weight of a galaxy, that a relative weight $W_g = 0.2$ for a faint galaxy was appropriate, and that their error for a single measure of a faint star was 2.0 microns. The relevance of the Lick results to the sub-micron world of milliarcsecond-class parallaxes can be questioned. A rough estimate given by van Altena (1987, private communication) is that a good galaxy has a positional uncertainty about twice as large as that for a star. The limited experience gained by the NOFS CCD program agrees with that estimate. On the other hand, King (1988) notes that "one QSO is worth twenty galaxies".

The conclusions are inescapable. With currently available CCDs, the probability of having the three QSOs needed for a minimum reference frame is extremely small. Field star reference frames will be the norm although they may be augmented by those field galaxies that have acceptably small measuring errors. Should a QSO be available, it will be treated more like a program star than a reference star. Minimization of the error in the correction from relative to absolute parallax is essential. A field star reference frame should be appropriately faint and spectrophotometric parallaxes will be needed for every member. A field galaxy reference frame must be rich enough to offset the decreased astrometric accuracy of each member. If the correction is determined by a single object then its error is limited by the measuring error for that object not the mean measuring error of the reference frame. In any case, an astrometric detector designed to observe the bright ($V < 10$) program stars at distances approaching 1000 pc must provide astrometric information for faint ($V > 20$) field stars, galaxies, and QSOs.

III. INSTRUMENTAL CONSTRAINTS

Of the two groups mentioned previously, neither has an ideal detector system. The MAP (Gatewood 1987) suffers from poor quantum efficiency and the NOFS CCD (Monet 1988b) suffers from a limited field of view. A general discussion of astrometric technologies and their advantages and limitations was presented by Monet (1988a) and various authors (*e.g.*, King 1983) have discussed theoretical limitations of astrometric accuracy. Assuming that astrometric accuracy in the focal plane is well understood and that the use of reference galaxies may occasionally alleviate some difficulties, the following three areas of detector technology limit the choice of possible program stars.

Limiting Magnitude: The previous section demonstrated that accurate absolute parallaxes require measuring faint ($V > 20$) reference objects. Here the CCD with its 80% quantum efficiency is clearly superior to the MAP whose quantum efficiency is limited to a few percent because it uses a half-blind, one-dimensional Ronchi ruling, fiber optics, and photomultiplier tubes.

Field Of View: Because QSOs and galaxies with the proper morphological properties are rare, the detector needs a field of view of a few square degrees not a few square arcminutes. Here the MAP is superior because its fibers can be repositioned anywhere in the focal plane. CCDs are single pieces of silicon and currently available devices are small, about 1.5 cm^2. The eagerly awaited Tektronix TK2048 is 30 cm^2. At some point, atmospheric effects should start degrading astrometric accuracy although both the MAP and CCD programs appear not to be limited by such effects yet.

Dynamic Range: The difference between the apparent magnitude of a bright program star and a faint reference star might be 15 magnitudes or more. Such a dynamic range is outside the normal MAP operating mode and exceeds the dynamic range of all known CCDs. Techniques are available for both the MAP and the CCD to selectively dim particular stars but only limited testing has been done to date.

IV. ASTROMETRIC DETECTORS AT NOFS

NOFS remains committed to the use of CCDs for astrometry for three reasons: their high quantum efficiency, their lack of moving parts, and their long term stability. Regarding this last point, the so-called Mark IV 800x800 CCD has been operating since 1979, has survived numerous accidents, and is still the standard astrometric CCD.

Various solutions to the inherent problems of CCDs have been proposed and some are being implemented:

Detector Size: A large field of view requires a large CCD. A thin, science grade TK2048 is on order and a thick, set-up grade device (Ables *et al.* 1988) has been implemented. Observing and simple data analysis software has been written but the large number of defective pixels on the set-up grade device appears to preclude serious astrometric testing.

Selective Dimming: One method for increasing the dynamic range is to selectively dim the program star. The photographic astrometry program (Dahn *et al.* 1988) routinely uses filters with neutral density spots up to $D = 4.9$ (12.25 magnitudes). To the limits tested by that program, these spots do not affect astrometric accuracy. Three factors must be evaluated before such spots can be used in the CCD astrometry program. First, the CCD is about a factor of ten more accurate than the photographic plate measuring machine used by Dahn *et al.* Do spots remain astrometrically acceptable in this regime? Second, the size of the photographic spots (approximately 1 cm^2) is too big for use with small CCDs. Can astrometrically acceptable spots be made smaller? Third, the photographic spotted filters are in close proximity to the plate but the CCD dewar prevents a similar configuration. Should spots be deposited directly on the CCD? If not, where should the spotted filter be placed?

A Mosaic Of CCDs: Another method of selective dimming is to use different exposure times for the bright star and the reference frame. Since currently available CCDs cannot be accessed on a pixel by pixel basis, an array of CCDs has been commissioned. Six front illuminated, unmounted CCDs will be attached to a silicon substrate by means of silicon-on-silicon bonding technology. Each device will be separately connected to package pins by means of circuitry on the substrate and a six-fold electronics package will provide simultaneous control. The CCDs are arranged in a 3x2 pattern with the columns aligned in the "3" direction. Five of the devices are Thomson-CSF TH7882 CCDs which have 23 micron square contiguous pixels in a format of 576 rows by 384 columns. The sixth device is a Fairchild CCD222 which uses interline transfer technology and has 12x18 micron pixels on 30x18 micron centers in a format of 488 rows by 380 columns. Because the first shift puts the charge in the opaque shift registers, the CCD222 is self-shuttering. While the TH7882's integration time is controlled by the camera shutter, the CCD222's integration time is determined by its readout rate. Therefore, the bright star will be imaged by the CCD222, whose charge will be continuously read by the host computer. At the end of the exposure the TH7882s will be read. Combining the dynamic range of the TH7882s and the short CCD222 integration time, the dynamic range of this mosaic may approach 13 magnitudes. Astrometric integrity is guaranteed by the substrate which is a single crystal of silicon. The sensor is being procured in 1988 but it may require several years of development before useful results are produced.

V. CONCLUSIONS

While the ideal astrometric detector has yet to be developed, many astrophysically important programs can be started with currently available detectors. Those underway at NOFS include the following. The CCD parallax program concentrates on intrinsically faint stars. Preliminary solutions have been done for more than 100 stars and errors of 1 milliarcsecond are typical. Well observed stars have errors as small as 0.5 milliarcsecond. The photographic parallax program uses fine grain *IIIa* emulsions and bright stars such as RR Lyrae are being observed. Preliminary tests show that these plates can yield accuracies approaching 1 milliarcsecond. A program to identify QSOs behind globular clusters is in progress and first epoch CCD observations of a few clusters have been made. Absolute globular cluster proper motions should be detectable in 5-10 years given currently available astrometric accuracy.

Ground-based astrometry will not be rendered obsolete by the current generation of satellites with enhanced astrometric capabilities. One hopes that *Hubble Space Telescope* and *HIPPARCOS* have survived their long storage without degradation and that the current projected launch dates in 1989 do not slip further. The Fine Guidance Sensors of *HST* offer the tremendous dynamic range needed to do relative astrometry of bright stars with respect to the brighter QSOs. The Baum spot of the Wide Field Camera offers the possibility of astrometry of extremely faint objects and selective dimming of 8 magnitudes. *HIPPARCOS* promises extremely accurate absolute parallaxes for brighter stars (1-2 milliarcseconds; Perryman 1985).

Until the next generation of astrometric detectors becomes available, astrometry of distance scale calibrators such a Cepheid and RR Lyrae variables will not provide definitive results. The MAP is limited by the uncertainty of the correction from relative to absolute parallax and the CCD is limited by its dynamic range. However, current astrometric detectors are producing many exciting scientific results as well as providing a sound basis for future work.

ACKNOWLEDGMENTS

Hugh Harris contributed significantly to this work, particularly in the evaluation of program star availability, and Bill van Altena's comments on this manuscript were appreciated. NOFS continues to thank the Hubble Space Telescope Wide Field and Planetary Camera team, especially Jim Westphal, for the continued availability of the Mark IV CCD. Astrometry at the Flagstaff Station is a task shared by all. Their assistance and insight forms the foundation that supports this work.

REFERENCES

Ables, H. D., Harris, H. C., and Monet, D. G. 1988, *Le Monde Des Galaxies* conference proceedings, Paris: Institute d' Astrophysique, *(in press)*.

Dahn, C. C., Harrington, R. S., Kallarakal, V. V., Guetter, H. H., Luginbuhl, C. B., Riepe, B. Y., Walker, R. L., Pier, J. R., Vrba, F. J., Monet, D. G., and Ables, H. D. 1988, A. J., 95, 237.

Gatewood, G. D. 1987, A.J., 94, 213.

Gatewood, G. D., de Jonge, J. K., Castelaz, M., Han, I., Persinger, T., Prosser, J., Reiland, T., and Stein, J. 1988, Ap. J., *(in press)*.

King, I. R. 1983, P. A. S. P., 95, 163.

King, I. R. 1988, in *Globular Cluster Systems In Galaxies*, ed. J. E. Grindlay and A. G. D. Philip (Dordrecht: Kluwer), p. 60.

Klemola, A. R., Jones, B. F., and Hanson, R. B. 1987, A. J., 94, 501.

Koo, D. C., Kron, R. G., and Cudworth, K. M. 1986, P. A. S. P., 98, 285.

Monet, D. G. 1988a, Ann. Rev. Astron. Astrophys., *(in press)*.

Monet, D. G. 1988b, in *Instrumentation for Ground-Based Optical Astronomy*, ed. L. B. Robinson (New York: Springer), p. 493.

Monet, D. G., Dahn, C. C., Harris, H. C., Ables, H. D., Luginbuhl, C. B., Pier, J. R., and Vrba, F. J. 1987, B. A. A. S., 19, 641.

Perryman, M. A. C. 1985, Highlights Of Astronomy, 7, 673.

van Altena, W. F. 1974, A. J., 79, 826.

DISCUSSION

J. GRAHAM: Do you have any problems with handling the data volume from your CCD astrometry?

D. MONET: Yes! Fortunately, a routine program like the astrometry program allows many "batch" routines to do most of the processing without human intervention. We still back-up all frames to magnetic tape but this may change in the future.

T. LUTZ: What is the brightest V magnitude for which you can measure a parallax?

D. MONET: In theory, a neutral density filter would allow work on an arbitrarily bright star. The key is to get enough reference stars on the same CCD frame as the program star. With the NOFS 61-inch telescope and a 800 × 800 CCD, the practical limit is near 15th V magnitude.

CALIBRATION OF THE CEPHEID PERIOD-LUMINOSITY RELATION

ALISTAIR R. WALKER
Cerro Tololo Inter-American Observatory, National Optical Astronomy Observatories, Casilla 603, La Serena, Chile.

ABSTRACT: For seventy-five years the absolute calibration of the Cepheid Period-Luminosity (PL) relation has been of vital importance since Cepheids are the primary distance indicator for nearby galaxies, whose distances are in turn crucially important for fixing the distance scale out to cosmologically interesting distances.

The PL relation, with slope fixed from Cepheids in the Magellanic Clouds and zeropoint primarily from Cepheids in galactic clusters now appears to be able to predict the absolute magnitudes of Cepheids to an accuracy of ~0.1 mag. A significant part of the remaining error is due to the uncertainty of the distance to the Pleiades.

INTRODUCTION

In the past five years there have been considerable advances in our knowledge of the absolute magnitudes of Cepheid variables, and a concomitant gain in confidence regarding their utility as standard candles. The most recent review on the use of Cepheids as distance indicators is that of Feast and Walker (1987) (FW87), who comprehensively investigate the Cepheid distance scale and reference earlier reviews. They point out that due largely to the advent of improved instrumentation Cepheids can be discovered using ground based telescopes out to beyond 5 Mpc while distances in general can be found with much greater accuracy than in the era of photographic plates due to the elimination of systematic errors in the photometry. The use of red and infrared passbands has lessened the importance of reddening while application of infrared arrays will improve and extend this work. Efforts are underway to discover and measure Cepheids in more distant galaxies, and to redo and extend the earlier photographic work on those in nearby galaxies. With the Hubble Space Telescope we can expect that there will be efforts to find Cepheids as distant as the Virgo cluster.

Cepheid variables remain the prime distance indicator for nearby galaxies, and have been so since Hertzsprung (1913) first presented a calibration of the PL relation. The early history of the PL relation is described in the fascinating review by Fernie (1969) and it is apparent that some of the suspicion with which the PL zeropoint calibration is regarded even today is due to the 40 year period in which it was incorrect by about 1.5 magnitudes.

This review will discuss the calibration of the Cepheid PL relation from observations of galactic Cepheids. Although traditionally Cepheid distance scale work has involved the use of B and V band photometry only, the advent of III-V photocathodes, CCDs and improved IR detectors has extended the wavelength coverage to redder passbands which have now assumed great importance for observations of extragalactic Cepheids. As work on the galactic calibrators (primarily those in clusters and associations) mainly involves observations in the B and V passbands we will concentrate on the calibration of the V band PL relation (PLV). The zeropoint of the V, B-V form of the Period-Luminosity-Color relation (PLCV) will also be found. The reliability of the calibrations can be checked in a number of ways, most importantly by the comparison of various distance indicators in the Magellanic Clouds (Feast 1988).

CEPHEID LUMINOSITY RELATIONS

The form of the PL and PLC relations can best be found from observations of Cepheids in the LMC. The LMC geometry is reasonably well understood while the Cepheids are bright enough so that high precision CCD, photoelectric and infrared photometry is possible.

Some recent determinations of the slope of the PLV relation are collected in Table I.

TABLE I Slope of the PLV Relation

Galaxy	Slope	Error	Nstars	Ref
LMC, SMC	-2.78		136	CC86
LMC	-2.91	0.15	26	CC86
SMC	-2.95	0.16	44	CC86
LMC	-3.02	0.10	43	Walker (1987c)
Milky Way	-2.90	0.08	28	FW87
Milky Way	-2.92	0.12	52	Gieren (1988)

Note: The first entry in the table is for a combined SMC, LMC solution using "statistical" reddenings, whereas the next two entries use stars with individual reddenings.

FW87 adopted a slope of -2.78 but there now seems to be reasonable evidence that the slope is slightly more negative than this, here we will adopt

$$\langle M_V \rangle = -2.91 \log P + \varphi(PLV).$$

In practice, small changes in slope are compensated by a corresponding change in the zeropoint. Slope differences such as those in Table I will typically produce $\langle M_V \rangle$ differences for Cepheids at the long and short ends of the period range of order 0.1 mag. For an ensemble of Cepheids with a range in periods the effect should be negligible.

The slope of PLV relations in the galaxy, LMC and SMC are expected to be the same on theoretical grounds (Stothers 1988) and this appears to be confirmed from the observational values in Table I. Possible different slopes found for PLB in some galaxies (eg IC1613, Sandage 1971) can be attributed either to calibration errors of the faint photographic photometry, or a selection effect arising from the finite width of the instability strip. This means that at close to the limiting magnitude, for any given period, only the brighter Cepheids will be found.

The zeropoint of the bolometric PL relation is expected to be insensitive to abundance effects (Becker, Iben and Tuggle 1977, Stothers 1988) and the same is true of the PLV relation. Although the bolometric correction is a weak function of metallicity (Laney and Stobie 1986), for the V passband the bolometric corrections are very small. Stothers (1983) assumed that metal poor RR Lyraes in the Magellanic Clouds have the same absolute magnitudes and then showed that the Cepheid PL relation zeropoints were the same to within ±0.10 mag despite $\Delta Z = 0.009$, while theoretical calculations of the size of the metallicity dependence have been made by Caldwell and Coulson (1987) and Stothers (1988), the latter finds that the PL zeropoint is a weak function of helium abundance (Y) and metal abundance (Z)

$$\Delta M_V = 0.5\ \Delta Y - 2.8\ \Delta Z.$$

It appears likely from observations of HII regions in galaxies that Y and Z are correlated, approximately (Peimbert 1986)

$$\Delta Y = 3.5\ \Delta Z$$

hence

$$\Delta M_V = -1.05\ \Delta Z.$$

Therefore even for large changes in Z the PL zeropoint hardly alters, e.g. from the galaxy to the SMC ($\Delta Z = 0.015$) ΔM_V =

0.016 mag, therefore for any comparison involving galactic Cepheids alone the zeropoint differences on PLV will be totally negligible. Calculations by Caldwell and Coulson (1987) show that the same is true for R and I, but at B and in the infrared metallicity corrections may be necessary.

There are however two potential problems associated with the PL relation. The first is that the finite width of the instability strip means that the PL relation is best used with sufficient Cepheids so that it can be assumed that the strip is uniformly filled. The second problem is that even with a large number of Cepheids, the filling of the strip may depend on metallicity, i.e. exactly how the evolutionary tracks cross the strip and the rate at which the stars evolve along these tracks (Madore 1985a). The importance of these effects is not presently known with any degree of accuracy, both can be minimized by observing in the infrared.

PLC relations have been extensively used for finding Cepheid distances following the simple theoretical demonstration of the need for such a relation by Sandage (1972), and particularly since the Martin, Warren and Feast (1979) (MWF79) derivation of the PLCV relation for LMC Cepheids. PLC relations have also been established for V magnitude and V-I color (Caldwell and Coulson 1986, CC86) and for H magnitude and J-K color by Laney and Stobie (1986). Determination of the color term for the PLCV relation is difficult due to the necessity for disentangling the effects of differential reddening. This has led to considerable controversy regarding the size of the color coefficient, however Laney and Stobie (1986) find from their JHK observations of MC Cepheids that their H, J-K PLC relation can be used to predict the color term in the PLCV relation which they find to be similar to that found from BVI data by MWF79 and CC86.

PLC relations appear to be nearly dispersionless (MWF79, CC86) and so compensate for any uneven filling of the instability strip (in M_V the strip width is approx 0.6 mag at log P = 1.2). In addition, moduli calculated from the PLC relation are less sensitive to reddening than those calculated from the PL relation (MWF79), but unfortunately the PLC relation is sensitive to metallicity. Most recently this has been quantified semi-empirically from observations of MC Cepheids by CC86 and purely theoretically by Stothers (1988), the latter finds

$$\Delta M_V(\mathrm{PLC}) = 2.7\ \Delta Y - 36.9\ \Delta Z$$
$$= -27.5\ \Delta Z$$

assuming the same Y and Z correlation as above. This is double the dependence found by Iben and Tuggle (1975) and predicts an offset from our galaxy (Z = 0.02) to the SMC (Z = 0.005) of 0.41 mag, similar to the 0.51 mag found by CC86. The PLCV relation formulated using V-I color shows about half the sensitivity to changes in metal abundance (CC86) but the color coefficient is larger. We conclude that if the metallicity of a Cepheid is known, and it has been measured in several passbands in order to find the reddening, then the use of the PLC relation may well be preferable to the PL relation. This is particularly true in nearby galaxies such as the MC where accurate photometry is possible. FW87 discuss the MWF79 and CC86 PLCV relations in some detail and establish distance scales for both, but since the results obtained are almost identical we will adopt the CC86 result, which for metal normal Cepheids is

$$\langle M_V \rangle = -3.53 \log P + 2.13 (\langle B_o \rangle - \langle V_o \rangle) + \varphi(\mathrm{PLCV}).$$

Short-cut methods, such as use of Feinheit functions, have been discussed by Madore (1985b), Moffett and Barnes (1986), Freedman (1988) and Walker (1988). Their use is somewhat controversial and they will not be covered here. They can probably be best formulated from LMC Cepheid data with zeropoint from some assumed LMC modulus.

CALIBRATING THE DISTANCE SCALE ZEROPOINT

There are potentially six different methods available for direct calibration of the PL relation zeropoint. These are:

(1) Cepheids in clusters and associations. This is considered to be the most accurate method and is covered in detail below.

(2) Geometric methods, from observations of light variations in gas or dust clouds surrounding a Cepheid. Only two stars, SU Cas and RS Pup, are in this category. There is some question as to whether or not SU Cas is a first overtone pulsator (Evans and Ferro 1986). The photographic work of Havlen (1972) on the light variations of the dust shells around RS Pup could profitably be repeated using a CCD. These two stars are included in Table II.

(3) Baade-Wesselink methods, using radial velocities and photometry (preferably obtained simultaneously) to determine a Cepheid's radius and hence distance. Fernie (1984) summarized earlier work and showed there was poor agreement between various investigators. Recent work appears to give more consistent results, and is described below.

(4) Statistical parallaxes. FY87 find that the statistical parallax work (eg Wielen 1974, Clube and Dawe 1980, Karimova and Pavlovskaya 1981) give zeropoints reasonably consistent with that found here but with large errors (at least ±0.3 mag), consequently this method will not be considered further.

(5) Binaries. We include here α UMi, a visual binary. Analyses of companions detected in the ultraviolet by IUE have been made by Böhm-Vitense (1985), Evans and Ferro (1986) and Ferro and Madore (1985). The results are consistent with the zeropoint here but with very large scatter. Since analyses of the same data by different investigators can give differences in M_V of up to 0.8 mag the method does not appear to be useful. Feast (1986) pointed out that systematic errors could arise if the companions were evolved.

(6) Theoretical estimates, from combining evolution and pulsation theory. Theory finds Cepheids to be 0.5 mag brighter than observations would expect (Iben and Renzini 1984, FW87), although see Carson and Stothers (1988) for a possible reconciliation of theory with observation. The theoretical predictions will not be considered further here.

We will restrict ourselves now to an evaluation of the presently most accurate Cepheid calibrating methods, Cepheids in clusters and associations, and Baade Wesselink techniques.

Cepheids in Clusters and Associations

In the last few years there have been a number of investigations aimed at greatly improving the photometric data for a number of galactic clusters containing Cepheids. This work has most recently been evaluated by FW87 and Walker (1987a), where the cluster zero age main sequences (ZAMS) are matched to that of the Pleiades, and for the remaining stars their distances are referenced to the Pleiades as well.

Most previous discussions of the cluster and association data (eg Fernie and McGonegal 1983, Caldwell 1983) referenced distances to the Hyades, using a ZAMS (eg Blaauw 1963) at least partially based on the Hyades. This has caused much confusion in the literature since both the distance to the Hyades and the size of the metallicity correction needed to account for its greater than solar metallicity are controversial, although the models of VandenBerg and Bridges (1984) can now be used to predict the size of the metallicity correction needed. Averaging the results of various authors in order to get a 'best' calibration is incorrect unless the results are reduced to a common Hyades distance and metal abundance. Further, the Hyades is a much older cluster than the typical clusters containing Cepheids consequently more of the Hyades main sequence is evolved (to as far as $M_V = +2$). These problems can all be bypassed by referencing to the Pleiades rather than the Hyades (Walker 1985a). The Pleiades is thought to have solar metallicity and its distance is non-controversial, Van Leeuwen (1983) finds from extensive Walraven photometry of the Pleiades main sequence which he compares to field stars with known parallaxes that the Pleiades modulus is 5.57 ± 0.08 mag, which we adopt. The Cepheid distance scale therefore is calibrated directly from the parallaxes of solar neighborhood F and G dwarfs. Eggen (1986), from Strömgren and R,I photometry finds a Pleiades modulus of 5.55 mag assuming solar metallicity but prefers to adopt [Fe/H] = 0.05 in which case the corresponding modulus is 5.65 (Eggen) or 5.61 mag (using VandenBerg and Bridges (1984) models).

Comparison of the Pleiades main sequence to those of other clusters involves two assumptions, firstly that in the CMD the main sequence shapes of the clusters containing the Cepheids is the same as the Pleiades, and secondly that the cluster metallicities are equal to that for the Pleiades.

The first assumption is generally not important since, at least for the photoelectric studies of the nearer clusters (eg M25, NGC 6087) and the CCD studies (Lyngå 6, Ruprecht 79, anon. van den Bergh, NGC 6067) the comparison is made with the unevolved A and F stars rather than with the B stars where the position of the upper main sequence may vary greatly due to differing cluster ages. Pel (1985) finds using Walraven photometry that the differential evolutionary corrections needed to adjust his moduli of M25 and NGC 6087 to the Pleiades are very small (-0.03, -0.02 mag respectively).

The second assumption, that of solar metallicity, is potentially very important. VandenBerg and Bridges (1984) show how the position of the ZAMS shifts as a function of [Fe/H], they calculate that Δ[Fe/H] = 0.2 causes $\Delta M_V = -0.23$ mag, averaging from B-V = 0.0 to 0.6. The question of the abundances of galactic Cepheids in the solar neighborhood has been reviewed by Feast (1987). He finds that there is no evidence for wide variations in abundance. The photometric methods that have been applied to this problem (eg Harris 1981, Eggen 1985a,b,c) all appear to require corrections for temperature and/or luminosity effects, and therefore should be regarded as giving upper limits to the variations in metallicity. Caldwell and Coulson (1987), in a study of the rotation and distance to the galactic center using Cepheids, adopt from the spectroscopic Cepheid [Fe/H] measurements tabulated by Giridhar (1986) a galactic abundance gradient of Δ[Fe/H] =-0.07 /kpc and estimate the intrinsic local dispersion to be about 0.11 in [Fe/H]. Since both the position of the ZAMS and the Cepheid move in the CMD as a function of metallicity the total dependence is reduced in the usual V, B-V CMD (Table V of Caldwell and Coulson 1987). The changes to the luminosities of the Cepheid calibrators caused by adopting the above abundance gradient is insignificant.

A list of Cepheids in clusters and associations is given by FW87, with extra information in Walker (1987a). Similar recent lists are those by Caldwell and Coulson (1987) and Gieren (1988). In the last few years considerable attention has been paid to improving the basic data for the clusters.

Unfortunately close, rich galactic clusters containing Cepheids are few indeed (M25, NGC 6067, NGC 6087) and most of the clusters are distant, sparse, and at low galactic latitude where both the field star density and the reddening are high. The Cepheids are often in the cluster 'coronae' and membership becomes problematical, particularly if the absorption in the vicinity is patchy, sometimes relying on the circular argument of whether the derived luminosity, assuming membership, agrees with the PL calibration. For several clusters, radial velocity and proper motion information is available while BVI photometry for the Cepheid can be used to find its reddening which then can be compared to that for the cluster OB stars.

CCD CMD diagrams for some of the clusters are now available (Walker 1985a, 1985b, 1987b, Walker and Laney 1987). These together with new photoelectric studies (Pel 1985, Turner 1986) and programs underway on Cepheids in cluster coronae and on associations possibly containing longer period Cepheids (D. Turner, private communication) have considerably improved the observational situation and will continue to do so. ZAMS fitting, as pointed out by Walker and Laney (1987), needs good quality data ($\sigma(B-V) < 0.05$ mag) in order to be able to remove binaries and rapid rotators which lie above the main sequence and so systematically influence the fit. Mermilliod, Mayor and Burki (1987) have measured very accurate radial velocities for individual stars in several clusters containing Cepheids and further radial velocity work on northern clusters is in progress (D. Turner, private communication).

The data in Table II summarize the cluster and association members available for the calibration of PLV and PLCV. Changes from FW87 are the inclusion of CS Vel (Walker 1987b), deletion of S Vul (Walker 1987a) and a different period for SZ Tau (D. Turner, private communication). The data can be grouped or weighted in various ways (eg see FW87, CC86), here we have chosen to group the association data separately and then divide the cluster data into two groups depending on reliability. We weight the groups (A:B:C:D) = (3:1:1:1) which gives PLV and metal normal PLCV relations

$$\langle M_V \rangle = -2.91 \log P - 1.21$$

$$\langle M_V \rangle = -3.53 \log P + 2.13(\langle B_o \rangle - \langle V_o \rangle) - 2.11.$$

Any other reasonable weighting scheme gives zeropoints within a few 0.01 mag of these. The zeropoints correspond to an LMC distance of 18.48 mag (PLV) and 18.44 mag (PLCV) using the MC BVI Cepheid photometry as analyzed by CC86 and Walker (1987c). If we conservatively assume that the zeropoints above are known to ±0.1 mag, and the Pleiades distance to ±0.08 mag, then the Cepheid luminosity scale is defined to

TABLE II Zeropoints of the PLV and PLCV relations from observations of Cepheids in clusters, etc.

A. Stars with reasonably definitive $\langle M_V \rangle$.

Star	Period (d)	$\langle M_V \rangle$	φ(PLV)	φ(PLCV)
CEb Cas	4.48	-3.33	-1.43	-2.14
CF Cas	4.87	-3.17	-1.17	-2.00
QZ Nor	5.10	-3.31	-1.25	-2.06
CEa Cas	5.14	-3.39	-1.32	-2.18
CV Mon	5.38	-3.27	-1.14	-1.91
CS Vel	5.90	-3.24	-0.99	-1.81
V367 Sct	6.29	-3.80	-1.47	-2.21
U Sgr	6.74	-3.77	-1.36	-2.25
DL Cas	8.00	-3.86	-1.23	-2.08
S Nor	9.75	-3.95	-1.07	-2.09
TW Nor	10.79	-3.79	-0.78	-1.81
V340 Nor	11.22	-3.80	-0.74	-1.93

B. Stars lacking definitive data or in very sparse clusters(?)

Star	Period (d)	$\langle M_V \rangle$	φ(PLV)	φ(PLCV)
EV Sct	3.09	-2.53	-1.11	-2.09
SZ Tau	4.52	-3.43	-1.52	-2.13
UY Per	5.37	-3.38	-1.25	-1.99
V Cen	5.49	-3.24	-1.09	-1.89
RU Sct	19.70	-5.16	-1.39	-2.05
WZ Sgr	21.83	-4.86	-0.96	-2.02

C. Stars probably in associations.

Star	Period (d)	$\langle M_V \rangle$	φ(PLV)	φ(PLCV)
VY Car	18.93	-4.95	-1.24	-2.38
RZ Vel	20.40	-5.14	-1.33	-2.25
SW Vel	23.47	-5.21	-1.22	-2.03
T Mon	27.02	-5.55	-1.38	-2.61
KQ Sco	28.69	-5.47	-1.23	-2.57
SV Vul	45.01	-6.12	-1.30	-2.44
GY Sge	51.07	-6.22	-1.25	-2.50

D. Others.

Star	Period (d)	$\langle M_V \rangle$	φ(PLV)	φ(PLCV)
SU Cas	1.95	-1.98	-1.14	-1.87
α UMi	3.97	-3.20	-1.46	-2.34
RS Pup	41.39	-5.96	-1.26	-2.21

±0.13 mag.

The luminosity of Cepheids in clusters and associations can also be found from Strömgren-beta photometry of the B stars using a suitable luminosity calibration. This was done for eight clusters by Schmidt (1984), and appeared to give a distance scale 0.4 mag shorter than the then current ZAMS fitting methods (eg Caldwell 1983). However, if the recalibration of the Strömgren-beta B star luminosity scale by Balona and Shobbrook (1984) is used and the distances are referred to a Pleiades modulus of 5.57 mag then the Schmidt (1984) Cepheid luminosity scale agrees with that given by FW87 using the ZAMS fitting technique. There is considerable disagreement about the Strömgren-beta B star luminosities (Eggen 1986, Mermilliod 1987) amounting to about a magnitude of uncertainty. This, together with questions about the need and size of corrections for evolution and rotation (Turner 1986) as well as the requirement of extreme accuracy in the beta photometry (better than 0.005 mag external error according to Balona and Shobbrook 1984), makes the technique of dubious value. Far more useful would be Strömgren photometry of the unevolved A or F stars. This is well within the capabilities of a moderate sized telescope equipped with a CCD.

In the infrared, C.D. Laney (private communication) has embarked on a program to measure the cluster and association Cepheids, and in addition to determine the ratio of total to selective absorption R and the reddening law as accurately as possible. This is important since the infrared Cepheid photometry has to be compared to optical photometry of the other cluster stars. He finds that the infrared PL zeropoints so obtained are consistent with those found here.

Baade Wesselink Methods

Baade Wesselink (BW) methods for determining Cepheid luminosities have the important property of giving a distance scale which is totally independent of that found from the cluster and association stars. Fernie (1984) summarized the BW results available at that time and showed that there was a great deal of discrepancy between the various studies. More recent programs have shown considerable improvement. The surface brightness (Barnes Evans) version of the BW method uses V-R (Johnson) color rather than B-V as a temperature indicator (e.g. Moffett and Barnes 1985, 1987), while Coulson, Caldwell and Gieren (1986) use V-I in a classical BW analysis. These studies, and those of Caccin et al. (1981, 1985) using the multicolor CORS method, give consistent results.

Gieren (1988), in a Barnes Evans analysis using the simultaneous photometry and radial velocities for 52 southern Cepheids from Coulson and Caldwell (1985) and Gieren (1986), together with zeropoint from Barnes and Moffett (1985), derives a PLV relation

$$M_V = -2.92 \log P - 1.23$$

which is almost identical to that found above from the cluster and association data. The internal error can be reduced to very small values by observing sufficient Cepheids. However there are still uncertainties which may affect the surface brightness luminosity scale by about ±0.3 mag. The zeropoint b in the relation

$$F_V = b + m(V-R)_o$$

where F_V is the surface brightness relies on the Cepheid angular diameters found from fitting model atmospheres to observed fluxes (Parsons and Bouw 1971), since Cepheid angular diameters cannot be determined directly. In addition, the projection factor p required to convert from observed to pulsational velocity is p = 1.31 (Parsons 1972) or alternatively p = 1.34 - 1.36 depending on period (Hindsley and Bell 1986). Together these give an error of ±0.3 mag according to Gieren (1988).

Therefore the most recent BW analyses give a PL relation agreeing with the cluster and association results but with a larger error. However Moffet and Barnes (1987) suggest that improvements can be made to the surface brightness method which may help to reduce the size of the errors. Use of the classical BW technique but with colors other than B-V also appears promising (Coulson, Caldwell and Gieren 1986).

CONCLUSIONS

Distances to Cepheids in clusters and associations define the Cepheid distance scale to ±0.13 mag of which ±0.08 mag is due to the uncertainty in measuring the distance to the Pleiades. Recent Baade Wesselink results agree with this scale, which corresponds to a V magnitude Period-Luminosity LMC distance modulus of 18.48 mag.

ACKNOWLEDGEMENTS

I would like to thank the many astronomers who provided preprints, reprints and reports of their work.

REFERENCES

Balona, L.A., and Shobbrook, R.R. 1984, M.N.R.A.S., 211, 375.
Barnes, T.G., and Moffett, T.J. 1985, in "Cepheids: Theory and Observations", Proceedings of IAU Colloquium 82, ed. B.F. Madore, (Cambridge: Cambridge Univ. Press), p. 53.
Becker, S.A., Iben, I., and Tuggle, R.S. 1977, Ap. J., 218, 653.
Blaauw, A. 1963, in "Basic Astronomical Data", ed. K.A. Strand, (Chicago: Univ. Chicago Press), p. 383.
Böhm-Vitense, E. 1985, Ap. J., 296, 169.
Caccin, B., Onnembo, A., Russo, G., and Sollazzo, C. 1981, Astr. Ap., 97, 104.
Caccin, B., Buonaura, B., Onnembo, A., Russo, G., Sambuco, A.M., and Sollazzo, C. 1985, in "Cepheids: Theory and Observations", Proceedings of IAU Colloquium 82, ed. B.F. Madore, (Cambridge: Cambridge Univ. Press), p. 43.
Caldwell, J.A.R. 1983, Observatory, 103, 244.
Caldwell, J.A.R., and Coulson, I.M. 1986, M.N.R.A.S., 218, 223. (CC86).
Caldwell, J.A.R., and Coulson, I.M. 1987, A.J., 93, 1090.
Carson, T.R., and Stothers, R.B. 1988, Ap. J., 328, 196.
Clube, S.V.M., and Dawe, J.A. 1980, M.N.R.A.S., 190, 591.
Coulson, I.M., and Caldwell, J.A.R. 1985, SAAO Circ., 9, 5.
Coulson, I.M., Caldwell, J.A.R., and Gieren, W.P. 1986, Ap. J., 303, 273.
Eggen, O.J. 1985a, A.J., 90, 1260.
Eggen, O.J. 1985b, A.J., 90, 1278.
Eggen, O.J. 1985c, A.J., 90, 1297.
Eggen, O.J. 1986, Pub. A.S.P., 98, 755.
Evans, N.R., and Ferro, A.A. 1986, in "New Insights in Astrophysics (Eight Years of Ultraviolet Astronomy with IUE)", ESA SP-263, p. 403 (Paris: ESA).
Feast, M.W. 1986, in "Galaxy Distances and Deviations from Universal Expansion", ed. B.F. Madore, R.B. Tully, (Dordrecht: Reidel), p. 7.
Feast, M.W. 1987, in "The Galaxy", ed. G. Gilmore, R. Carswell (Dordrecht: Reidel) p. 1.
Feast, M.W. 1988, this conference.
Feast, M.W., and Walker, A.R. 1987, Ann. Rev. Astr. Ap., 25, 345. (FW87).
Fernie, J.D. 1969, Pub. A.S.P., 81, 707.
Fernie, J.D., and McGonegal, R. 1983, Ap. J., 275, 732.
Fernie, J.D. 1984, Ap. J., 282, 641.
Ferro, A.A., and Madore, B.F. 1985, Observatory, 105, 207.
Freedman, W.L. 1988, Ap. J., 326, 691.
Gieren, W.P. 1986, Ap. J., 306, 25.
Gieren, W.P. 1988, Ap. J., in press.
Giridhar, S. 1986, J. Ap. Astr., 7, 83.
Harris, H.C. 1981, A.J., 86, 707.

Havlen, R.J. 1972, Astr. Ap., 16, 252.
Hertzsprung, E. 1913, A.N., 196, 201.
Hindsley, R., and Bell, R.A. 1986, Pub. A.S.P., 98, 881
Iben, I., and Renzini, A. 1984, Phys. Rep., 105, 329.
Iben, I., and Tuggle, R.S. 1975, Ap. J., 197, 39.
Karimova, D.K., and Pavlovskaya, E.D. 1981, Sov. Astr., 25, 561.
Laney, C.D., and Stobie, R.S. 1986, M.N.R.A.S., 222, 449.
Madore, B.F. 1985a, in "Cepheids: Theory and Observations", Proceedings of IAU Colloquium 82, ed. B.F. Madore, (Cambridge: Cambridge Univ. Press), p. 166.
Madore, B.F. 1985b, Ap. J., 298, 340.
Martin, W.L., Warren, P.R., and Feast, M.W. 1979, M.N.R.A.S., 188, 139. (MWF79).
Mermilliod, J.C. 1987, Bull. Inform. CDS, 33, 41.
Mermilliod, J.C., Mayer, M., Burki, G. 1987, Astr. Ap. Suppl., 70, 389.
Moffett, T.J., and Barnes, T.G. 1985, in "Stellar Pulsation", ed. A.N. Cox, W.M. Sparks and S.G. Starrfield, (Berlin: Springer-Verlag) p. 169.
Moffett, T.J., and Barnes, T.G. 1986, Ap. J., 304, 604.
Moffett, T.J., and Barnes, T.G. 1987, Ap. J., 323, 280.
Parsons, S.B., and Bouw, G.D. 1971, M.N.R.A.S., 152, 133.
Parsons, S.B., 1972, Ap. J., 174, 57.
Peimbert, M. 1986, Pub. A.S.P., 98, 1057.
Pel, J.N. 1985, in "Cepheids: Theory and Observations", Proceedings of IAU Colloquium 82, ed. B.F Madore, (Cambridge: Cambridge Univ. Press), p. 1.
Sandage, A.R. 1971, Ap. J., 166, 13.
Sandage, A.R. 1972, Q.J.R. Astr. Soc., 13, 202.
Schmidt, E.G. 1984, Ap. J., 285, 501.
Stothers, R.B. 1983, Ap. J., 274, 20.
Stothers, R.B. 1988, preprint.
Turner, D.G., 1986, A.J., 92, 111.
Van Leeuwen, F. 1983, PhD Thesis. Leiden University, Netherlands.
VandenBergh, D.A., and Bridges, R.B. 1984, Ap. J., 278, 679.
Walker, A.R. 1985a, M.N.R.A.S., 213, 889.
Walker, A.R. 1985b, M.N.R.A.S., 214, 45.
Walker, A.R. 1987a, SAAO Circ., 11, 131.
Walker, A.R. 1987b, M.N.R.A.S., 229, 31.
Walker, A.R. 1987c, M.N.R.A.S., 225, 627.
Walker, A.R. 1988, Ap. J. (Letters), submitted.
Walker, A.R., and Laney, C.D. 1987, M.N.R.A.S., 224, 61.
Wielen, R. 1974, Astr. Ap. Suppl., 15, 1.

DISCUSSION

E. MILONE: Has anyone succeeded in measuring the phase lag between SU Cas and its nebulosity, and also RS Pup and its nebulosity?

A. WALKER: Havlen succeeded in measuring the phase lag between RS Pup and its nebulosity at Mt. Stromlo some 20 years ago. It may be possible to improve on this work by using CCDs, but it will be difficult due to the low contrast.

D. TURNER: It was pointed out to me by Bob Havlen many years ago that it is impossible to use the geometric distance method on SU Cas. The combination of its small amplitude and short period make light pulses in its nebula impossible to detect at its distance, even using CCD imaging at good seeing sites.

D. WELCH: If I may comment on RS Pup, it will require special handling because the Cepheid is very much brighter than the nebulosity and will saturate the CCD if care is not taken.

J. MOULD: One criticism of your composite P-L relation would be that, for periods exceeding 10 days (which are the only Cepheids seen in extragalactic samples), the calibration rests entirely on Cepheids in associations.

A. WALKER: I agree - it is important to measure Cepheids of all periods in the Magellanic Clouds in order to provide a firm link between the Galactic calibrators and the longer period Cepheids seen in extragalactic systems.

B. MADORE: If one should avoid the (B-V) colour in the Baade-Wesselink technique, should the (B-V) colour also be avoided in the PLC?

A. WALKER: Possibly; however, the coefficient of the colour term in the (V-I) version of the PLC is some 50% larger than it is in the (B-V) version and this is a disadvantage.

THE BRIGHTEST STARS AS DISTANCE INDICATORS

ROBERTA M. HUMPHREYS
University of Minnesota and Landessternwarte Heidelberg

ABSTRACT The luminosities of the brightest early-type supergiants have a well known dependence on the luminosity of the galaxy. In addition, many of the brightest blue "stars" are actually compact HII regions, clusters, and composite images. This is a serious problem in increasingly distant galaxies. Therefore they are not recommended as distance indicators. The brightest M supergiants have an upper bound to their bolometric luminosities due to the stability limit of the most massive evolved stars, making the red supergiants potentially powerful distance indicators.

I. INTRODUCTION

Hubble (1936) was the first to emphasize the potential usefulness of the brightest stars as distance indicators; although, many of Hubble's brightest blue stars were later shown to be HII regions and clusters (Sandage 1958). Using observations of the resolved brightest blue stars in nearby galaxies, Sandage and Tammann (1974a) showed the dependence of their luminosities on the luminosity of the parent galaxy; i. e. the brightest blue stars are found in the brightest and largest galaxies. This well known result is attributed to statistical effects such that the largest galaxies have the largest population sample and therefore more of the brightest stars. They also found evidence that the brightest red stars may have a nearly constant visual luminosity, independent of galaxy type.

My own work on this subject was motivated by my interest in stellar populations and massive star evolution in galaxies of different types. Any study of stellar evolution depends critically on knowing the distance to the galaxy, and good stellar distance indicators which can be reliably and easily used for resolved galaxies will be especially valuable. With modern photographic and electronic imaging, spectroscopy, and visual and infrared photometry it is possible to not only identify the brightest supergiants but separate them from foreground stars and non-stellar objects and accurately measure their luminosities.

The results of the early spectroscopy of the brightest stars in Local Group galaxies was summarized by Humphreys (1983). More recent observations in more distant galaxies plus use of infrared photometry for the M supergiants is in several more recent papers (Elias et al. 1981, 1985, Elias and Frogel 1985, Humphreys et al. 1986, Humphreys and Aaronson 1987a, b, Humphreys et al. 1988).

II. RESULTS OF THE LUMINOSITY CALIBRATION

In this discussion I will distinguish between the absolute visual magnitude, M_V, and the bolometric luminosity, M_{Bol}. M_{Bol} is the most physically meaningful and is what we require for understanding the physics of the stars' evolution. However, M_V is the most practical. It is what we observe; although with increased use of the infrared for the late-type stars, M_K (2.2 μ) and M_{Bol} are also directly observed.

Spectra of the visually brightest blue or early-type stars show that they are all late B or early A-type supergiants. The spectroscopic results for Local Group Sc type spiral and irrregular galaxies confirm the dependence on galaxy luminosity (Figure 1, also see Humphreys 1987). Recent observations of the brightest stars in M31 (in preparation), an Sb spiral, show that its brightest blue supergiants are somewhat less luminous than would be expected for a galaxy of its mass and luminosity. Our observations of the early-type supergiant candidates in NGC 2403, M81, and M101 (Humphreys and Aaronson 1987a, b) revealed additional problems with their use as distance indicators.

Many of the presumed single stars in these more distant galaxies (Sandage 1983, 1984a, b) were found to be compact HII regions, young clusters and composite systems. The problem of blended images for blue stars obviously becomes more serious at increasing distances. This contamination with nonstellar objects severely limits their usefulness as distance indicators. I do not recommend them as distance indicators unless spectra for classification are available, and then only in conjunction with other indicators.

The luminous blue variables (LBV's), a group of hot and luminous, but highly unstable stars have also been considered as possible distance indicators (Sandage 1983, 1984a, b). However, these stars are not necessarily the brightest and they cover a two magnitude range of visual and bolometric luminosities in their respective galaxies. Primarily because of the range in their luminosities these unstable stars are not good distance indicators.

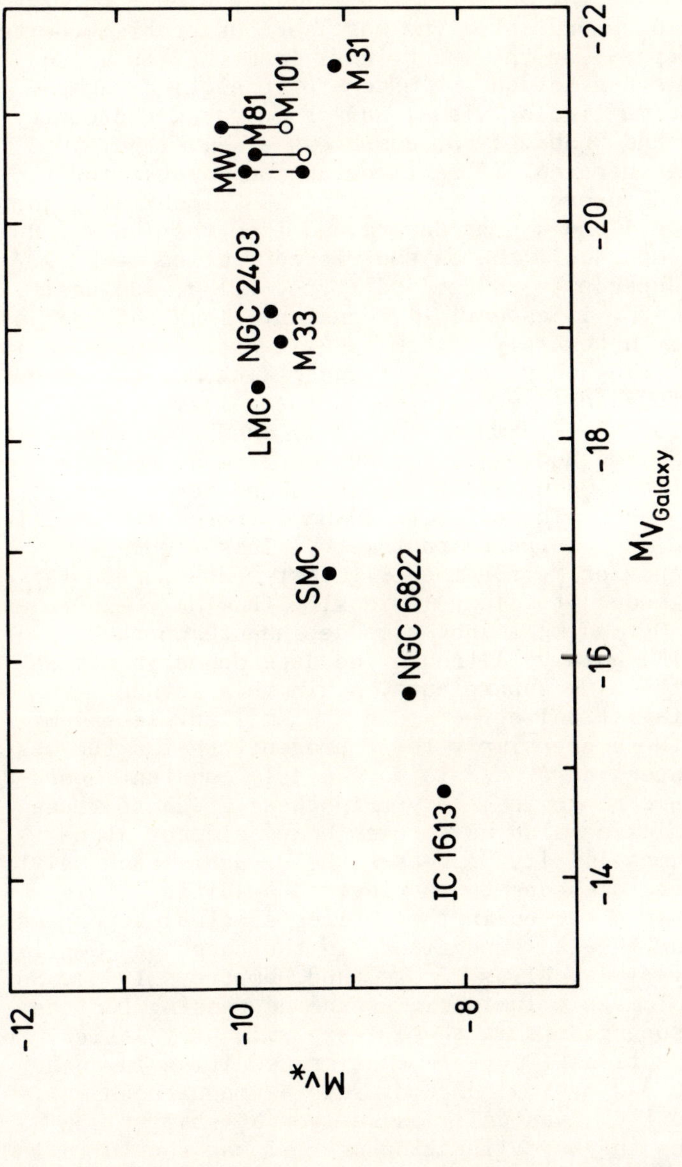

Figure 1 - The visual luminosity calibration for the brightest blue supergiants in Local Group spirals and irregulars plus N2403, M81, and M101. The filled circle for M81 is the only confirmed single star, a very luminous F-type supergiant and for M101 two A-Type supergiants whose spectra are probably contaminated by hotter stars. The open circles are the next three brightest blue stars in each galaxy.

In contrast to the blue supergiants, the most luminous late-type supergiants (spectral types F, G, K, and M) have an essentially constant upper luminosity near $M_{Bol} \simeq -9.5$ mag. The most luminous M supergiants have the additional advantage of being identified by their very red color; therefore, they are not as affected by contamination. Most overlapping images will be blue, so contaminated images simply wouldn't be listed as candidate red supergiants. The most serious problem for the red stars is separating the supergiants from the foreground red dwarfs. This can be done by moderate resolution far-red spectroscopy or multicolor visual and/or infrared photometry (see Humphreys and Graham 1986, Humphreys et al. 1986, 1988).

Use of the infrared offers important advantages for the study of the red supergiants. The extinction is normally quite small ($A_K \simeq .1\ A_V$) and can be determined from the J-H vs. H-K diagram. Most important, the bolometric correction at K (2.2 μ) has little dependence on spectral type, color, luminosity, or metallicity. BC_K is essentially constant. Thus both M_K and M_{Bol} can be very accurately determined given that the star is an M supergiant. The only source of uncertainty is the error in the K magnitude and the distance to the galaxy.

Figure 2 (also see Humphreys (1987)) shows the luminosity calibrations (M_V, M_K and M_{Bol}) for the confirmed red supergiants in the Local Group Sc-type spirals and irregulars plus NGC 2403 and NGC 300. The bolometric and infrared luminosities are all derived from infrared photometry. This figure shows the near constancy of M_V for the brightest red supergiants over five magnitudes of galaxy luminosity from $M_V = -15$ to -20.5 mag. Both M_K and M_{Bol} show some dependence on the luminosity of the galaxy; although the dependence is not as strong as for the blue supergiants. Again this is due primarily to statistical effects; in the smaller, less luminous galaxies there are simply fewer progenitors for the most luminous red supergiants. M_V is more nearly constant near -8 mag over the same range in galaxy brightness, because these same smaller galaxies also have lower heavy element abundances. As the metallicity decreases the Hayashi track shifts to warmer temperatures due to the lowered opacities. This effect is clearly observed in the earlier spectral types and bluer colors for the red supergiants in the Large and Small Cloud (Humphreys 1979, Elias, Frogel and Humphreys 1985). So, although the bolometric luminosities are decreasing for the brightest red supergiants in these less luminous galaxies, the bolometric corrections (BC_V) are smaller resulting in an M_V that is near -8 mag over a range of several magnitudes in galactic luminosity. Eventually as we look at smaller and smaller galaxies the M_V calibration near -8 mag should fail because of the decreasing population sample. We may be seeing such a trend for the irregular galaxies with M_V below -16 mag.

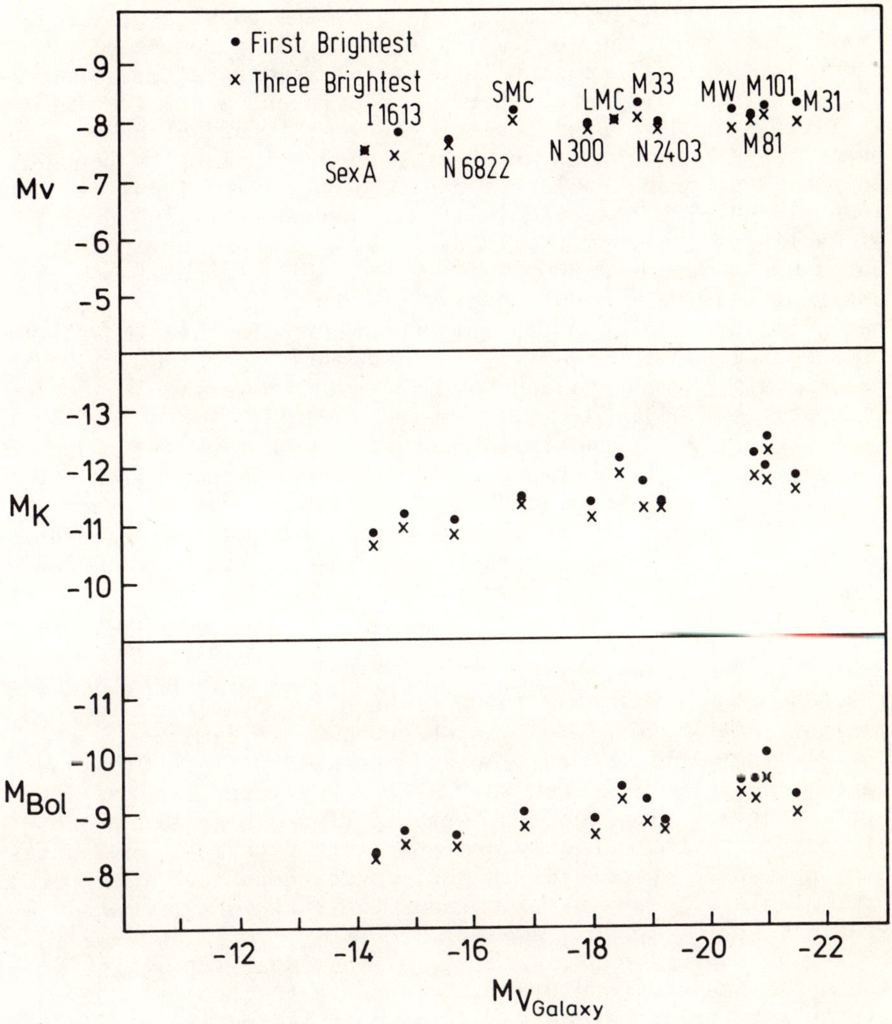

Figure 2 - The visual, infrared (K) and bolometric luminosity calibrations for the first brightest and three brightest red supergiants in Local Group spirals and irregulars plus N 300, N 2403, M81 and M101.

The visual luminosities of the brightest red supergiants at $M_V \simeq -8$ mag is due to an upper bound to the luminosities of the evolved cooler hypergiants near $M_{Bol} \simeq -9.5$ mag. This upper luminosity limit is explained by the physics governing the evolution of the most massive stars. As a result of numerous observational and theoretical studies of massive stars in the Galaxy and others, a consistent model for their evolution has developed. One of the most important results has been the recognition of an observed upper luminosity boundary in the HR diagram (Humphreys and Davidson 1979, 1984, de Jager 1980, 1984). The observed luminosity boundary is an envelope of declining luminosity with decreasing temperature for the hot stars which near 8000 - 10000 K turns over and becomes constant near $M_{Bol} \simeq -9.5$ mag. It is due to instabilities from radiation pressure and possibly turbulence in the atmospheres of the evolved most massive stars (see de Jager 1980, 1984, Maeder 1983, Humphreys and Davidson 1984, Lamers and Fitzpatrick 1988).

The origin of the instability in the hot stars is due to increasing radiation pressure (Lamers 1986, Appenzeller 1986). As the stars more than 40-50 $M_\odot$ evolve away from the main sequence, the opacity in their expanded atmospheres increases and they approach the Eddington limit for their mass and temperature. Lamers and Fitzpatrick (1988) have calculated the location of the Eddington limit in the HR diagram with the effects of increasing opacity and find reasonably good agreement with the observations. They especially note a turnover in the Eddington limit near 10000 K where the opacities reach a maximum which agrees with the observed upper limit.

In the most luminous cooler supergiants ($<$ 8000 K), de Jager (1984, 1987) has shown that the pressure gradient due to turbulence may also lead to instabilities. These stars are already known to be highly unstable with high mass loss rates and supersonic velocities in their atmospheres. If these cool hypergiants were any more luminous the turbulent pressure gradient would exceed gravity.

Thus radiation pressure, i. e. the Eddington limit, in the hot stars and possibly turbulence in the cooler supergiants set limits not only on the upper luminosities of stars, but also establish an upper limit to the mass, near 40-50 $M_\odot$, at which stars can evolve to the red supergiant stage. This physical limitation explains why there is an upper bound to the bolometric luminosities of the evolved, cool supergiants and why in combination with the effects of lower metallicity, M_V for the brightest red supergiants is nearly constant over several magnitudes of galaxy luminosity.

In addition to these results, observations have also been obtained of the candidate red supergiants in M81 and M101 (Humphreys et al. 1986) and most recently for the nearby Sb-type spiral M31 (Humphreys et al. 1988). The results for these three spirals are shown in Figure 1 together with the other galaxies. M81 now has a preliminary distance from Cepheids (Freedman and Madore 1988), but the distance to M101 is still controversial (Sandage and Tammann 1974b, Cook et al. 1986, de Vaucouleurs 1978, 1979, Humphreys et al. 1986). At the larger suggested distance modulus, the candidate red supergiants exceed the stability limit by a magnitude in M_{Bol}. At these luminosities they exceed the Eddington limit for their blue star progenitors. It is therefore highly unlikely that evolved red supergiants can exist at such high luminosities, consequently, I have adopted the nearer distance modulus of 28.4 mag to derive the luminosities in Figure 2 and Table I. Confirming spectra of several of its most luminous red stars are definitely needed. I still consider the characteristics of the massive star population in M101 highly uncertain.

The red supergiants in M31, a giant spiral, are especially important. Although M81 is also an Sb-type spiral, the data for its red stars are sparse; M31 is the first Sb-type spiral for which we have a good representative sample. There are visually luminous red supergiants in M31 with $M_V \simeq -8.0$ mag but there are few stars with the brightest infrared and bolometric luminosities. For a galaxy of the size and mass of M31 we would have expected to find several of the most luminous red supergiants in comparison with M33, the Galaxy and LMC. It is possible that in an Sb spiral the rate of massive star formation is lower producing fewer of the most massive stars per unit volume.

III. CONCLUDING REMARKS

The brightest blue stars are seriously compromised as distance indicators first by the well known dependence of their luminosities on the luminosity of the galaxy and by their confusion with compact HII regions, blue clusters, and close stars. I would not recommend their use as distance indicators unless confirming spectra are available, and only if they are used in conjunction with other distance indicators.

The brightest red supergiants are good distance indicators. Their maximum visual luminosity near $M_V \simeq -8$ mag is empirical. It is due to the fundamental limit to the stability of the atmospheres of evolved stars > 40-50 $M_\odot$ caused by radiation pressure and turbulence. This sets an upper boundary to the luminosities of the M supergiants which in combination with lower metallicity in less luminous galaxies has produced a visual luminosity calibration which is essentially constant over several magnitudes of galactic luminosity.

Finally I want to point out a group of stars that have been ignored as potential distance indicators. These are the

TABLE I Summary of the Luminosities of the Brightest M Supergiants in Local Group and Nearby Galaxies

GALAXY	M31	Milky Way	M33	LMC[1]	SMC	N6822	IC1613	M101	M81	N2403[2]	N300
Distance Modulus	24.26	-	24.2	18.5	19.0	23.2	24.3	28.4:	27.5	27.5	26.1
Type	Sb I-II	Sbc II:	Sc III	Im	Im	Im	Im	Sc I	Sb I-II	Sc III	Sc III
M_V	-21.5	-20.5:	-18.9	-18.5	-16.8	-15.7	-14.8	-20.8:	-20.6	-19.0	-18.0
Luminosities of M Supergiants											
M_V	- 8.5:	- 8.2	- 8.2	- 8.0	- 8.2	- 7.9	- 8.1	- 8.2(2)	- 8.1	- 7.8	- 8.0
	- 8.0	- 7.6	- 8.1	- 7.9	- 7.7(2)	- 7.8	- 7.6	- 8.1	- 8.0(2)	- 7.7	- 7.8
	- 7.8(2)	- 7.5(2)	- 7.9	- 7.6	- 7.5	- 7.3	- 7.0	- 8.0	-	- 7.6	- 7.7(2)
M_K	-11.7	-12.2	-11.8	-12.0	-11.5	-11.2	-11.3	-12.6	-12.0	-11.1	-11.4
	-11.4	-11.8	-11.2	-11.8	-11.3	-10.9	-11.2	-12.4	-11.5(2)	-11.0	-11.0(2)
	-11.2	-11.6	-11.1	-11.6	-11.2(4)	-10.8	-10.2	-12.1(3)	-	-10.8	-10.8
M_{Bol}	- 9.2	- 9.6	- 9.3	- 9.3	- 9.0	- 8.7	- 8.8	- 9.9	- 9.5	- 8.5	- 9.2
	- 8.9	- 9.2	- 8.7	- 9.1	- 8.8	- 8.4	- 8.7	- 9.8	- 9.0(2)	- 8.4(2)	- 8.7
	- 8.7(2)	- 9.0	- 8.6	- 9.0	- 8.7(4)	- 8.3	- 7.7	- 9.4(4)	-	-	- 8.5

1. Two IRAS sources, not included here, which are probable M supergiants with circumstellar dust shells have M_{Bol} of -9.6 and -9.3 mag (Elias, Frogel, and Schwering 1986).

2. The survey for red supergiants in NGC 2403 is incomplete. Only the three brightest red variables have been observed.

very high luminosity F, G and K-type supergiants near the upper luminosity boundary with $M_V \simeq -9$ to -9.5 mag! They have generally not been considered as distance indicators because of confusion with foreground yellow dwarfs. But at the very large distances that can be reached by Space Telescope, contamination by foreground dwarfs will be minimal. These yellow hypergiants are a magnitude brighter than the brightest red supergiants and three magnitudes brighter than the Cepheids. It is now time to take a closer look at these hypergiants as potential distance indicators.

ACKNOWLEDGEMENTS

This paper was written while R. M. H. was a visitor at the Landessternwarte in Heidelberg. R. M. H. gratefully acknowledges the suport of a senior U.S. scientist award from the Alexander von Humboldt Foundation.

REFERENCES

Appenzeller, I. 1986 in IAU Symposium 116, Luminous Stars and Associations in Galaxies, ed. C. W. H. de Loore et al. (Dordrecht: Reidel), p. 139.
Cook, K. N., Aaronson, M. and Illingworth, G. 1986, Ap. J. Letters, 301, L45.
de Jager, C. 1980, The Brightest Stars (Dordrecht: Reidel), p. 11.
de Jager, C. 1984, Astr. Ap., 138, 246.
de Jager, C. and Nieuwenhuijzen, H., 1987, in Workshop on Instabilities in Luminous Hot Stars.
de Vaucouleurs, G. 1978, Ap. J., 224, 710.
de Vaucoulerus, G. 1979, Astr. Ap., 79, 274.
Elias, J. H. and Frogel, J. A. 1985, Ap. J., 289, 141.
Elias, J. H., Frogel, J. A. and Humphreys, R. M. 1981, Ap. J., 249, L55.
Elias, J. H., Frogel, J. A. and Humphreys, R. M. 1985, Ap. J. Suppl., 57, 91.
Elias, J. H., Frogel, J. A. and Schwering, P. B. W. 1986, Ap. J., 302, 675.
Freedman, W. L. and Madore, B. F. 1988, preprint.
Hubble, E. P. 1936, Ap. J., 84, 279.
Humphreys, R. M. 1979, Ap. J., 231, 384.
Humphreys, R. M. 1983, Ap. J., 269, 335.
Humphreys, R. M. 1987, P.A.S.P., 99, 5.
Humphreys, R. M., Aaronson, M., Lebofsky, M., McAlary, C. W., Strom, S. E., and Capps, R. W. 1986, A. J., 91, 808.
Humphreys, R. M. and Aaronson, M. 1987a, Ap. J., 318, L69.
Humphreys, R. M. and Aaronson, M. 1987b, Ap. J., 94, 1156.
Humphreys, R. M. and Davidson, K. 1979, Ap. J., 232, 409.

Humphreys, R. M. and Davidson, K. 1984, Science, 223, 243.
Humphreys, R. M. and Graham, J. A. 1986, A. J., 91, 522.
Humphreys, R. M., Pennington, R. L., Jones, T. J., and Ghigo, F. D. 1988, A. J., submitted.
Lamers, H.J.G.L.M. 1986, in IAU Symposium 116, Luminous Stars and Associations in Galaxies, ed. C. W. H de Loore et al. (Dordrecht: Reidel), p. 157.
Lamers, H.J.G.L.M. and Fitzpatrick, E. 1988, Ap. J., 324, 279.
Maeder, A. 1983, Astr. Ap., 120, 113.
Sandage, A. 1958, Ap. J., 127, 513.
Sandage, A. 1983, A. J., 88, 1569.
Sandage, A. 1984a, A. J., 89, 621.
Sandage, A. 1984b, A. J., 89, 630.
Sandage, A. and Tammann, G. A. 1974a, Ap. J., 191, 603.
Sandage, A. and Tammann, G. A. 1974b, Ap. J., 194, 223.

DISCUSSION

S. DJORGOVSKI: Sampling the brightest, or even the three brightest stars may be tricky, because of the steep slope of the IMF, patchy extinction, etc. Did you consider taking, say, the 20 brightest stars, ignore the brightest 3, and average the rest? Wouldn't that be much more robust?

R. HUMPHREYS: The physical justification for using the red supergiants is that the upper boundary to their luminosities is determined by instabilities in the most massive evolved stars, i.e. the Eddington limit. Therefore it is important to use the red supergiants closest to that limit, that is the brightest.

I suspect that the average M_v or M_B of the 20 (?) brightest M supergiants might show a lot of variation depending on the galaxy. There are only three galaxies in which that many M supergiants have been confirmed: the Milky Way, the LMC and the SMC.

W. WALKER: Could you please discuss your results on M101 in the context of the other differing results?

R. HUMPHREYS: We (Humphreys et al. 1986) have identified probable red supergiants in M101 using JHK photometry. At the larger distance modulus, 29.2 mag, the brightest red stars would exceed the upper luminosity boundary by more than a magnitude. Their bolometric luminosities would be near -11 and their blue stellar progenitors would have to violate the Eddington limit for their masses.

It is for this reason that I favour the closer distance modulus for M101. I am using this luminosity boundary to set a limit to the distance to M101. This is sufficiently important that we need to get confirming spectra for these red stars in M101.

NOVA EXPANSION PARALLAXES

JUDITH G. COHEN
Palomar Observatory

ABSTRACT

A review of the observational parameters and assumptions necessary to derive the calibration of the maximum luminosity-rate of decline rationationship for novae from expansion parallaxes of galactic novae is given. The sample as it is known today consists of 11 objects with high quality data which define the zero-point, and a total of 21 objects which define the relationship itself. I review the prospects for enlarging this sample from ground-based observations and with the Hubble Space Telescope. A doubling of the sample of objects with high quality data seems possible within a decade.

The critical issue of how close real nova shells come to the idealized uniformly expanding spheres is discussed. A detailed model of DQ Her 1934 based on spectra with high spatial and spectral resolution is presented; its shell is not spherical, but by chance the final galactic calibration is unaffected. The zero-point of the galactic calibration must be treated with caution until the sample of spatially resolved shells can be enlarged and those objects known not to be spherical can be modelled in detail.

It is suggested that the spatial non-uniformities in emission line ratios within nova shells reflect non-uniformities in density and/or ionizing radiation, rather than real variations in the chemical abundances within the ejected shell.

I. INTRODUCTION

The detection of a shell around a nova offers a unique opportunity to directly determine the distance of the object. We need to find a shell that is spatially resolved, so that its diameter can be measured. If one knows the expansion velocity of the shell, and if one assumes that the shells are round and are now and have been expanding with a constant velocity since the initial explosion, it is possible to determine the distance to the nova via an expansion parallax. We then look for some property of the nova, preferably involving the period near maximum light when the object is brightest, which offers a chance of behaving in a predictable way and which yields distances with a suitably small uncertainty. Whatever this property may turn out to be, if it is well behaved and can be measured from observations of some nova in a distant galaxy which is so far away that we cannot expect to ever resolve the nova shell, we can still determine the distance to the galaxy. Behavior which can be understood theoretically, so one could anticipate the magnitude of its dependence on metallicity or other parameters which might vary from galaxy to galaxy, or even within the same galaxy, would be ideal.

McLaughlin realized in the mid-1940s (see his review of 1960) that classical novae appear to obey a luminosity-rate of decline relationship, where the rate of decline is usually parameterized as $t_n = \Delta t\,(\text{days})/n$ mag, or the time in days to decline n magnitudes below maximum light. n is usually taken as 2 or 3 mag, and for $n = 2$, t_2 ranges from 2.5 to 150 days. This relationship is established as a purely empirical one. There have been extensive searches for novae in M 31 by Arp (1956) and by Rosino and collaborators (see Rosino 1964), which have resulted in the discovery of at least 40 well-observed novae in that galaxy. These objects appear to obey the same relationship between brightness at maximum light and rate of decline from maximum as do the galactic novae.

Since novae near maximum light are very bright and have already been detected as far as the Virgo cluster (Pritchet and van den Bergh 1985, 1986) this offers the prospect of a direct distance determination to Virgo in one step using a technique with a good galactic calibration which is independent of any cluster distance moduli.

II. CALIBRATION OF THE MAXIMUM LUMINOSITY-RATE OF DECLINE RELATIONSHIP USING NOVA SHELLS

The observational material required to establish this relationship are the apparent magnitude (in some well-defined color) at maximum light, the light curve around maximum light, the bolometric correction or conversion among the various color systems that might have been used at various times, the interstellar absorption, the expansion velocity, and the diameter of the spatially resolved shell.

Let us consider each of these in turn. The discovery of classical novae in our galaxy, and the measurement of detailed light curves, is generally the purview of the amatuer astronomer, and the AAVSO has made great contributions in this area. The expansion velocity is best measured when the shell is spatially resolved, otherwise great care must be taken to avoid confusing emission lines arising in the accretion disk around the white dwarf with those from the expanding shell itself. Also, the interval just after maximum light should be avoided as, at that time, the spectra are very complex with multiple velocity systems present. The interstellar absorption is best measured from interstellar lines seen in the spectrum of the nova near maximum light, when high dispersion spectroscopy is most feasible. HI measurements, the mean reddening of external galaxies in the field, maps of absorption versus distance for stars in the field, etc. can also be used. Unfortunately many novae are found toward the galactic center in the heavily reddened regions of Sgr, and accurate absorption determinations for such objects are difficult. The diameter of the shell would seem straightforward, but all the easy bright large shells had been found by 1960, and number only eight objects. Most of the remaining known nova shells were found by myself in a recent survey (Cohen and Rosenthal 1983, Cohen 1985), and are still quite small with typical radii of a few arcsec. Thus, measurement errors induced by poor modeling, incorrect seeing profiles, etc. are serious for the smaller shells, but patience over a timescale of decades will presumably result in further expansion of these shells to the point where such errors become insignificant.

At the present time, there are 20 galactic novae with shells that have been spatially resolved, as well as one moderately well-observed nova in a galactic globular cluster, so potentially there are 21 objects to calibrate this relationship. In practice, however, two of these novae (BT Mon 1939 and RW UMi 1956) have light curves near maximum light which are so poorly observed that the necessary parameters cannot be derived. For example, RW UMi 1956, which was observed to reach sixth magnitude at maximum light, was not discovered until 1962, when it was noticed as bright on an old plate. This leaves only 19 objects to calibrate the relationship. Of these, four novae have extremely uncertain absorption determinations. If we restrict ourselves to objects with A_V less than 1.5 mag, distance less than 1.5 kpc, and date of outburst later than 1900 (to ensure good photometry around maximum light on a well defined color system), there are 11 objects with high quality data to define the calibration.

From these 11 objects, I derived via a least square fit (Cohen 1985) the relationship

$$M_V\,(\max) = -10.66\,(\pm 0.33) \;+\; 2.31\,(\pm 0.26)\log(t_2).$$

The fit is shown in Figure 1 (from my 1985 paper), where the data for the full sample of 19 objects are presented. The dispersion of the high quality sample around the fitted line is 0.52 mag. As we will see below, there are good reasons to suspect that most of this scatter is due to problems with the observational parameters rather than an intrinsic dispersion in the properties of novae near maximum light.

Novae are cataclysmic binaries which consist of a short period system containing a white dwarf and a low mass secondary filling its Roche lobe. The secondary is transferring mass onto the white dwarf via an accretion disk around the white dwarf. For accretion rates of about $10^{-7} M_\odot$/yr, Paczynski and Zytkow (1978) found that hydrogen burning in a shell is stable and just keeps up with the accreted matter, while for even larger accretion rates, the white dwarf expands and effectively becomes a red giant again. For lower accretion rates, the initiation of nuclear burning in a degenerate gas leads to a thermonuclear runaway (i.e., a nova outburst). Outbursts in recurrent novae are also believed to be due to a thermonuclear runaway within the accreted hydrogen rich material (see, for example, Starrfield, Sparks, and Truran 1985), but for these stars the secondary is a red giant rather than a main sequence dwarf. Consequently, the mass transfer

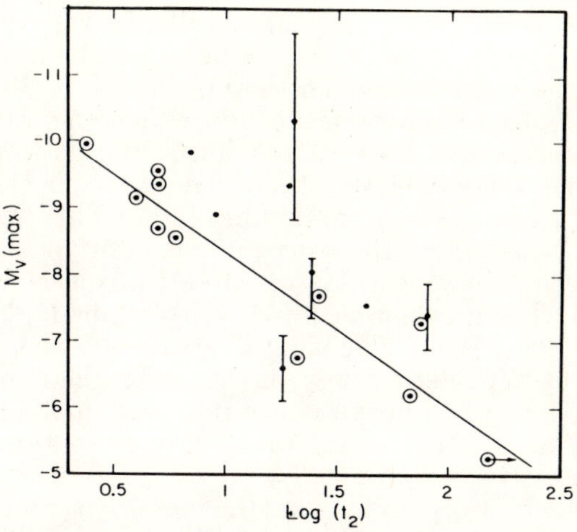

Figure 1. The absolute V-magnitude-rate of decline relationship for novae with well-determined distances. The solid line is the least squares fit for the high-quality sample (*circled points*). Error bars are shown where the uncertainty in interstellar absorption is particularly large.

rates are high and the accreted envelope can be replenished in the necessary timescale of a few decades. Dwarf nova outbursts are believed to be the result of a thermal instability in the accretion disk itself (Meyer and Meyer-Hofmeister 1984, Smak 1984).

Since we believe we understand in some detail the mechanism for producing nova outbursts, it should be possible to derive the absolute magnitude-rate of decline relationship from first principles, including properties of models of stellar structure, nuclear burning, etc. This has, in fact, been carried out in a series of papers by Bath and Shaviv (1976); Hartwick and Hutchings (1978); Prialnik, Shara, and Shaviv (1978); and Shara (1981a). The assumptions introduced by the first of these are that the nova luminosity near maximum light is constant and is the Eddington luminosity, while the apparent decline from maximum seen in the visual bandpass is in fact a consequence of increasing $T_{\rm eff}$ producing increasing bolometric corrections, and hence declining visual brightness, and that the opacity in the ejected material is dominated by electron scattering. Shara (1981a) added a calculation of the minimum accreted envelope mass necessary for a thermonuclear runaway, combined that with the white dwarf mass-radius relationship, added a definition of t_3 as the time at which enough of the envelope mass has been ejected to shut off the outburst, and produced from theoretical models a relationship extremely close to the observed one. This gives us great confidence that we understand the origin of the nova phenomenon. Since it involves thermonuclear burning of hydrogen rich material accreted onto white dwarfs, which are so evolved that their present chemical composition and physical parameters are largely independent of their original metallicity, we expect that to first order the behavior of novae during outburst will be independent of the properties of the galaxy in which they reside.

There is another useful relationship which was first derived from data on the novae in the Magellanic Clouds by Buscombe and de Vaucouleurs (1955). They suggested that all novae reach the same absolute magnitude M_B 15 days after maximum light. This would be particularly useful in cases where the light curve is poorly sampled, and there may be an uncertainty of a day or so as to exactly when maximum light occurred. The mean value of $M_V(15)$ for the high quality sample from Cohen (1985) is -5.60 mag. The theoretical models described above can be used to derive the time at which the expected spread in the visual luminosity of novae is a minimum; Shara (1981b) derived the time of 18 days after maximum light, in good agreement with the available data.

We can, therefore, assume that in fact $M_V(15)$ is constant, and ascribe all departures from this value to errors in the observational parameters. With these corrections, the adjusted M_V (max)-rate of decline relationship is shown in Figure 2 (from Cohen 1985). We see a dramatic improvement in the scatter about the mean line, which strongly suggests that $M_V(15)$ is indeed approximately constant. Fig. 2 is the best available calibration of the M_V (max)-rate of decline relationship for the galactic novae; the least squares fit is

$$M_V\,(\text{max, corr}) = -10.70\,(\pm 0.30) + 2.41\,(\pm 0.23)\log\,(t_2).$$

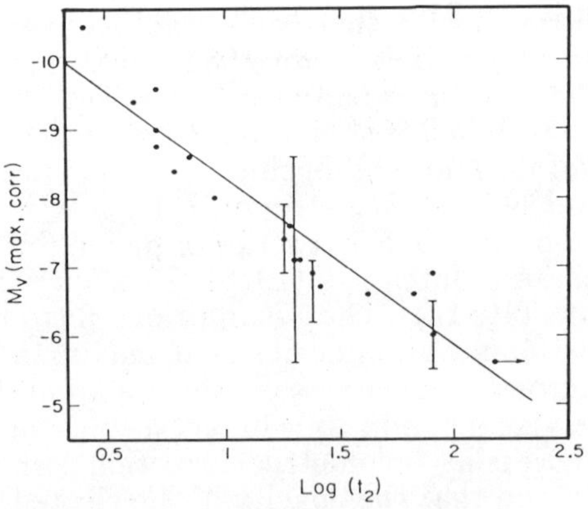

Figure 2. $M_V(15)$ is assumed to be constant at -5.60 mag, and a corrected value of M_V(max) is derived and shown as a function of log t_2. The least squares fit using all the points is indicated by a solid line.

III. PROSPECTS FOR THE FUTURE

What are the prospects for improving this by discovering more spatially resolved nova shells? I have been looking intermittently and have discovered one more probable shell, around nova DN Gem 1912, but it is very faint and needs confirmation. Figure 3 summarizes the situation; it is a histogram of the brightness at maximum light for novae from the catalog of Duerbeck (1987). Those novae with spatially resolved shells are indicated by filled squares. Objects which had maxima prior to 1915 are indicated by the letter o inside the square. Those objects with maxima subsequent to 1975 are indicated by the letter r inside the square. There are no known novae with spatially resolved shells that were fainter at maximum light than mag 7.5. BT Mon 1939 (the faintest point in the figure which does have a detected shell), probably actually reached as bright as mag 5; its light curve is very poorly sampled as it was discovered at least two months after maximum light.

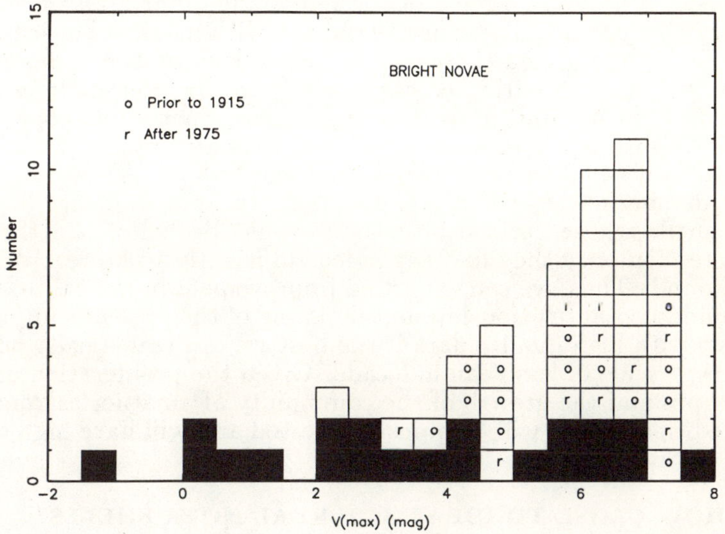

Figure 3. Histogram of the apparent magnitude of novae of maximum light. Filled squares denote objects with spatially resolved shells, "o" indicates novae with maxima prior to 1915, and "r" indicates novae with maxima after 1975.

Thus, it is only the novae that appear on this figure which are likely candidates. Fainter novae are probably too far away; the shell would become too faint to be detected before it expands enough to be resolved. (Recall that the brightness of the shell in the Balmer lines is given by standard recombination formulae and is proportional to $n(p) \times n(e)$, or density squared.) Novae with maxima prior to 1915 are likely to have poorly observed maxima, and are also likely to have very faint shells, as they have escaped detection thus far in spite of searches by several observers. There is also the concern, discussed further in §IV, that the oldest shells will have had time to interact with the ambient interstellar medium, and hence their shells may be distorted in shape.

Novae with outbursts after 1975 have probably not yet expanded enough to be detectable. Assuming the novae obey the calibration relationship determined earlier, the time for the expanding shell to reach 1 arc-sec in radius (t_1) is given by

$$t_1\,(\text{years}) = \frac{4.5\times 10^{0.2\,(m_V - A_V)}}{(t_2/10\ \text{days})^{0.48}\ (v_{\text{exp}}/500\ \text{km s}^{-1})}$$

which is 28 years for a nova with $m_V - A_V = 4$ mag and with the standard values for t_2 and v_{exp}.

For ground based searches, the best objects are those represented in fig. 3 by unmarked squares, of which there are 16. Unfortunately, six of the 16 are in Sgr, and their fields are very crowded and highly reddened. I have tried most of the remaining 10 at least once, and I will try them again. There are 10 objects with maxima subsequent to 1975 in the figure, and these should be resolved by the Hubble Space Telescope, or, with patience, eventually from the ground. In addition, depending on its limiting surface brightness, Space Telescope should be able to resolve the shells of somewhat more distant novae while they are still too small to be resolved from the ground. These somewhat more distant novae, which are too faint to appear in fig. 3, would have shells whose surface brightness would be below the threshold for detection once they had expanded sufficiently to be resolved from the ground. Thus, we can expect an improvement in the calibration of the fundamental relationship, but doubling of the present sample of 11 objects with high quality data is the best we can realistically hope for on a time scale of less than a decade. Given the proliferation of small telescopes and the growth of the community of amatuer astronomers, future bright novae will be quickly detected and will have high quality light curves.

IV. HOW CLOSE TO IDEAL ARE REAL NOVA SHELLS?

The principal assumption made above is that nova shells expand uniformly and hence are spherical. This assumption is known to be invalid for several novae; the notes in Cohen and Rosenthal (1983) list five novae with shells known to be non-circular. Among those novae with well resolved shells, one which at present shows considerable asymmetry is that of GK Per 1901. It's shell is extremely clumpy and assymetrical, appearing to have run into a brick wall on one side of nova. Yet Ritchey (1918) reported that the shell was circular with a diameter of 17 arc-seconds shortly after its discovery in 1917.

We can calculate the timescale for an expanding shell to sweep up a mass of interstellar material equal to its own mass, taking as a typical value for nova shells 5×10^{-5} $M_\odot$, which is

$$\tau = 160 \text{ years} \left(\frac{1 \text{ cm}^{-3}}{n}\right)^{0.33} \left(\frac{500 \text{ km s}^{-1}}{v_{\text{exp}}}\right)^{0.33} \left(\frac{M_{\text{shell}}}{5\times10^{-5}\, M_\odot}\right).$$

Thus, the shells of novae with ages exceeding 100 years may be decelerated by interaction with the local ISM and deformed in appearance if the local ISM is not uniform. GK Per 1901 has $v_{\text{exp}} \approx$ 1100 km s^{-1}, hence $\tau \approx$ 70 years, so its current deformed appearance should not surprise us.

More serious is the fact that, in their early development, nova shells tend to appear to consist of what are commonly denoted "equatorial rings" and "polar condensations". This is well illustrated in the time series of photographs of the shell of nova DQ Her 1934 by Baade (shown in Mustel and Boyarchuk 1970). Initially, the shell had extremely non-symmetric emission concentrated into two "polar condensations" 180° apart in position angle, and an "equatorial ring". Yet since about 1960, the shell emission, at least in Hα, has appeared to be clumpy but essentially azimuthally uniform. It is an interesting question how shells transform themselves from the structure described above, where emission depends strongly on position angle, to more or less uniformly emitting spheres (or spheroids), but we shall not address this point further.

The projection of the shell of DQ Her 1934 onto the plane of the sky is somewhat elliptical, with the ratio of major to minor axes being about 1.3; the major axis coincides with the position angle of the "polar condensations". Mustel and Boyarchuk point out that the shell of v603 Aql 1918 showed similar features. Hutchings (1972) has produced models which attempt to relate the multiple peaks in the emission line profiles of several recent novae, including FH Ser 1970, to the possible existence of the "polar condensations" and the "equatorial ring". He suggests that there may be an interaction between the expanding shell and the orbit of the secondary star plus the accretion disk which would affect the shape of the shell. Weaver (1974) analyzed an extensive set of spectra of v603 Aql 1918 to produce a complex model involving these types of structures, while Solf (1983) did the same for HR Del 1967.

These observations and their interpretation give rise to concern on two fronts. The first is the meaning of measurements of the "diameters" if many shells are in fact not round. The second is the expansion velocities; for non-sperical shells, the expansion velocity cannot be deduced without more information about the shell geometry and its orientation with respect to the plane of the sky than one normally has available.

In some cases, this concern may not be serious. For example, a photograph of the shell of FH Ser 1970 taken earlier this year through a wide Hα filter shows a circular shell with brightness enhancements 180° apart in position angle.

In view of this concern, A. Rosenthal's thesis, which was never completed, was to be a study of the geometry and kinematics of a few of the brightest and best resolved nova shells. Long slit spectra with high spectral and spatial resolution were obtained in 1982 by

Cohen and Rosenthal for DQ Her 1934 with the goal of determining whether the shell is expanding as a whole, or whether the structures seen by Baade represent real velocity perturbations. A 2 arc-sec slit was used with the Double Spectrograph (Oke and Gunn 1982) on the Hale telescope. The spectral resolution near Hα was 1.62Å (74 km s^{-1}) for two pixels, and 1.65 pixel/arc-sec along the slit. The spectra were taken with the slit positioned along the major axis of the shell, in steps of 2 arc-sec along the major axis. Figure 4 shows a montage of the resulting spectra around Hα. Figure 5 shows a few of the blue spectra, which had somewhat less spectral resolution. While there are obvious clumps of emission, it is clear that the shell expansion is a smoothly varying function of position. The shell, at least in its kinematic structure, is a single coherent entity, not some combination of distinct components, such as an "equatorial belt" and "polar condensations". Cuts both along and perpendicular to the spectral dispersion in the central part of the shell show that the shell has a width which does not exceed either the 2 arc-sec slit width or the spectral resolution as inferred from the widths of emission lines from a noble gas lamp, so the shell must be very thin. The centroid in wavelength of the Hα emission shifts by less than one pixel across the entire set of spectra. Since we already know that the shell is not spherical, this suggests that the major axis of the spheroid must be close to the plane of the sky.

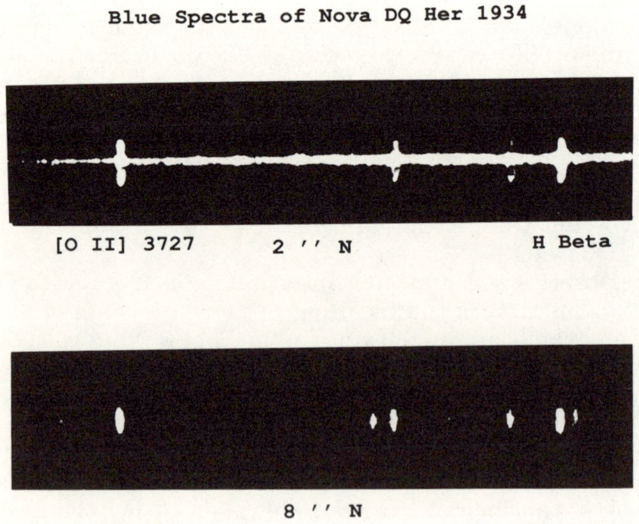

Figure 5. Two of the spectra from the blue side of the Hale telescope Double Spectrograph which are paired with those in Figure 4.

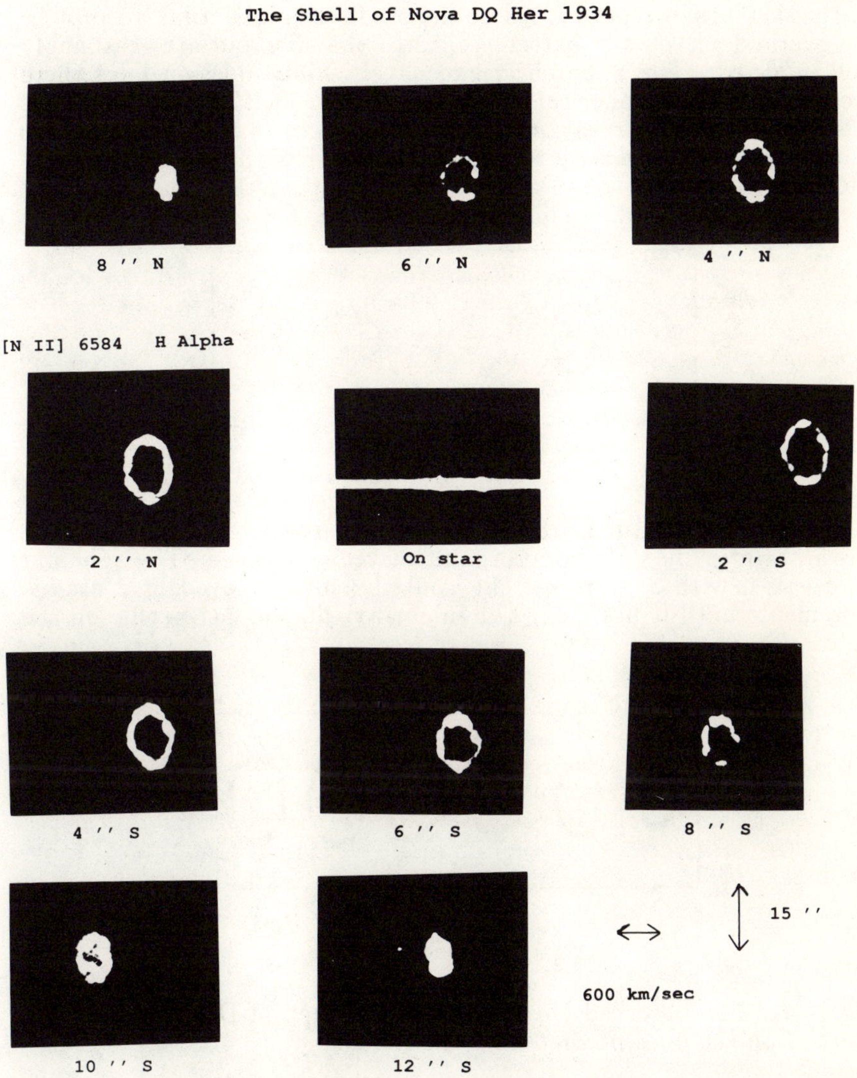

Figure 4. Series of spectra of the shell of nova DQ Her 1934 in the region of Hα taken with a 2 arc-sec slit in steps of 2 arc-sec.

A model of the expanding shell was constructed which consists of a prolate spheroid whose long axis is in the plane of the sky. Only the surface of the spheroid is assumed to radiate. The spheroid is expanding homologously with a maximum expansion velocity of

480 km sec^{-1}. The resulting predicted emission line profiles for a 2 arc-sec slit, drawn to the same relative scales as the observed Hα profiles, are shown in Figure 6. These provide an extremely good match with the observed profiles. The quality of the fit is illustrated by the sketches in Figure 7, where the predicted and actual Hα profiles are overlaid. Even the extreme ends of the spheroid are reasonably well modeled. The apparently excessively wide emission lines there are due to the non-spherical shape and of the wide (2 arc-sec) slit that was used for the observations. If the major axis of the spheroid were substantially inclined with respect to the plane of the sky, the velocity widths of spectra at the extreme ends of the projected shell would be even larger than they are observed to be.

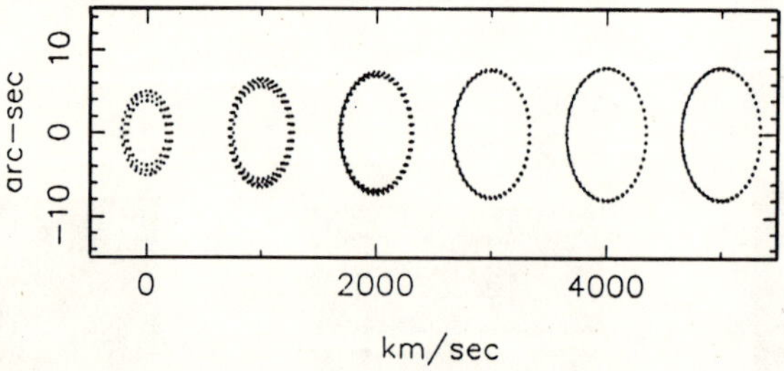

Figure 6. Predicted Hα profiles for the prolate spheroid model. Each has an arbitrarily shift in zero point in velocity space. The spheroid is observed with a 2 arc-sec slit, and the slit is moved in 2 arc-sec increments until it lies along the minor axis in the last profile on the right.

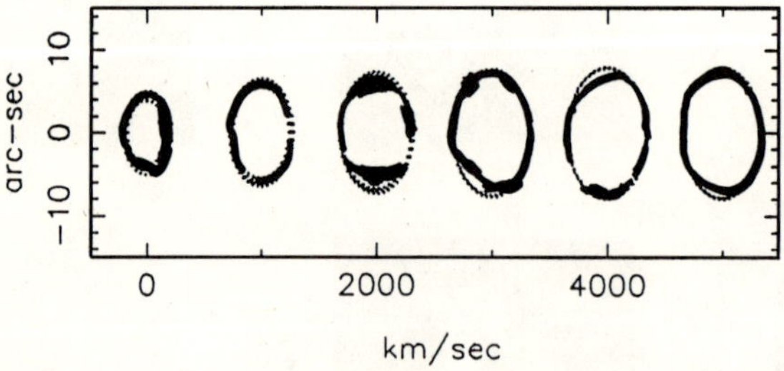

Figure 7. The observed Hα profiles of the shell of DQ Her 1934 are overlaid on the ones predicted by the model.

This detailed model of the shell of DQ Her 1934 gives a distance modulus which is 0.3 mag larger than the value in Cohen (1985), where in fact the "diameter" adapted was close to that of the major axis, with a value of v_{exp} appropriate to the minor axis.

We must consider the effect of the non-sphericity of observed nova shells on the calibration given in §II. We can only hope that by observing enough objects the statistical effects of varying angles of inclination of the major axis with respect to the plane of the sky will average out to zero, as one expects them to do eventually. When one corrects for the one detailed model available thus far, namely that of DQ Her 1934 described above, in fact the galactic calibration is essentially unchanged. This is because the absolute visual magnitude of DQ Her 1934 15 days after maximum light was deduced as −5.2 in Cohen (1985). The correction to the mean for the high quality sample of −5.60 is in fact almost the same as the correction derived above for the geometrical effects ignored in my earlier paper. (We ignore here the change of −0.03 mag in the mean value of $M_V(15)$.) Thus, for this case only, the galactic calibration given above is unaffected. However, with the present sample of only 11 objects with high quality data to define $M_V(15)$, and 19 points to define the final relationship, one cannot feel comfortable with the $M_V(15)$ argument that statistics conquers all sins. Hopefully, in the next few years, the sample of known shells can be enlarged and a few additional cases can be modeled in detail.

V. EMISSION LINE RATIOS

Another interesting question is the possible spectral differences of the "equatorial ring" gas as compared to the "polar condensations". Williams *et al.* (1978) found that for the shell of DQ Her 1934 the regions which correspond to the equatorial ring (i.e., the major axis of the shell) have spectra which show unusually strong permitted recombination lines of ions of C, N, and O, and that these lines appear stronger with respect to Hα or the usual forbidden lines of N and O along the major axis of the shell than along the minor axis.

Interpretation of any spectra to derive abundances and physical conditions in nova shells is difficult. The ionization is a strongly varying function of time, and we have little information on the spectral distribution of the ionizing flux from the post-nova star plus its gradually restored accretion disk as a function of time after eruption. At one time, it was believed that the ionization in the older nova shells represented a non-equilibrium frozen-in ionization, but it now seems clear (see Ferland *et al.* 1984) that even the relatively old shell of DQ Her 1934 is in ionization equilibrium, albeit a rather peculiar equilibrium.

While the new high dispersion spectral observations described below confirm the difference in relative emission line strengths between the major and minor axes of the shell of DQ Her 1934, such differences could easily be due to ionization or density effects and may not reflect real differences in the metallicity of the ejecta. The spectra of the shell of DQ Her 1934, particularly the presence of Balmer continuum emission, as discussed by Williams *et al.* indicate the presence of cold gas, with temperature close to 500 K, probably ionized by radiation from the accretion disk and the post-nova star. As pointed out by Ferland *et al.* (1984), the high abundances with respect to hydrogen in the nova ejecta, particularly of oxygen, lead to very efficient cooling by the infrared fine structure lines. But for a given radiation field, the temperature of the gas (i.e., the efficiency of the cooling in the fine structure lines of [OIII], which can be collisionally de-excited if the density is high) is a very strong function of the density, rising from 600 to 3000 K when the density is increased by a factor of 10 from 100 atoms cm^{-3} to 1000 atoms cm^{-3}. Since the emission in the Balmer lines is a less sensitive function of temperature than that in the permitted recombination lines from the trace elements, which at low temperatures are presumably formed by recombination but at higher temperatures are collisionally excited, we can easily imagine small variations in density producing exaggerated variations in emission line ratios. Another possibility is absorption of the ionizing radiation from the post-nova star by the accretion disk itself, which could produce the observed azimuthal dependence of the relative strength of the C, N, and O emission lines.

ACKNOWLEDGEMENTS

I am grateful to H. Ford and R. Ciardullo for drawing my attention to two interesting references. Financial support was provided by Caltech and by the Caltech Recycling Center.

VI. REFERENCES

Arp, H. C. (1956), *A.J.*, **61**, 15.
Bath, G. T., and Shaviv, G. (1976), *M.N.R.A.S.*, **175**, 305.
Buscombe, W., and de Vaucouleurs (1955), *Observatory*, **75**, 170.
Cohen, J. G., (1985), *Ap.J.*, **292**, 90.
Cohen, J. G., and Rosenthal, A. J. (1983), *Ap.J.*, **268**, 689.
Duerbeck, H. W. (1987), *Space Science Reviews*, **45**, 1.
Ferland, G. J., Williams R. E., Lambert, D. L., Shields, G. A., Slovak, M., Gondhalekar, P. M., and Truran, J. W. (1984), *Ap.J.*, **281**, 194.
Hartwick, F. D. A., and Hutchings, J. B. (1978), *Ap.J.*, **226**, 203.
Hutchings, J. B. (1972), *M.N.R.A.S.*, **158**, 177.
McLaughlin, D. B. (1960), in *Stars and Stellar Systems*, Vol. **6**, *Stellar Atmospheres*, ed. J. L. Greenstein (University of Chicago Press).
Meyer, F., and Meyer-Hofmeister, E. (1984), *Astr. Ap.*, **132**, 143.
Mustel, E. R., and Boyarchuk, A. A. (1970), *Ap. Space Sci.*, **6**, 183.
Oke, J. B., and Gunn, J. E. (1982), *P.A.S.P.*, **94**, 586.
Paczynski, J. and Zytkow, A. N. (1978), *Ap.J.*, **222**, 604.
Prialnik, D., Shara, M. M., and Shaviv, G. (1978), *Astr. Ap.*, **62**, 339.
Prichet, C. J., and van den Bergh, S. (1985), *Ap.J.*, **288**, L41.
Prichet, C. J., and van den Bergh, S. (1986), *Ap.J.*, **318**, 507.
Ritchey, G. W., (1918), *P.A.S.P.*, **30**, 163.
Rosino, L. (1964), *Ann. d'Ap.*, **27**, 498.
Shara, M. M. (1981a), *Ap.J.*, **243**, 926.
Shara, M. M. (1981b), *Ap.J.*, **243**, 268.
Smak, F. (1984), *P.A.S.P.*, **96**, 5.
Solf, J. (1983), *Ap.J.*, **273**, 647.

Starrfield, S. A., Sparks, W. M., and Truran, J. W. (1985), *Ap.J.*, **291**, 136.

Weaver, H. (1974), *Highlights of Astronomy*, **3**, 509 (ed. G. Gontopoulos).

Williams, R. E., Woolf, N. J., Hege, E. K., Moore, R. L., and Kopriva, D. A. (1978), *Ap.J.*, **224**, 171.

DISCUSSION

D. HANES: Is it correct that Mike Shara's theory of nova outbursts yields the downward concavity observed in your relationship between magnitude at 15 days after maximum light and decay rate?

J. COHEN: Yes, Shara predicts a slight downward concavity in the relationship between $M_B(15)$ and decay rate, but one can trust neither the theory nor the observations at that level of precision.

NOVAE AS DISTANCE INDICATORS

Holland C. Ford
Department of Astronomy
University of Michigan, Ann Arbor, MI 48109
Space Telescope Science Institute
Homewood Campus, Baltimore, MD 21218

Robin Ciardullo
Kitt Peak National Observatory, Tucson, AZ 85726
Space Telescope Science Institute
Homewood Campus, Baltimore, MD 21218

ABSTRACT

Nova shells are characteristically prolate with equatorial bands and polar caps. Failure to account for the geometry can lead to large errors in expansion parallaxes for individual novae. When simple prescriptions are used for deriving expansion parallaxes from an ensemble of randomly oriented prolate spheroids, the average distance will be too small by factors of 10% to 15%. The absolute magnitudes of the novae will be underestimated and the resulting distance scale will be too small by the same factors. If observations of partially resolved nova shells select for large inclinations, the systematic error in the resulting distance scale could easily be 20% to 30%.

Extinction by dust in the bulge of M31 may broaden and shift the intrinsic distribution of maximum nova magnitudes versus decay rates. We investigated this possibility by projecting Arp's and Rosino's novae onto a composite $B - \lambda 6200$ color map of M31's bulge. Thirty two of the 86 novae projected onto a smooth background with no underlying structure due to the presence of a dust cloud along the line of sight. The distribution of maximum magnitudes versus fade rates for these "unreddened" novae is indistinguishable from the distribution for the entire set of novae. We conclude that novae suffer very little extinction from the filamentary and patchy distribution of dust seen in the bulge of M31.

Time averaged B and Hα nova luminosity functions are potentially powerful new ways to use novae as standard candles. We analyzed our modern CCD observations and the photographic light curves of M31 novae found during the last 60 years to show that these functions are power laws. Consequently, unless the eruption times for novae are known, the data cannot be used to obtain distances.

I. INTRODUCTION

The photographic discovery of novae in "spiral nebulae" by Ritchey in 1917 was one of the earliest keys to correctly sizing the Universe. After finding a previously unrecorded star in NGC 6946, Ritchey (1917) searched a series of plates taken in 1909 and found two objects in M31 with nova-type light-curves. Systematic observations over the next two years resulted in the detection of fourteen additional novae in M31. Quoting from Hubble (1936) "Curtis immediately pointed out that the apparent faintness of novae in spirals indicated large distances, averaging at least one hundred times greater than the mean distance of galactic novae".

Novae proved to be good distance indicators because they were bright and easily identifiable. The first distance measurements were made by comparing the mean maximum magnitudes of novae in nearby galaxies to the mean absolute magnitude of galactic novae. In 1945 McLaughlin deduced a relationship between the "duration" of a nova and its absolute magnitude. Eleven years later in a landmark paper Arp (1956) used the light curves of 24 novae in M31 to quantify the relationship between absolute magnitude and the time to fade 2 magnitudes from maximum.

Novae continue in good standing as distance indicators because they are bright, easily identifiable, and because their light curves can be used to determine their absolute magnitudes. De Vaucouleurs (1978) lists novae as one of four primary distance indicators, the other three being Cepheids, RR Lyrae and HB stars, and AB Supergiants and Eclipsing binaries. Cohen and Rosenthal (1983; CR) and Cohen (1985) used expansion parallaxes of galactic novae to calibrate the dependence of absolute magnitude on fate rate. Van den Bergh and Pritchet (1986) recently discussed the advantages of novae as distance indicators, and subsequently (Pritchet and van den Bergh 1987; PvdB) used observations of 9 novae in NGC 4472 and NGC 4365 to estimate the distance to the Virgo cluster.

In spite of their good standing, there are problems in calibrating novae as standard candles. Expansion parallaxes, which were suggested by McLaughlin as early as 1942, are in principle an easy and accurate way to measure the distances to galactic novae, and thus calibrate their absolute magnitudes. However, rather than being spherical, nova shells typically have a prolate geometry consisting of polar caps and an equatorial ring. Failure to account for the geometry can lead to large systematic errors in the derived distances.

Payne-Gaposchkin (1957) was one of the first authors to suggest using observations of novae in M31 and M33 (and the Magellanic Clouds) to calibrate novae relative to Cepheids and RR Lyrae stars. Although this approach demotes novae from primary to secondary standard candles, it enables us to extend the Cepheid and RR Lyrae calibration to early type galaxies which do not have Population I Cepheids or are too distant for detecting RR Lyrae stars. A potential problem with calibrating novae in M31 is internal extinction in M31's bulge. If extinction is present, it will broaden and shift the magnitude fade-rate calibration.

In the following sections we take a critical look at these problems, and then discuss a new approach, randomly sampled B and Hα luminosity functions, for using novae as standard candles.

II. CALIBRATION PROBLEMS AND SELECTION EFFECTS

A) Shell Morphology

We can easily see that a nonspherical shell can lead to a systematic error in an expansion parallax by considering an oblate spheroid. The maximum projected dimension of the spheroid always will be the major axis, whereas the observed expansion velocity will always be less than or equal to the velocity in the equatorial plane. Consequently, the derived distance will always be less than the true distance. In order to determine whether or not the geometry of nova shells should be included in the derivation of expansion parallaxes, we must first establish the nature of the geometry.

As early as 1941 Payne-Gaposchkin had recognized that the four-peaked emission line profiles seen in some novae showed that the shells are not spherically symmetrical. This conclusion is confirmed by detailed studies of individual novae. V 603 Aql (nova Aql 1918), a typical fast nova, is a spectroscopic binary with a period of 3 hours and 20 minutes and a very small velocity amplitude, suggesting that the orbital plane is viewed nearly face on. Mustel and Boyarchuk (1970; MB) concluded that V 603 Aql's shell has a central equatorial belt, a pair of tropical/temperate belts, and polar caps at each end of the major axis. The belts and polar caps are caused by large density enhancements relative to other latitudes on the shell. MB's morphologic description was verified by Weaver's (1974) exhaustive analysis of the spectral development of the nova from outburst in 1919 through 1922. Weaver described the shell as a series of "truncated cones" with the axis of the cones inclined by less than a degree to the line of sight. If this is correct, the planes of the equatorial band and the belts are parallel to the plane of the binary.

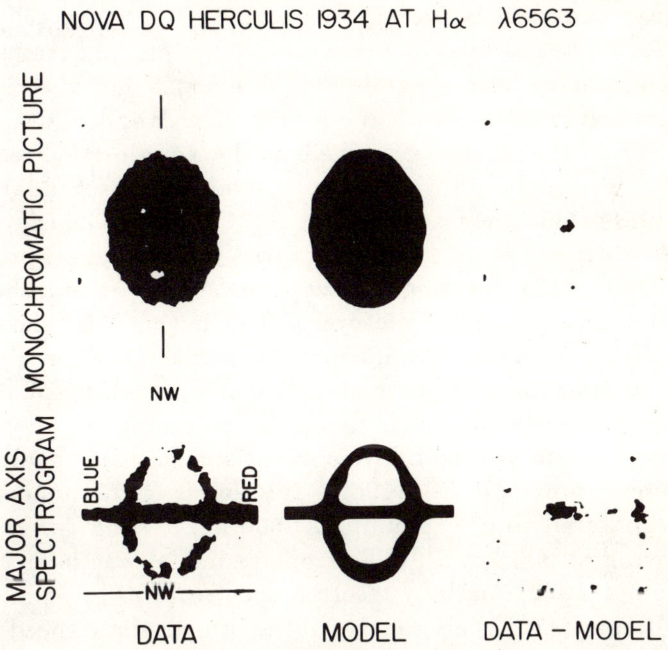

Fig. 1 The top left hand panel shows an $H\alpha + [N\ II]$ image of DQ Her taken at the prime focus of the Shane 3-m telescope using a single stage Westinghouse image intensifier; the lower left hand panel shows an Hα spectrogram taken at the Coude focus of the Shane 3-m telescope using a single stage ITT image intensifier. The two middle panels show Jenner's (1978) model which best fit the observations. The model was used to derive the distance given in the text. The right hand panels show the difference between the model and the observations.

Weaver did not give a value for the ratio of the equatorial radius to polar axis. Because the shell is seen almost end on, there is no feasible way to determine the ratio of the major axis (the polar axis) to minor axis (the equatorial radius). *Consequently, the shell is "geometrically degenerate"; a parallax cannot be determined from the angular size and expansion velocity.* The only way to derive a parallax is to assume or guess the ratio of the major and minor axes.

MB and Weaver described the morphological development of the slow nova DQ Her as an expanding shell with an equatorial band and two polar caps. The bright equatorial band seen in early photographs has evolved into the three bands seen in Figure 1 and in a picture taken by Williams *et al.* (1978). The preservation of the bands as the nova evolves shows that the shell is prolate. Since DQ Her is an eclipsing binary we are seeing the orbital plane almost edge on. The most physically intuitive geometry (also suggested by Weaver) is one wherein the central equatorial belt coincides with the accretion disk and orbital plane of the binary. The expansion of the shell in the equatorial direction is then slowed by an interaction with the mass-transfering secondary.

MB note that the fast nova T Aur (1891) is an eclipsing binary with a period of 4 hours and 54 minutes. Since the system eclipses it must be nearly edge on, and thus might be expected to have a morphology similar to DQ Her. Baade's 1956 picture of T Aur (*cf.* MB) shows an "oval" with a 1.5/1 axial ratio. From comparison with DQ Her it is reasonable to conclude that the nonspherical shell also is a prolate spheroid with the polar axis perpendicular to the binary's orbital plane.

Based on the spectral development and multiple peaks in the slow (and unique) nova HR Del (1967), Hutchings (1972) modeled the shell as two equatorial rings expanding at 500 km $s^{-1}$and two polar blobs or caps moving at 200 km s^{-1}. This oblate model was later confirmed by Soderblom's (1976) spatially resolved spectra.

Hutchings (1972) also analyzed the intermediate speed nova FH Ser (1970) and concluded that shortly after outburst the nova was slightly oblate with an equatorial ring and polar caps and was surrounded by a low density, high velocity spherical shell. Eventually the spherical shell and ring dissipated, leaving only the polar caps, *i.e.,* a prolate geometry. If there are only polar caps, the distance will be give by

$$D = 2 \times V_{obs} \times T \times tan(i)/\theta$$

where i is the inclination of the line connecting the polar caps to the line of sight, V_{obs} is the observed radial velocity ($V_{obs} = V_{true} \times cos(i)$), T is the time since outburst, and θ is the angular separation of the polar caps on the sky. *Since we do not know the inclination, we cannot determine D.*

The preceeding discussion is summarized in Table 1. In a nutshell, symmetrical nova shells appear to be composed of equatorial rings and a pair of polar caps, with maximum elongation along the polar direction.

The best novae for understanding the geometry are the eclipsing binaries like T Aur and DQ Her where the shell is seen from the equatorial plane.

Table 1. Geometry of Nova Shells

Nova	Type	Morphology	Geometry
V 603 Aql 1918	Fast	Equatorial Rings Polar Caps	Prolate or Oblate
T Aur 1891	Fast	Oval	Prolate
HR Del 1967	Slow	Equatorial Rings Polar Caps	Oblate
DQ Her 1934	Slow	Equatorial Rings Polar Caps	Prolate
FH Ser 1970	Intermediate	Polar Caps	Prolate

B) Distances to Individual Novae

The importance of properly including the geometry of the shell in expansion parallaxes can be illustrated by considering V 603 Aql, FH Ser, DQ Her, and T Aur. As we have seen, V 603 Aql and FH Ser are geometrically degenerate. The uncertainty in the distance to V 603 Aql is equal to the uncertainty in the ratio of the major axis to the minor axis. Since some nova shells may be spherical, whereas other nova shells such as T Aur have a major to minor axis ratio $a/b = 1.5$, the uncertainty in the distance can be $\sim 50\%$, and is bounded only by our beliefs about the permissible range of expansion velocities for fast novae. Consequently, the errors in the distance could be a factor of 2 or more. In spite of these large uncertainties, the published parallaxes have not included qualifications about their derived distances.

The nearly edge-on, prolate geometry of DQ Her's shell should provide one of the most accurate expansion parallaxes. Nonetheless, published parallaxes range from 320 pc (MB) and 302 pc (CR) to 485 pc ($\pm$ 50 pc) (Jenner 1978,1988). Figure 1 shows the high spatial resolution image of the shell and the high dispersion Hα velocity ellipsoid which Jenner used to model DQ Her. The difference between the data and the model is shown in the right hand panels of Figure 1. The sharpness of the bands and the small degree of tilt in the velocity ellipsoid confirm that the prolate shell is seen nearly edge on. Jenner derived an axial ratio $a/b = 1.4$, an inclination of 87 degrees between the line of sight and the major axis, a maximum expansion velocity (*i.e.*, along the major axis) of 550 km s^{-1}, and a distance of 485 pc. The primary difference between Jenner's and MB's parallax stems from the different minor axis expansion velocities which were used (393 km $s^{-1}$and 290 km s^{-1}, respectively). The difference between Jenner's and CR's parallax is due again to the assumed tangential expansion velocities (CR used 315 km s^{-1}) and the assignment of the *observed* expansion velocity to the minor axis (Jenner) and the major axis (CR).

CR's distance to T Aur (1320 pc) was derived by assigning the observed expansion velocity to the major axis. Since T Aur is an eclipsing binary (Kraft 1964), the major axis of the shell is most likely nearly perpendicular to the line of sight. If this is correct, the ratio of the true distance to CR's distance could be as large as $a/b = 1.5$.

Based on the preceding discussion, we conclude that the errors in expansion parallaxes to individual novae may be much larger than is generally recognized. In the next section we investigate the question of whether or not the errors average to zero when a large number of randomly oriented shells are observed.

C) Expansion Parallaxes for Randomly Oriented Spheroids

In one of the earliest papers on expansion parallaxes McLaughlin (1945) stated the following. "Involved in these results is the assumption that the velocity of ejection was equal in all directions. We know this is not strictly true, for some of the shells have visibly elliptical cross section. However, we can expect the orientation of ellipsoidal shells to be random relative to the line of sight." McLaughlin assumed that the errors in the parallaxes would average to zero for randomly oriented spheroids. This obviously cannot be true if the shells are oblate, since a parallax based on projected dimensions and the observed radial velocity will always be less than the true parallax. However, the result of observing an ensemble of randomly oriented prolate spheroids is not so obvious, since the derived distance can be too large or too small, depending on the orientation of the spheroid.

We will use two limiting geometries to show that the errors in the parallaxes do not average to zero when observing randomly oriented prolate spheroids. In the first case we simplify the characteristic geometry of novae by assuming a thin equatorial band and two small polar caps. We then use CR's prescription of combining the observed expansion velocity with the maximum angular radius of the shell to derive the "observed distance." In Table 2 we tabulate the ratio of the observed distance to the true distance, $D_{obs}/D_{true} = f(i)$, as a function of the inclination between the line of sight and the major axis of the spheroid.

Table 2. Dependence of Observed Distance on Inclination

Inclination	D_{obs}/D_{true}
$i < arctan(b/a)$	a/b
$arctan(b/a) < i < arctan(a/b)$	$cot(i)$
$arctan(a/b) < i$	b/a

The distribution of inclinations will be given by sin (i). The average ratio of distances will be given by

$$< D_{obs}/D_{true} >= \frac{\int_0^{\frac{\pi}{2}} f(i) sin(i) di}{\int_0^{\frac{\pi}{2}} sin(i) di}$$

Using $f(i)$ from Table 2, we obtain

$$< D_{obs}/D_{true} >= a/b + (1 - a/b) \times cos(\theta) - (1 - b/a) \times sin(\theta)$$

where $\theta = arctan(b/a)$.

The ratio of distances is very flat for values of b/a between 0.3 and 0.7, with a value of ~ 0.90. Thus, there is a 10% systematic error in the distances and an 0.1 mag error in the absolute magnitudes for the simple case we have considered.

For the second geometry we assume that the shell is a thin prolate spheroid with uniform surface brightness and an axial ratio b/a. The ratio of the observed distance to the true distance is then given by

$$D_{obs}/D_{true} = \sqrt{\frac{a^2 + b^2 \times tan^2(i)}{b^2 + a^2 \times tan^2(i)}}$$

Numerical integration of $D_{obs}/D_{true} \times sin(i)$ for different values of b/a gives the results tabulated in Table 3.

Table 3. Systematic Errors for Randomly Oriented Prolate Spheroides

b/a	$< D_{obs}/D_{true} >$
0.0	1.00
0.2	0.88
0.4	0.86
0.6	0.89
0.8	0.94
1.0	1.00

The fact that there is very little difference between the average systematic errors for the two geometries shows that the error is rather insensitive to details of the shell structure.

In both cases the ratio of distances is very flat for values of b/a between 0.3 and 0.7, with a value ranging from ~ 0.86 to ~ 0.90. Thus, there is a 10% to 14% systematic error in the distances and a 0.10 mag to 0.14 mag error in the absolute magnitudes for these simple cases.

One additional point should be made. When observing partially resolved shells, there may be a preferred selection of novae with large major axis inclinations relative to the line of sight. In this case, the systematic error approaches b/a. We conclude that the systematic errors in the published calibrations of absolute magnitude versus fade rate may be as large as 0.2 mag to 0.3 mag. This effect may explain the fact that Cohen's (1985) nova calibration gives a distance modulus to M31 which is 0.3 magnitudes smaller than the distance modulus derived from RR Lyrae stars (PvdB) and infrared observations of Cepheids (Welch *et al.* 1986), and 0.4 magnitudes smaller than the distance derived from the Population II giant branch (Mould and Kristian 1986).

D) Recommendations for Measuring Expansion Parallaxes

In view of the potential problems in measuring expansion parallaxes, we make the following recommendations.

1) High quality, homogenous, moderate resolution spectra should be obtained systematically during the first two or three years after outburst of novae in the Galaxy and the Magellanic Clouds. The development of the absorption/emission line profiles can then be used to infer the geometry of the shell following the methods used by Hutchings (1972) and Weaver (1974).

2) Eclipsing binaries such as DQ Her and T Aur should receive special attention since the nearly edge on geometry reduces the uncertainty in the expansion parallax. A high resolution, spatially resolved spectrum of T Aur should be obtained in order to improve the parallax of this nova.

3) Expansion parallaxes should be derived by combining spatial and spectral models with high quality spatial/spectral observations.

E) Internal Reddening in M31

Many authors have noted that there is scattered dust throughout the bulge of M31 (Johnson and Hanna 1972; Hodge 1980; Gallagher and Hunter 1981; McElroy 1983; Kent 1983; and Ciardullo *et al.* 1988). Extinction by this dust could in principle account for a large fraction of the width in the distribution of maximum magnitudes versus rates of decline observed in M31 (*cf.* Rosino 1964 and PvdB). We investigated this possibility by converting the nebular coordinates of Arp's and Rosino's novae into standard coordinates and then projecting the positions onto a television image of Ciardullo *et al.* 's (1988) composite $B - \lambda6200$ color map of M31's bulge. The position of each nova was inspected and the nova was classified as unreddened if it projected onto a smooth background with no underlying structure due to the presence of a dust cloud along the line of sight. Thirty two of the 86 novae which project into the boundaries of the composite picture fall in regions which do not show any evidence for dust. The distribution of maximum magnitudes versus fade rates for these "unreddened" novae is indistinguishable from the distribution for the entire set of novae. Consequently, we conclude that there is very little extinction from the filamentary and patchy distribution of dust seen in the bulge of M31.

The apparent absence of extinction in the magnitudes of novae in the bulge of M31 lends confidence to PvdB's calibration of novae via RR Lyrae stars in M31. The surprising absence of significant extinction can be explained if the dust clouds and filaments are optically thick with a relatively small filling factor. Novae are then either in front of the dust and unobscured, or they are behind the dust and are undetected. The variation in absolute magnitude at a given fade rate which is seen in Arp's and Rosino's light curves may then be due to a combination of observational errors in the zero points of the photographic magnitudes and intrinsic scatter in the characteristics of the outbursts (Shara 1981).

III. TIME AVERAGED NOVA LUMINOSITY FUNCTIONS

One of the biggest, if not the biggest difficulty in using nova light curves for extragalactic distance estimates is logistical rather than scientific. The problem is illustrated by PvdB's heroic observations of novae in the Virgo cluster. With 15 half nights on the Canada-France-Hawaii 3.6-m telescope they were able to detect 8 novae in NGC 4472 and 1 nova in NGC 4365. Five of the novae had maximum B magnitudes fainter than 24.0, and thus were close enough to the limiting magnitude to make determination of their decay times difficult. Between the problems of faintness and incomplete coverage, they were able to measure fade rates for six of the novae. Given the fierce competition for time on large telescopes and the common practice of assigning time to observers in small blocks of whole nights, PvdB's observations will not be easily repeated or surpassed.

On the other hand, observing extragalactic novae *at some phase* of their evolution is much easier. In the broadband B filter, fast novae in Virgo will be detectable for about a week; novae of moderate speed should remain visible for a month. Three novae were detected in three fields on the first night of PvdB's observing run in March. Five novae were detectable in 3 fields on the first night of PvdB's observations in April. Thus, the detection rate will be approximately 1 nova per field for a deep CCD picture of a bright elliptical in Virgo. A large sample of novae will be collected if a series of frames are taken over a time span of years. Taken together, the magnitudes of these novae, observed at random phases, form a time averaged nova luminosity function. Because this type of data is relatively easy to obtain, it is important to see if it can be used to measure distances.

In order to investigate this possibility, we created two time averaged broadband luminosity functions using the observations of M31 novae found in the literature. For the first curve, we inverted and differentiated Arp's (1956) homogeneous set of complete B light curves to create the Δt *vs.* m relation of each nova. We then found the theoretical $N(m)$ *vs.* m function by summing these curves. Unfortunately, although this works rather well, Arp only observed 30 novae, and the sample may not be truly representative. However, since the turn of the century, over 200 novae have been discovered in M31—the great majority of which were observed in the B or photographic passband (Alksnis and Sharov 1969; Arp 1956; Ciardullo *et al.* 1987; Duncan 1928; Grubissich and Rosino 1957a,b; Hubble 1929; Rosino 1973; Rosino and Grubissich 1955; Sharov 1972; and Sharov and Alksnis 1969, 1970). We therefore created a second, empirically derived luminosity function by randomly sampling these light curves. Although incompleteness is a major problem for the histogram of historical novae, the two curves agree extremely well. However, as Figure 2 shows, when plotted logarithmically, the relation is a line. Unless the outburst times for a sample of novae are known, broadband observations cannot be used to derive distances.

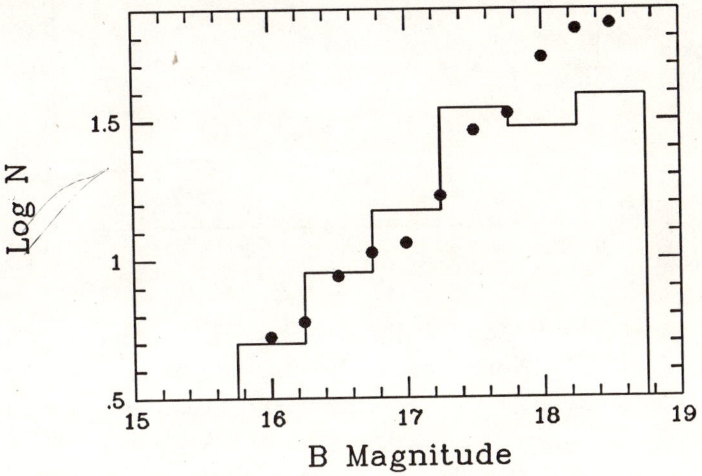

Fig. 2 The time averaged M31 nova B luminosity function derived from historical novae.

As discouraging as this result is, it is not necessarily fatal. Observations aimed at detecting novae past maximum light should be performed in the narrow band Hα filter, not in B. To see why, we compare the Hα and B light curves of Nova Cyg 1975 and a typical nova we recently observed in M31 (Figures 3 and 4). The broadband data for Nova Cyg 1975 is that of Young *et al.* (1976) and Williamon (1977). The Hα curve for Nova Cyg 1975 was derived from the spectrophotometric measurements of Ferland, Lambert, and Woodson (1986). The zero point of the Hα magnitude scale is referred to Vega, with $m_{\mathrm{H}\alpha} = 0$ corresponding to a total flux of 1.5×10^{-7} ergs cm^{-2} s^{-1} through a 75 Åfilter. All novae begin as continuum sources, but after maximum, the Hα emission continues to increase for a period of several days (the precise length of time for this brightening depends on the speed of the nova). Thereafter, while the continuum is undergoing the phase labeled by McLaughlin (1960) as "early decline", the Hα light fades more slowly, in an exponential decay

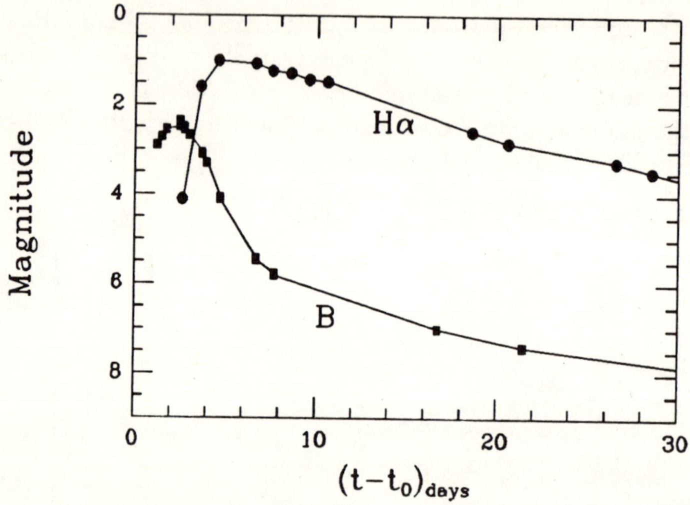

Fig. 3 The Hα and B light curves for the first month of Nova Cyg 1975. Note that the Hα flux continues to rise after the maximum in B, and remains brighter than B as the nova fades.

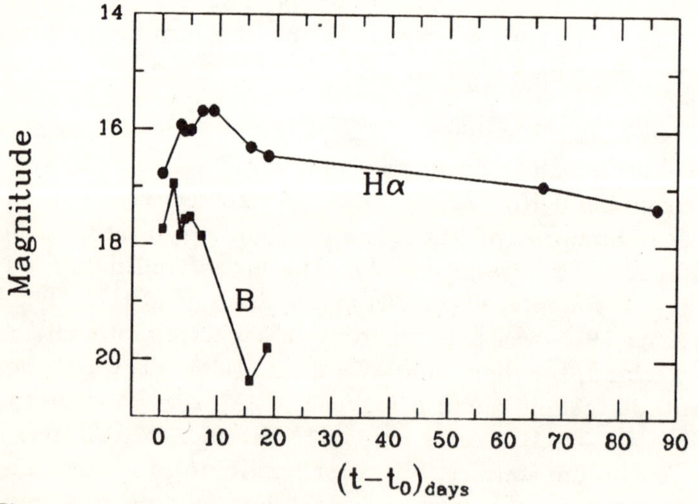

Fig. 4 The B and Hα light curves for a typical nova in M31 (CFJN 31). This characteristic light curve shows that shortly after outburst the nova is much more easily detected in Hα than in B.

rate labeled by McLaughlin as "final decline." As a result, for most of a nova's life, the flux of photons emitted in a 75 Å Hα bandpass is

30% to 60% of that radiated in the entire 1000 Å wide B filter. When a nova is at maximum, it is easy to observe in the broadband filter since it is a continuum source; once past maximum, the high contrast of the emission line over the background galaxy makes it far easier to detect in Hα. This is true for novae of all classes and speeds, and is especially true for the slowest novae, which may have a 3 mag decay time of well over a year. Therefore, if one wants to detect large numbers of extragalactic novae regardless of their evolution phase, whether for the purposes of studying their underlying population, identifying objects for spectroscopic study, or making distance estimates, the survey should be done in the narrow Hα line, rather than with a broadband B filter.

For the past few years, we have been performing such an Hα survey in several nearby galaxies, in order to investigate the underlying nova-producing stellar population via the observed nova rate and distribution. As a by-product, we have also been building up an estimate of the Hα time averaged luminosity function for novae. Figure 5 displays this relation derived from 35 novae in M31. Unfortunately, once again, the distribution is very suggestive of a power law, thus precluding its use as a distance indicator.

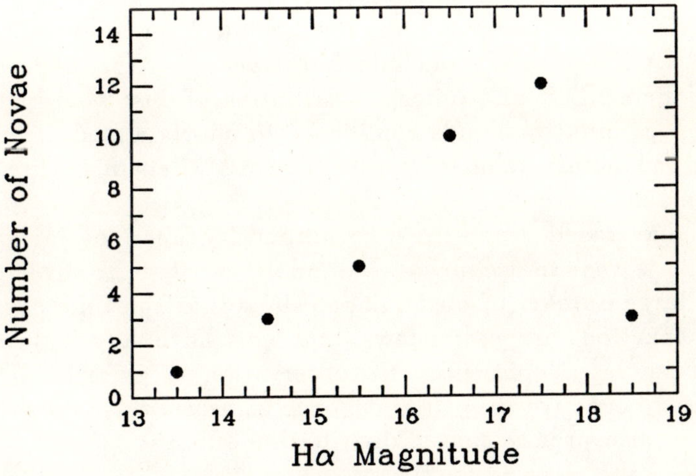

Fig. 5 The time averaged M31 nova Hα luminosity function derived from modern CCD observations of M31.

IV. SUMMARY AND CONCLUSIONS

We have shown that nova shells are characteristically prolate with equatorial bands and polar caps. Failure to take the geometry into consideration can lead to large errors in expansion parallaxes for individual novae. When simple prescriptions are used for deriving expansion parallaxes from an ensemble of randomly oriented prolate spheroids, the resulting absolute magnitudes will be underestimated and the distance scale will be too small by factors of 10% to 15%. If observations of partially resolved nova shells select for large inclinations, the systematic error in the resulting distance scale could easily be 20% to 30%. Although we make recommendations on how to avoid systematic errors in expansion parallaxes, we think the better approach is to invest effort in observations of M31 to improve and understand the relationship between maximum magnitude and fade rate, and to tie the absolute magnitudes of novae to RR Lyrae stars and Cepheids.

We investigated extinction in the bulge of M31 by projecting Arp's and Rosino's novae onto a composite $B - \lambda 6200$ color map of M31's bulge. Thirty two of the 86 novae projected onto a smooth background with no underlying structure due to the presence of a dust cloud along the line of sight. The distribution of maximum magnitudes versus fade rates for these "unreddened" novae is indistinguishable from the distribution for the entire set of novae. We conclude that novae suffer very little extinction from the filamentary and patchy distribution of dust seen in the bulge of M31. We think this lends confidence to efforts aimed at calibrating novae (and planetary nebulae) in the bulge of M31 against other standard candles in M31.

Time averaged B and Hα nova luminosity functions are potentially powerful new ways to use novae as standard candles. However, our analysis of a large number of observations show that the time averaged luminosity functions are power laws, and thus have very little utility as standard candles. Nonetheless, Hα observations of galaxies will provide an extremely effective way to find novae in order to establish the nova rate and to measure the spatial distribution of novae.

V. ACKNOWLEDGEMENTS

The authors were Visiting Astronomers at Kitt Peak National Observatory, which is operated by the Association of Universities for Research in Astronomy, Inc., under contract with the National Science

Foundation. We are grateful to Xiaohui Hui for providing the analytical solution for D_{obs}/D_{true} for a prolate spheroid with uniform surface brightness, and for help in preparing the camera ready manuscript. This work was supported in part by NASA grants NAS 529293 and NAGW-421.

REFERENCES

Alksnis, A.K., and Sharov, A.S. 1969, *Astron. Tsirk.*, **507,** 1.

Arp, H.C. 1956, *A. J.*, **61,** 15.

Ciardullo, R., Ford, H.C., Neill, J.D., Jacoby, G.H., and Shafter, A.W. 1987, *Ap. J.*, **3**18, 520.

Ciardullo, R., Rubin, V.C., Jacoby, G.H, Ford, H.C., and Ford, W.K. Jr. 1988, *A. J.*, **95**, 438.

Ciardullo, R., Jacoby, G.H., Ford, H.C., and Neill, J.D. 1988b, *Ap. J.*, (*submitted for publication*).

Cohen, J.G., and Rosenthal, A.J. 1983, *Ap. J.*, **268,** 689.

Cohen, J.G. 1985, *Ap. J.*, **292,** 90.

de Vaucouleurs, G. 1978, *Ap. J.*, **223,** 351.

Duncan, J.C. 1928, *Pub. A.S.P.*, **40,** 347.

Ferland, G.J., Lambert, D.L., and Woodman, J.H. 1986, *Ap. J. Suppl.*, **60,** 375.

Gallagher, J.S., and Hunter, D.A. 1981, *A. J.*, **86**, 1312.

Grubissich, C., and Rosino, L. 1957a, *Con. Asiago* no. 77.

Grubissich, C., and Rosino, L. 1957b, *Con. Asiago* no. 93.

Hodge, P.W. *A. J.*, **85**, 376.

Hubble, E. 1929, *Ap. J.*, **69,** 103.

Hubble, E. 1936, *The Realm of the Nebulae* (New Haven and London: Yale University Press).

Hutchings, J.B. 1972, *M.N.R.A.S.*, **158,** 177.

Jenner, D.C. 1978, *B.A.A.S.*, **10,** 427.

Jenner, D.C. 1988, private communication

Johnson, H.M., and Hanna, M.M. 1972, *Ap. J. (Letters)*, **174**, L71.

Kent, S.M. 1983, *Ap. J.*, **266**, 562.

McElroy, D.B. 1983, *Ap. J.*, **270**, 485.

McLaughlin, D.B. 1942, *Pop. Astr.* 50, 233.

McLaughlin, D.B. 1945, *Pub. A.S.P.*, **57,** 69.

Mould, J. and Kristian, J. 1986, *Ap. J.*, **305**, 591.

Mustel, E.R., and Boyarchuk, A.A. 1970, *Astrophys. Space Sci.* **6**, 183.

Payne-Gaposchkin, C. 1941, in *XIII Colloq. Int. Astrophys., Novae and White Dwarfs,* A.J. Shaler ed., p.69.

Payne-Gaposchkin, C. 1957, *The Galactic Novae* (New York: Dover).

Pritchet, C.J., and van den Bergh, S. 1987, *Ap. J.,* **318**, 507.

Ritchey, 1917, *Pub. A.S.P.,* **29,** 210.

Rosino, L. 1964, *Ann. Astrophys.* **27**, 498.

Rosino, L. 1973, *Astr. Ap. Suppl.,* **9,** 347.

Rosino, L., and Grubissich, C., 1955, *Con. Asiago* no. 68.

Shara, M.M. 1981, *Ap. J.,* **243**, 926.

Sharov, A.S. 1972, *Soviet Ast.,* **15,** 1001.

Sharov, A.S., and Alksnis, A.K. 1969, *Astron. Tsirk.,* **514,** 1.

Sharov, A.S., and Alksnis, A.K. 1970, *Astron. Tsirk.,* **560,** 1.

Soderblom, D. 1976, *Pub. A.S.P.,* **88,** 517.

van den Bergh, S., and Pritchet, C.J. 1986, *Pub. A.S.P.,* **98,**, 110.

Weaver, H. 1974, in *Highlights of Astronomy, Vol. 3* G. Contopoulos ed. (Dordrecht: D. Reidel Publishing Co.), p. 509.

Welch, D.L., McAlary, C.W., McLaren, R.A., and Madore, B.F. 1986, *Ap. J.,* **305**, 583.

Williamon, R.M. 1977, *Pub. A.S.P.,* **89,** 44.

Williams, R.E., Woolf, N.J., Hege, K., Moore, R.L., and Kopriva, D.A. 1978, *Ap. J.,* **224,** 171.

Young, P.J., Corwin, H.G., Bryan, J., and de Vaucouleurs, G. 1976, *Ap. J.,* **209,** 882.

DISCUSSION

D. WELCH: What about the handful of overluminous novae on your graph? These will be the first objects seen at great distances.

H. FORD: Pritchet and van den Bergh used the distribution of MMRDs to derive the distance to the Virgo cluster. Since their novae appear to follow the form of the MMRD, it does not look as though they have selected anomalously bright novae.

S. VAN DEN BERGH: Is there a maximum magnitude versus rate of decline relation in Hα?

H. FORD: Because of the way our telescope time has been assigned we have observed only two or three novae well enough to have a clearly defined Hα maximum.

SUPERNOVAE AND THE HUBBLE CONSTANT

DAVID BRANCH
Department of Physics and Astronomy, University of Oklahoma, Norman, Oklahoma 73019 USA

ABSTRACT Five ways to use Type II and Type Ia supernovae to estimate the value of the Hubble constant give results ranging from 53 to 59 km s^{-1} Mpc^{-1}, and a mean of $H_o = 57 \pm 10$ km s^{-1} Mpc^{-1}.

INTRODUCTION

Supernovae offer an attractive alternative to galaxies as indicators of distances on cosmologically significant scales. The Proceedings of the Workshop on Supernovae as Distance Indicators (Bartel 1985) contain articles dealing with both optical and radio approaches to a supernova-based distance scale. I will argue here that as observations accumulate and understanding improves, supernovae continue to point toward a low value of the Hubble constant.

TYPE II SUPERNOVAE

As expected on the basis of theoretical considerations, neither the light-curve peaks nor tails of Type II supernovae (SNe II) are useful as standard candles (T. R. Young and D. Branch, in preparation), so if SNe II are to be used as distance indicators, they must be used as individuals. Distances to individual SNe II can be determined by Baade's method, and by VLBI.

The Baade Method

The angular radius of a supernova can be expresed in terms of the observed flux at the Earth and the absolute flux at the photosphere, $\theta = (f_\nu/F_\nu)^{\frac{1}{2}}$. In lieu of mature model atmospheres for supernovae, the angular radius can be obtained by appealing to equilibrium and taking the absolute flux to be the blackbody flux, $\pi B_\nu(T)$, where B_ν is the Planck function and T is the optical-infrared color temperature. From the angular radius one can obtain a distance-dependent photometric velocity at the photosphere, $V_p = \theta D/(t - t_o)$, where D is the distance and t_o is the time of explosion. An independent and distance-independent spectroscopic velocity at the photosphere, V_s, can be obtained from the widths and blueshifts of the spectral lines. Matching $V_p(t)$ and $V_s(t)$ yields the distance and the time of the explosion, which ordinarily also is unknown. Table I lists distances to SNe II that have been derived in this way. The value of H_o that is implied by each distance determination has been obtained by scaling the supernova distance against the parent-galaxy distance given in the Tully catalog (TC; Tully 1988). Distances in the Tully catalog are based on a Virgo-infall model with H_o = 75 km s^{-1} Mpc^{-1}.

TABLE I. The Baade Method Applied to Type II Supernovae

SN	Galaxy	D(SN) (Mpc)	D(TC) (Mpc)	H_o (km s^{-1} Mpc^{-1})
1969L	NGC 1058	12±4 (Ref. 1)	10.0	62
1970G	M101	7±2 (Ref. 1)	5.2	56
1979C	M100	23±3 (Ref. 2)	16.8 (Virgo)	55
1980K	NGC 6946	12±4 (Ref. 3)	5.1	32

Ref. 1: Kirshner and Kwan (1974)
Ref. 2: Branch et al. (1981)
Ref. 3: Kirshner (1985)

If the discrepant result for SN 1980K (as well as the Kirshner [1985] result of 21 ± 4 Mpc for SN 1979C) is ignored on the grounds that no details of the analysis have been published, the Baade method applied to SNe II gives

$$H_o = 57 \pm 4\ km\ s^{-1}\ Mpc^{-1},$$

with most of the weight being on SN 1979C in M100, in the Virgo cluster.

The internal error is small, but the dominant uncertainty is external - in the blackbody approximation. Let ϵ be the ratio of the true flux and the blackbody estimate, $\epsilon = F_\nu / \pi B_\nu(T)$. The effects of a scattering-dominated opacity tend to produce a dilute flux, $\epsilon < 1$, (Hershkowitz and Wagoner 1987, and references therein), while the effects of atmospheric extension tend to produce $\epsilon > 1$ (Branch 1979). To illustrate the effects of departures from a blackbody flux, suppose for simplicity that the method is applied at only two times. In the blackbody approximation the distance is given by

$$D_{BB} = \frac{V_{s1}V_{s2}(t_2 - t_1)}{\Theta_2 V_{s1} - \Theta_1 V_{s2}}$$

The true distance, however, would be given by

$$D = \frac{V_{s1}V_{s2}(t_2 - t_1)}{\Theta_2 V_{s1}/\epsilon_2^{\frac{1}{2}} - \Theta_1 V_{s2}/\epsilon_1^{\frac{1}{2}}}$$

The effects of various combinations of ϵ_1 and ϵ_2 on the distance to SN 1979C are shown in Table II (cf. Branch 1985):

TABLE II. Ratio of True Distance to Blackbody Distance

	ϵ_2=1.0	0.8	0.6	0.4
ϵ_1 = 1.0	1.0			
0.8	1.1	0.9		
0.6	1.3	1.0	0.8	
0.4	1.7	1.3	0.9	0.6

Because the effects of scattering may decrease with time as hydrogen recombines, while the effects of atmospheric extension may increase with time as the photosphere retreats into deeper layers of ejected matter where the density gradient is less steep than it is at the surface, only $\epsilon_1 < \epsilon_2$ is considered. In the case $\epsilon_1 = \epsilon_2$, we are on the diagonal in Table II and the blackbody Baade distance is an overestimate. But if ϵ increases with time, as expected, then the blackbody distance may be about right, or even an underestimate.

Supernova 1987A, being at known distance (by H_o standards), of known explosion time, and exceptionally well observed, provides us with an opportunity to check on the blackbody approximation. David Jeffery and I have compared photometric velocities inferred by the South African observers on the assumption that the distance to SN 1987A is 50 kpc (Menzies et al. 1987, Catchpole et al 1987), with spectroscopic velocities that we have derived from a synthetic-spectrum analysis of the Cerro Tololo optical spectra (Phillips et al. 1988). For five days after the explosion the photometric and spectroscopic velocities showed a significant difference which, if entirely due to departures from a blackbody flux, would correspond to $\epsilon = 0.4$. During the next fifteen days the velocity differences decreased and ϵ approached unity, where it remained until at least 100 days. After the first few days the blackbody approximation was a good one, so SN 1987A provides support for the previous blackbody distance estimates for SNe II.

Supernova 1987A also is stimulating detailed spectral modelling, with continuum radiative transfer, NLTE effects in the lines, and direct estimates of its distance by the Baade method. An analysis which takes the scattering dilution into account but neglects atmospheric extension has given a distance to SN 1987A of 43 ± 4 kpc (Chilukuri and Wagoner 1988), and one which treats both scattering and atmospheric extension but uses a questionable supernova density profile has given 46 ± 4 kpc (Höflich 1987). The attempts that have been made so far probably should be regarded as tests of the method rather than as competitive determinations of the distance to the LMC, but as atmosphere models develop in the next few years it will be important and worthwhile to obtain extensive photometry and spectroscopy of relatively faint SNe II, beyond the local supercluster, for accurate distance determinations.

The VLBI Method

Another way to derive distances to SNe II was introduced by Bartel et al. (1985). The angular radius of a radio-emitting supernova is measured by means of VLBI observations, and the distance is obtained from $\theta D/(t-t_o) = V_r$, where V_r is the expansion velocity of the radio emitting matter. The strength of this method is that there is no need to predict a flux. The weaknesses are that the inferred angular size depends on the assumed radio brightness distribution, and that V_r is not observed but must be inferred from optical expansion velocities and a model of the radio emission. The method now gives a distance to SN 1979C of 22^{+7}_{-6} Mpc, where all sources of error are taken into account, and a lower limit of 5 Mpc to the distance to SN 1980K (Bartel et al. 1988). Scaled against the Virgo- cluster distance of the Tully catalog, the VLBI distance to SN 1979C corresponds to

$$H_o = 57^{+25}_{-20}\ km\ s^{-1}\ Mpc^{-1},$$

and the lower limit for SN 1980K corresponds to $H_o < 76$ km s^{-1} Mpc^{-1}.

TYPE I SUPERNOVAE

Because the observational sample of Type Ib supernovae is small, and they are fainter than Type Ia, Type Ib will be ignored here. An important question about Type Ia is whether they all are practically the same. The debate has focussed mainly on photometric properties, but spectroscopic properties are perhaps cleaner for this purpose because they don't depend on distances, extinctions, and subtle photometric errors. It has become clear that when spectroscopic velocity at the photosphere is plotted against time for a sample of SNe Ia, the differences are not reconcilable with the notion that SNe Ia are all alike (Branch, Drucker, and Jeffery 1988). SNe Ia also appear to show real differences in their photometric properties, and they may obey a relation between peak brightness and rate of post-peak light-curve decline (in the opposite sense of novae — supernovae with fast light curves are faint, not bright). In any case, SNe Ia at peak brightness probably are not splendid standard candles.

Why, then, do remote SNe Ia show only small scatter in the Hubble diagram? From a sample of 16 SNe Ia in elliptical galaxies in the Virgo cluster and beyond (mostly beyond), Sandage and Tammann (1982) found $M_B = -18.24 \pm 0.24 + 5\log(H_o/100)$, while from six SNe Ia in ellipticals beyond 3000 km s^{-1}, Arnett, Branch, and Wheeler (ABW; 1985) found a similar relation: $M_B = -18.4 \pm 0.2 + 5\log(H_o/100)$. The key to the small scatter is that among remote supernovae there is a strong observational bias in favor of the brightest ones - but there is a rigorous upper limit to the brightness of SNe Ia. The SN Ia light curve is powered by the decay of radioactive ^{56}Ni through ^{56}Co to stable ^{56}Fe, the peak luminosity is proportional to the amount of ^{56}Ni ejected by the explosion, and that amount is guaranteed not to exceed 1.4 $M_\odot$.

We can proceed, therefore, to attempt to calibrate the relation between the absolute magnitude of SNe Ia and the value of H_o, but we must remember that the SNe Ia discovered and observed nearby will tend to be intrinsically fainter than the bright ones that are discovered and observed at large distances, so we run the risk of underestimating distances to the remote ones and overestimating the value of H_o.

The Distance to IC 4182 and SN 1937C

Sandage and Tammann (1982) determined the distances to SN 1937C in IC 4182 and SN 1954A in NGC 4214 by identifying the brightest red stars in IC 4182, NGC 4214, and NGC 4395, a member of the same group as NGC 4214. The assumption that the mean visual magnitude of the three brightest red stars in each galaxy, $\langle M_V(3) \rangle$, is -7.72 ± 0.06 mag., led to $H_o = 50 \pm 7$ km s^{-1} Mpc^{-1}. However, SN 1954A contained He I lines (McLaughlin 1963, Branch 1972), so it is now recognized to have been of Type Ib. If SN 1954A is ignored, the Sandage-Tammann result becomes $H_o = 49 \pm 8$ km s^{-1} Mpc^{-1}. For consistency with other results presented here, this value is transformed to the ABW relation mentioned above, to obtain

$$H_o = 53 \pm 8 \; km \; s^{-1} \; Mpc^{-1}.$$

SN 1937C was a well observed SN Ia with no known peculiarities, and Sandage and Tammann reported that IC 4182 was a particularly easy object regarding photometry of its brightest stars. The surface brightness of the background disk is low, and the spiral features and associations make the identification of the member stars by position quite certain. A spectroscopic check is needed to make sure that none of the brightest red stars are really foreground dwarfs (R. M. Humphreys, private communication), and an independent determination of the distance to IC 4182 via Cepheids would be worthwhile.

Possible Remnants of Type Ia Supernovae in the Galaxy

Four supernova remnants in the Galaxy – Tycho, Kepler, SN 185, and SN 1006 – are usually suspected of having come from Type I supernovae. An interesting attempt to put these SNRs on a common, self-consistent distance scale has been made by Strom (1988). The known ages and angular diameters, together with the assumption that the remnants are in the adiabatic phase, were first used to determine relative distances to 10 per cent accuracy, then various absolute distance determinations were considered to fix the scale (to 20 per cent). In Table III, the relative distances, absolute distances, and absolute magnitudes are from Strom. The value of H_o that would be obtained by using each individual remnant as the SNe Ia calibrator of the ABW relation also is listed.

TABLE III. Possible SNe Ia Remnants as Calibrators

SNR	D_{rel}	D (kpc)	M_V	H_o (km s^{-1} Mpc^{-1})
SN 185	1.0	1.2	-19.2 (±2)	69
SN 1006	1.17	1.4	-19.8 (±1)	52
Tycho	2.04	2.5	-17.6 (±0.5)	145
Kepler	3.43	4.2	-19.6 (±0.5)	58

In view of the faint peak brightness of Tycho relative to the others, Strom suggests that Tycho may have been a Type Ib. Green (1986) has made the same suggestion, for the same reason. Whether Tycho was a Type Ib or some other kind of relatively faint supernova, it would make no sense to use Tycho to calibrate the intrinsically bright SNe Ia seen at large distances. If Tycho is excluded, the weighted mean (dominated by Kepler) is

$$H_o = 58 \pm 10\ km\ s^{-1}\ Mpc^{-1}.$$

Nickel Radioactivity

There is strong evidence that the Type Ia light curve is powered by the radioactive decay of ^{56}Ni and ^{56}Co (Woosley and Weaver 1986, and references therein). A key result for the distance scale, first predicted by Arnett (1982) on the basis on analytical models and subsequently confirmed by numerical calculations (J. C. Wheeler, S. E. Woosley, private communication), is that at the light-curve peak the SN Ia luminosity simply equals the instantaneous luminosity of radioactive decay, i.e., $L = L(M_{Ni}, t_{max} - t_o) = M_{Ni} f(t_{max} - t_p)$. This relation has been used by Arnett, Branch, and Wheeler (1985) to calibrate the SNe Ia. An observational value for $t_{\mathrm{max}} - t_o$ of 17 days, a bolometric correction based on the observation that at maximum light the energy distribution of SNe Ia is well characterized by a 20,000 K blackbody curve in the optical and near infrared but truncated at 0.4 μ to represent the strong ultraviolet deficiency, and the ABW relation from the Hubble diagram leads to a relation between nickel mass and Hubble constant: $H_o = 46 \pm 4 M_{Ni}^{-1/2}$ $km\ s^{-1}\ Mpc^{-1}$.

The nickel mass is not yet firmly predicted by theory, but it can be estimated on the basis of the kinetic energy, to which it is closely coupled. Although the delayed injection of radioactivity energy keeps the supernova warm and produces the light curve, the kinetic energy of the explosion is the difference between the nuclear energy released by fusion reactions and the white dwarf binding energy. On the basis of detailed nucleosynthesis calculations (Nomoto, Thielemann, and Yokoi 1985, Woosley, Axelrod, and Weaver 1984) ABW found $E_K = 3.1 M_{Ni} - 0.5$, where energies are in 10^{51} ergs, the nickel mass is in solar masses, and the second term is the binding energy. Several possibilities for M_{Ni} and the associated kinetic energy are considered in Table IV.

TABLE IV. Nickel Radioactivity as Calibrator of SNe Ia

M_{Ni} ($M_\odot$)	E_K (10^{51} ergs)	H_o (km s^{-1} Mpc^{-1})	Comment
1.4	3.8	39	No intermediate-mass elements; E_K too high
1.0	2.6	46	
0.6	1.4	59	Best current estimate
0.4	0.7	73	
0.2	0.1	103	Too little Fe for spectrum; E_K too low

$M_{Ni} = 1.4$ is an upper limit. We know that intermediate-mass elements are also ejected, and that a kinetic energy of 3.8 produces a light curve that evolves too fast and a spectrum whose lines are too broad and too blueshifted. If Type Ia supernovae are powered by nickel-cobalt radioactivity, $H_o > 40$ km s^{-1} Mpc^{-1}. At the other extreme, $M_{Ni} = 0.2$ is too small to account for the Fe-dominated spectrum, and because the nuclear energy becomes comparable to the binding energy, the kinetic energy is small and both the light curve and the spectrum would be much too slow. Considering the speed of the light curve and the spectrum, as well as the direct evidence for Fe and Co in the spectrum, the best estimate is $M_{Ni} = 0.6$ (as in the Nomoto et al. [1983] model W7). The ABW result from this method is

$$H_o = 59 \pm 14 \; km \; s^{-1} \; Mpc^{-1}.$$

The Baade method also has been applied to SNe Ia, with results ranging from $H_o = 40$ (Branch and Patchett 1973) to $H_o = 59$ (Branch 1979). However, in view of the complexity of SNe Ia spectrum formation (cf. Harkness 1987), the nickel-radioactivity method appears to be the preferable astrophysical approach for SNe Ia. The great virtue of the nickel method is that the luminosity follows directly from the nickel mass and the rise time to maximum light; blackbody approximations, knowledge of the gas opacity, and an understanding of spectrum formation are not required.

SUMMARY

The five estimates of H_o that have been discussed above are listed in Table V.

TABLE V. Supernova Results for the Hubble Constant

Method	H_o (km s^{-1} Mpc^{-1})
SNe II - Baade Method	57 (±4)
SNe II - VLBI	57 (+25, -20)
SNe Ia - Brightest Stars in IC 4182	53 (±8)
SNe Ia - Galactic SNRs	58 (±10)
SNe Ia - Nickel Radioactivity	59 (±14)

No individual method is unassailable, and each needs to be refined, but the supernova results do show an interesting consistency. Because the listed errors are a mixture of internal and external ones, it may be best to summarize with an unweighted mean and an informal estimate of the uncertainty:

$$H_o = 57 \pm 10\ km\ s^{-1}\ Mpc^{-1}.$$

Recalling that results based on SNe Ia may be biassed towards high H_o, to the extent that SNe Ia are not standard candles, it is tempting to suggest that the debate about the Hubble constant may no longer be "Is it 50 - or is it 100?", but

"Is it 60 - or is it 50?".

ACKNOWLEDGEMENTS

My research on supernovae is supported by NSF grant AST 8620310.

REFERENCES

Arnett, W. D. 1982, Ap. J., 253, 785.

Arnett, W. D., Branch, D., and Wheeler, J. C. 1985 (ABW), Nature, 314, 337.

Bartel, N. (ed.) 1985, Supernovae as Distance Indicators (Berlin-Heidelberg, Springer-Verlag); Lecture Notes in Physics, Vol. 224.

Bartel, N., Rogers, A. E. E., Shapiro, I. I., Gorenstein, M. V., Gwinn, C. R., Marcaide, J. M., and Weiler, K. W. 1985, Nature, 318, 25.

Bartel, N. 1988, in Supernova Shells and their Birth Events, ed. W. Kundt (Berlin-Heidelberg, Springer-Verlag); Lecture Notes in Physics, in press.

Branch, D. 1972, Astr. Ap., 16, 247.

——. 1979, M.N.R.A.S., 186, 609.

——. 1985, in Supernovae as Distance Indicators, ed. N. Bartel (Berlin-Heidelberg, Springer-Verlag); Lecture Notes in Physics, 224, 138.

Branch, D., Drucker, W., and Jeffery, D. J. 1988, Ap. J. (Letters), 330, L117.

Branch, D., Falk, S. W., McCall, M., Rybski, P., Uomoto, A. K., and Wills, B. J. 1981, Ap. J., 244, 780.

Branch, D., and Patchett, B. 1973, M.N.R.A.S., 161, 71.

Catchpole, R. M., and 19 others, 1987, M.N.R.A.S., 229, 15P.

Chilukuri, M., and Wagoner, R. V. 1988, preprint.

Doggett, J. B., and Branch, D. 1985, A.J., 90, 2303.

Green, D. A. 1986, Observatory, 106, 165.
Harkness, R. P. 1987, in Radiation Transport and Hydrodynamics, ed. D. Mihalas and K. H. Winckler (Dordrecht, Reidel).
Hershkowitz, S., and Wagoner, R. V. 1987, Ap. J., 301, 220.
Höflich, P. 1987, in Proceedings of the 4th Workshop on Nuclear Astrophysics.
Kirshner, R. P. 1985, in Supernovae as Distance Indicators, ed. N. Bartel (Berlin-Heidelberg, Springer-Verlag); Lecture Notes in Physics, 224, 171.
Kirshner, R. P., and Kwan J. 1974, Ap. J., 193, 27.
McLaughlin, D. B. 1963, P.A.S.P., 75, 133.
Menzies, J. W., and 15 others, 1987, M.N.R.A.S., 227, 39P.
Nomoto, K., Thielemann, F.-K., and Yokoi, K. 1984, Ap. J., 268, 644.
Phillips, M. M., Heathcote, S. R., Hamuy, M., and Navarette, M. 1988, A.J., 95, 1087.
Sandage, A. R., and Tammann, G. A. 1982, Ap. J., 256, 339.
Strom, R. G. 1988, M.N.R.A.S., 230, 331.
Sutherland, P. G., and Wheeler, J. C. 1984, Ap. J., 280, 282.
Tully, R. B. 1988, Nearby Galaxies Catelog, (Cambridge University Press).
Woosley, S. E., Axelrod, T. S., and Weaver, T. A. 1984, in Stellar Nucleosynthesis, ed. C. Chiosi and A. Renzini (Dordrecht, Reidel), p. 263.
Woosley, S. E., and Weaver, T. A. 1986, Ann. Rev. Astr. Ap., 24, 205.

DISCUSSION

J. MOULD: Getting a Cepheid distance for IC 4182 requires a big investment in telescope time. The usefulness of the result depends on the quality of the 1937 photometry. How good is the light curve? Does the photometry require significant corrections?

D. BRANCH: As I recall, the peak photographic magnitude was well determined at 8.2 mag; there was no problem with background contamination, and no reason to suspect an extinction problem. Whether significant corrections to Baade's magnitudes might be required - I am not the one to say.

R. NOLTHENIUS: Have you looked at how these derived values for H_o might converge to a global value when corrected for observed large scale non-Hubble motions?

D. BRANCH: We've only looked out to ~ 3000 km/sec - no corrections have been made.

N. BARTEL: You mentioned that at present you do not see anything wrong with the Baade method. However, the recent speckle interferometric results obtained by the U of A group give us cause for worry. They found that SN 1987A is significantly elongated, by 30% or so. Could that mean that your good agreement between the photospheric velocity and the spectroscopic velocity of SN 1987A at late times is partly coincidental and that SN 1987A is not as reliable for a check on the Baade method as one might hope?

D. BRANCH: Worse yet, spectropolarimetry showed that the supernova was also asymmetric at the earlier epochs that are of concern for the Baade method. But I think the effects of asymmetry on distance determinations are likely to come in at only the 10% level.

S. VAN DEN BERGH: As you know, Bandiera has recently suggested that Kepler's SN was not a SNIa. It might therefore not be wise to use it as a Type Ia calibrator.

D. BRANCH: Indeed it's quite possible that Kepler's SN was not a Type Ia. But the remote Type Ia's that we're trying to calibrate appear to have the highest luminosity of all observed SNE. If Kepler's SN was not a Ia, it was almost certainly less luminous; this would lead to an overestimate of H_o!

GIANT HII REGIONS AS DISTANCE INDICATORS

Jorge Melnick
European Southern Observatory

ABSTRACT The application of giant HII regions as distance indicators is reviewed and new results obtained using giant HII region-like dwarf galaxies (HII galaxies) are presented.

1. INTRODUCTION

In the early fifties Gum and deVaucouleurs (1953) showed that large ring-like HII regions could be used as extragalactic distance indicators. This motivated Sérsic (1959) to investigate the usefulness of large HII regions in general as distance indicators. Later in 1959, Sérsic had the opportunity of using plates taken by Hubble and by Sandage to establish firmly the existence of a strong correlation between the linear sizes of HII regions and the morphological types of their parent galaxies (Sérsic, 1960). This correlation was later developed by Sandage (1962) but its definitive consecration as a "standard" method to determine distances to late type galaxies did not come until 1974 when Sandage and Tammann (1974; ST) published the first paper of their classical series on the determination of the Hubble constant.

Today, largely owing to the discovery of other powerful techniques to determine distances to galaxies, giant HII regions no longer play the crucial role in the calibration of the Hubble law they once played in the Sandage-Tammann distance ladder. But, as I will show in this presentation, giant HII regions *together* with HII region-like dwarf galaxies (HII galaxies) may still provide an important tool for the calibration of the extragalactic distance scale.

2. THE DIAMETER-LUMINOSITY CLASS RELATION

The original discovery by Sérsic was that the mean sizes of HII regions increased from morphological types Sa to Sc and then decreased from Sc to Irr. On the basis of Sérsic's results, Sandage and Tammann excluded Sa and Sb galaxies from their analysis and found that for later types (Scd to Irr) there was an excellent correlation between the mean diameters of the three largest HII regions, $< D >_3$, and the absolute magnitudes of their parent galaxies M_B. Because both the absolute magnitudes and the linear diameters depend on distance, the slope of the relation between $< D >_3$ and M_B must be significantly different from -0.2 if this correlation is to be useful as a distance indicator. The slope found by Sandage and Tammann (and indeed by all subsequent workers) was ~ -0.14, too close (given the scatter of the correlation) to the "null" value of -0.2. ST circumvented this problem by replacing the absolute magnitudes by DDO luminosity classes (which *in principle* are distance independent) and defined the by now classical correlation between $< D >_3$ and DDO luminosity class (LC) as the "giant HII region step" in their cosmic ladder.

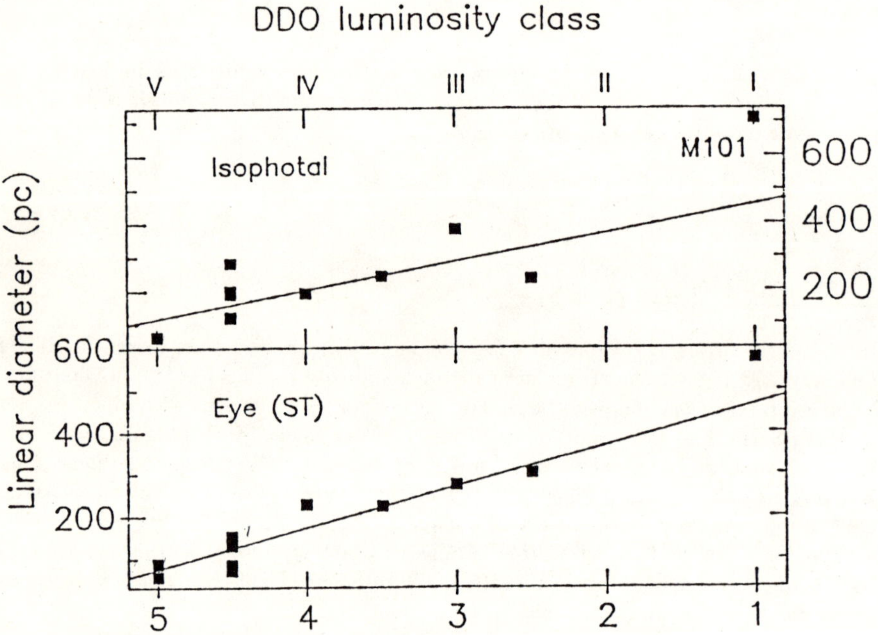

FIGURE 1: Relation between HII region diameters and parent galaxy luminosity classes for the galaxies in the ST sample. The eye measurements are the core-halo diameters of ST; the isophotal values are from Kennicutt (1979).

The next and last step in the ST ladder was to use giant HII regions to determine the absolute magnitudes of distant late-type field galaxies. However, as illustrated in Figure 1, the only ScI galaxy in the calibrating sample, M101, does not seem to fit the correlation. Thus, either the ($< D >_3 - LC$) relation is curved, or the ST distance to M101 is wrong, or M101 has abnormally large HII regions. ST adhered to the last hypothesis and extrapolated the calibration of the lower luminosity classes to ScI galaxies.

Besides the possible dependence of DDO type on distance and the large extrapolation to ScI's, the Sandage and Tammann method has been criticized on the ground that the boundaries of giant HII regions are extremely ill-defined and, therefore, that the diameters are very difficult to define and measure. This fact, illustrated in figure 2, was recognized by ST who explicitly stated that a "personal" system must be defined before their results can be applied to any independent sample of galaxies. A related

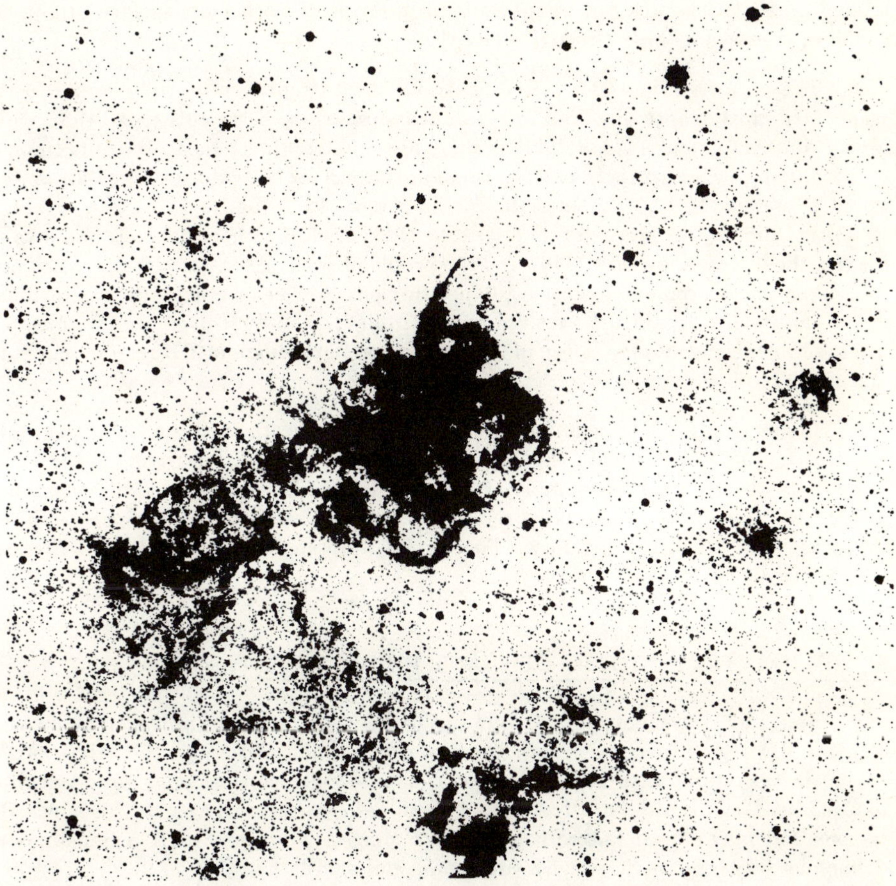

FIGURE 2: This R photograph of the 30 Doradus nebula illustrates some of the difficulties to reproduce the ST "core-halo" diameters of giant HII regions. The giant HII region (located near the center of the image) is embedded in a large super-association. The ST "halo" diameter encompasses the whole super-association (approximately the whole area covered by this photograph) while the ST "core" dimension corresponds to the diameter of the giant HII region. The overexposed area in the center of the HII region traces the location of the ionising cluster and corresponds to physical core of the nebula. On heavily exposed large scale photographs, such as that used by ST, the 30 Doradus nebula blends with the low emission-measure nebulosity that pervades the super-association and the whole complex appears as a single gigantic nebula. Other giant HII regions, some of those in M101 for example, are also parts of large super-associations.

difficulty was that they introduced a "core-halo" scheme to define their diameters which could not be fully reproduced by subsequent workers using independent plate material (Kennicutt, 1979). My own opinion, based on surface photometry of two nearby HII regions, is that the cores can be interpreted in terms of isothermal surface brightness profiles (at least for these two HII regions).

These caveats motivated Kennicutt (1979) to repeat the study of ST using *isophotal* diameters instead of eye core-halo measurements. Kennicutt found that, while for distant field galaxies isophotal diameters are superior to eye measurements (mostly by permitting seeing corrections), his isophotal values for the calibrating galaxies were essentially equivalent to the "core-halo" diameters of ST (figure 1). Thus, although Kennicutt did apply his results to determine the distances to field galaxies and a value for Ho, the discrepancy with M101 and the problems with the extrapolation to brighter luminosity classes remained. In fact, depending on whether or not M101 was included in the calibration, Kennicutt's results ranged from Ho=59 to Ho=84 $km\ s^{-1} Mpc^{-1}$. These problems invalidate the use of the correlation between giant HII region diameters and DDO luminosity classes to determine distances.

Before going on to a different topic, I would like to mention an attempt by Kennicutt (1981) to correlate the $H\alpha$ luminosities of giant HII regions ($L(H\alpha)$ with LC. On the basis of a large sample of galaxies in the Virgo cluster, Kennicutt found a large *intrinsic* scatter in this correlation which rendered this approach unsuitable for measuring (accurate) distances. The results I present below suggest that this is a metallicity effect.

3. THE VELOCITY WIDTHS OF GIANT HII REGIONS

Smith and Weedman (1972) found that the velocity widths of the emission lines of giant HII regions largely exceed the speed of sound. Although the origin of the supersonic gas motions in giant HII regions is still controversial, *supersonic* velocities may be used to discriminate truly giant nebulae from large aggregates of small HII regions that may appear unresolved on direct images. Thus, *giant* HII regions may be defined as very large nebulae with supersonic emission-line velocity widths.

In the course of an investigation of the nature of the supersonic motions observed in giant HII regions, I discovered a narrow correlation between the velocity dispersion of the gas (σ) and the linear diameters (D) as measured by ST (Melnick, 1977). This correlation, established on the basis of slightly more than a dozen data points, was criticized by Gallagher and Hunter (1983) who failed to reproduce it using a sample of HII regions in Irr galaxies. However, most of the objects in their sample have subsonic velocity widths and therefore are not *giant* HII regions (Melnick *et al.* 1987a).

Recently, the correlation between σ and D has been independently confirmed by two groups working with ST diameters (Melnick *et al.* 1987a) and with Kennicutt's isophotal measurements (Roy *et al.* 1986). Although the slope of the correlation is still somewhat controversial, these studies, based on large samples, confirm the reality of the correlation but with a larger scatter than my earlier result (Melnick, 1977). Thus, the $(D - \sigma)$ relation is not a suitable distance indicator.

Since the velocity widths are distance independent, an obvious thing to try is to correlate the mean velocity widths of the three largest HII regions, $< \sigma >_3$, with the absolute luminosities of the parent galaxies. In 1978 I used this approach to obtain a distance for NGC 300 (Melnick, 1978); later deVaucouleurs (1979), and Roy and Arsenault (1986) have demonstrated its potential usefulness as a distance indicator but the method has thus far not been applied to distant field galaxies. Figure 3 shows a plot of $< \sigma >_3$ versus absolute blue magnitude for the sample of giant HII regions in nearby galaxies studied by Melnick *et al.* (1987a).

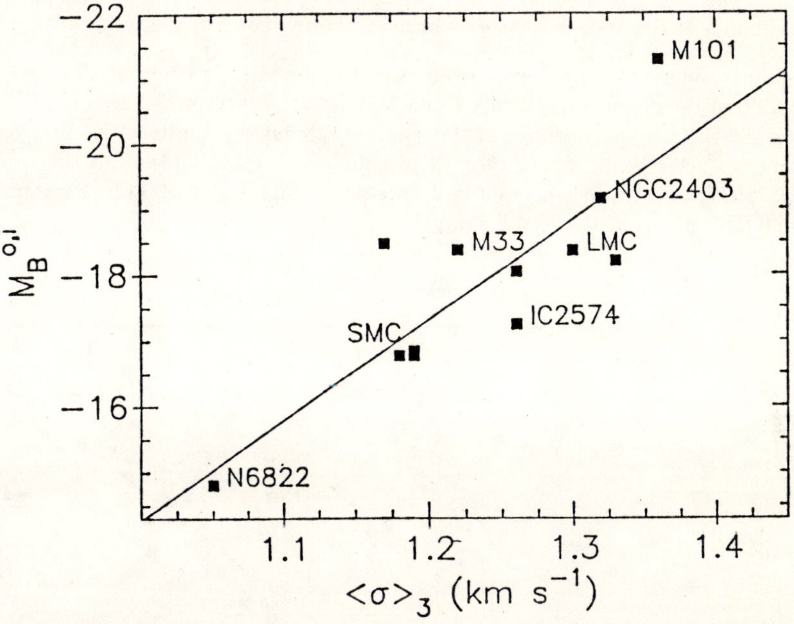

FIGURE 3: Plot of the mean velocity dispersion of the three largest HII regions, $< \sigma >_3$, versus the absolute blue magnitude of the parent galaxies corrected for extinction and inclination effects, $M_B^{o,i}$ (taken from the *Second Reference Catalogue*).

4. THE EMISSION LINE LUMINOSITIES OF GIANT HII REGIONS

Another possibility of avoiding the use of the controversial HII region diameters is to correlate σ with the emission line *luminosities* which are, at least in principle, straightforward to determine observationally. A first attempt based on a small sample of nebulae led me to the conclusion that the scatter of the correlation between H_β luminosity, $L(H_\beta)$, and σ was too large to be useful as distance indicator (Melnick, 1979).

However, Terlevich and Melnick (1981; TM) combined giant HII regions with HII galaxies and found that the scatter in the $(L(H_\beta) - \sigma)$ relation was correlated with the nebular abundance of oxygen (O/H) in the sense that, for a given value of σ, metal poor HII regions are systematically more luminous than metal rich ones. TM concluded that the correlation between $L(H_\beta)$, σ, and O/H was a potentially powerful distance indicator especially since HII galaxies can be observed out to very large distances.

Later, studies by Hippelein (1986) and Melnick *et al.* (1987a) failed to confirm the strong dependence on O/H found by TM, but the results of Melnick *et al.* showed that, even without abundance corrections, the scatter in the $(L(H_\beta) - \sigma)$ relation for a significant sample of nebulae with homogeneously determined velocity widths and luminosities is comparable to that of most modern methods to determine extragalactic distances. The results of Melnick *et al.* (1987a; MMT) form the basis for the application of HII galaxies as distance indicators which will be reviewed in the following section.

MMT undertook a photometric and spectrophotometric study of all the nebulae used by ST in their pioneering study (for which accurate distances based on Cepheids are available), with the purpose of determining global H_β luminosities and oxygen abundances. Velocity dispersions for these nebulae were available from the works of Melnick (1977) and Hippelein (1986). Figure 4 presents a plot of $L(H_\beta)$ versus σ for the giant HII regions in the MMT sample.

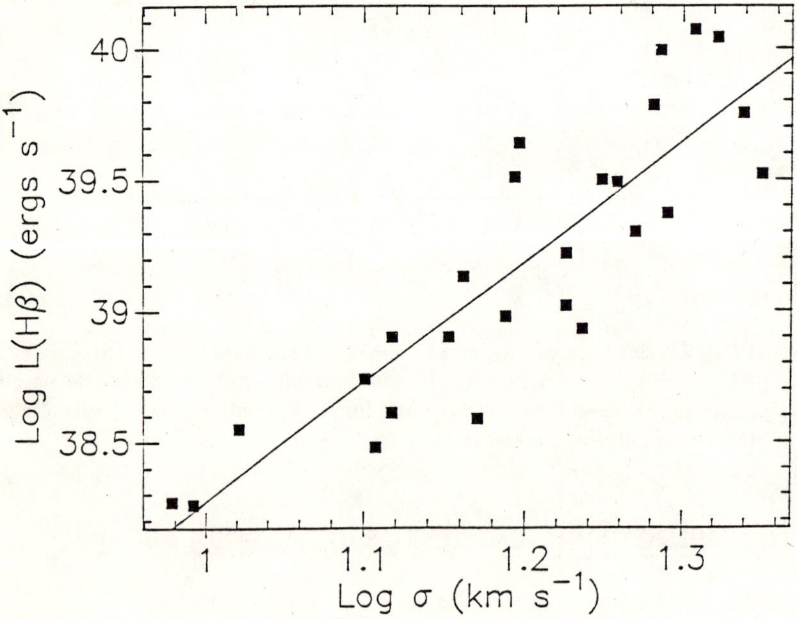

FIGURE 4: Relation between integrated H_β luminosity ($L(H_\beta)$) and velocity width (σ) for the giant HII regions in the ST sample.

The solid line represents a least squares fit to the data of the form $L(H_\beta) \sim \sigma^4$. (A Maximum Likelihood fit, taking errors into account gives $L(H_\beta) \sim \sigma^5$; MMT). The distances to the parent galaxies adopted by MMT are listed in Table 1 together with the old (ST74) and the new (TA87) Sandage-Tammann scales (Tammann, 1987).

TABLE 1
Distance moduli to the zero point calibrators

Galaxy	ST74	TA87	Here
LMC	18.59	18.50	18.50
SMC	19.27	18.85	18.89
NGC 6822	23.95	23.4	23.3
M33	24.56	24.4	24.17
NGC 2366	27.56	27.8	27.5
NGC 2403	27.56	27.8	27.5
Holmberg II	27.56	27.8	27.5
IC 2574	27.56	27.8	27.5
NGC 4236	27.56	27.8	27.5
M101	29.2	29.2	29.2

MMT tested the dependence of $L(H_\beta)$ on other parameters such as D or O/H using *Principal Component Analysis* (PCA) techniques. They found a weaker dependence on O/H than the previous investigation by TM but one which is strong enough to influence the application of the $(L(H_\beta) - \sigma)$ correlation to HII galaxies.

Because of the uncertainties involved in determining diameters and abundances for giant HII regions in distant galaxies, MMT did not attempt to apply their results directly to giant nebulae but used HII galaxies instead.

5. HII GALAXIES

In the course of a spectroscopic survey of Markarian and Zwicky compact galaxies Sargent (1970), found that a significant fraction of these galaxies had narrow emission line spectra resembling those of (giant) HII regions. In a classical paper, Sargent and Searle (1970) obtained detailed observations of two of these emission line compact dwarfs and introduced the term "isolated extragalactic HII regions" to describe the new class of dwarf galaxies identified by Sargent. Because some of these galaxies are not truly isolated, we have preferred to use the term "HII galaxies" instead of the original Sargent-Searle designation.

HII galaxies are easily found on objective prism surveys and Terlevich *et al.* (1988) have catalogued spectrophotometric observations of some 400 objects of this class, the majority of which were discovered on Southern Hemisphere objective prism surveys originally aimed at finding QSO's. Systematic studies of galaxies in the Terlevich *et al.* survey confirm the conclusions of Sargent (1970) and Sargent and Searle (1970) that HII galaxies are ionized by massive young star clusters formed in short lived episodes or "starbursts" (Melnick, 1988). A morphological study of a selected sub-sample of HII galaxies using deep CCD images (Melnick, 1987), showed that, while virtually all HII galaxies are very compact dwarf systems, about 50% of the objects surveyed are starlike. The remainder appear to be either parts of interacting systems, or dwarf galaxies with multiple giant HII regions which (as in all other types) completely dominate the luminous output of their parent galaxies.

About 70 bright HII galaxies from the Terlevich *et al.* catalogue were selected for a study of their emission line widths, H_β luminosities, and oxygen abundances with the aim of comparing HII galaxies with the sample of giant HII regions studied by MMT. Details of the observations and the sample selection are given in Melnick *et al.* (1987b; MTM). Figure 5 shows the plot of $L(H_\beta)$ versus σ for giant HII regions and HII galaxies.

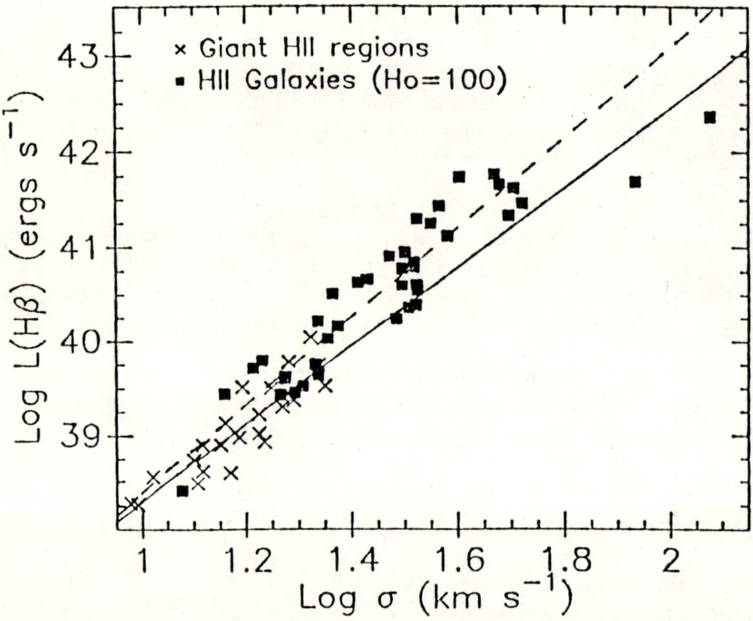

FIGURE 5: The $(L(H_\beta) - \sigma)$ relationship for giant HII regions and HII galaxies. The solid line is the mean relation for giant HII regions. The dashed line represents a fit to the HII galaxies alone. $Ho = 100$ was assumed to compute HII galaxy luminosities.

The offset between giant HII regions and HII galaxies seen in this figure is due to the fact that HII galaxies *as a class* are systematically metal poorer. This difference, easily understood in terms of a mass-metallicity relation (Skillman *et al.* 1988), implies that the luminosities *must* be corrected for abundance effects if giant HII regions are to be used to calibrate the zero point of the $(L(H_\beta) - \sigma)$ relation.

Using PCA techniques, MTM showed that the H_β luminosities of giant HII regions and of HII galaxies depend in essentially the same way on σ and O/H: namely $L(H_\beta) \sim \sigma^5 (O/H)^{-1}$. This lends firm observational support to the conclusion that HII galaxies are dwarf systems with one or more dominating giant HII region. The scatter in the relationship between $L(H_\beta)$ and $M_Z \equiv \sigma^5(O/H)^{-1}$ is about 0.5 magnitudes, comparable to that of the Tully-Fisher relation, confirming the potential power of giant HII regions and HII galaxies as distance indicators. Figure 6 shows the final correlation between $L(H_\beta)$ and M_Z for giant HII regions and HII galaxies. In order to reduce age effects, the sample was restricted to neAbulae with H_β equivalent widths, $W(H_\beta)$ (an age indicator for giant HII regions; Copetti *et al.* 1987), larger than 30Å.

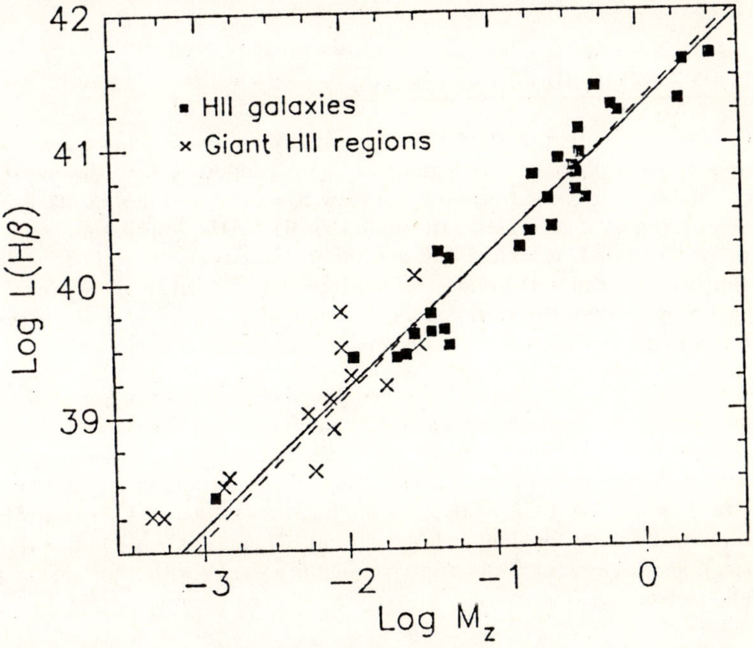

FIGURE 6: The distance indicator for HII galaxies. $M_Z \equiv \sigma^5(O/H)^{-1}$. The slope of the least-squares line is 1.0. The giant HII regions used to calibrate the zero point are also shown.

6. THE HUBBLE CONSTANT

There are 12 galaxies in the sample of MTM with radial velocities larger than 5000 $km\,s^{-1}$. Assuming that the Hubble flow beyond $cz = 5000$ $km\,s^{-1}$ is not perturbed by large scale streaming motions, the calibration shown in figure 6 yields a value of $Ho = 105 \pm 7$ $km\,s^{-1}Mpc^{-1}$. Other values for different redshift cuts range between $94 < Ho < 105$ $km\,s^{-1}Mpc^{-1}$. These values must be corrected for Malmquist bias but *a priori* the corrections are not expected to be very large because the scatter in the distance indicator is relatively small, and because the HII galaxies used in the determination are among the brightest but *not* among the most luminous in the Terlevich *et al.* (1988) catalogue.

After correcting the luminosities for Malmquist bias (using the formalism developed by Giraud, 1987) and the redshifts for the motion of the Local Group relative to the Cosmic Microwave Background, MTM obtained a value of $Ho = 87 \pm 10$ $km\ s^{-1} Mpc^{-1}$. Also, from 4 HII galaxies in the southern extension of the Virgo cluster MTM derive a distance of 15±2 Mpc to Virgo that yields a similar value for Ho. The comparison of these results with other determinations in the literature is the subject of this conference and will not be discussed here. However, in order to have confidence in the present results it is important to understand the physics of the $(L(H_\beta - \sigma)$ relation for giant HII regions and HII galaxies and the systematic effects that may bias our results.

7. THE PHYSICS OF HII GALAXIES

The ionising sources of giant HII regions are massive star clusters or *starbursts*. Since, for ionisation bounded HII regions $L(H_\beta)$ is proportional to the masses of the ionising clusters, a correlation between velocity dispersion and luminosity is expected if turbulence is mostly generated (and maintained) by the ionising stars (either via gravity or stellar winds). It is clear that age and metallicity effects will be the dominant factors that will determine the scatter of this correlation. But, if there are no sources of turbulence other than the young stars, for giant HII regions *of a given age* $L(H_\beta)$ should correlate tightly with velocity dispersion and chemical composition.

In principle this should also hold for HII galaxies but, since here the giant HII region components may extend over a significant fraction of the parent galaxies, part of the observed line broadening may be due to the gravity of the underlying old stellar population. I do not think that this effect is very important in the present sample because, as shown by the PCA analysis in table 2, the correlations between $L(H\beta)$, σ, and O/H are very similar for giant HII regions and HII galaxies (both in functional form and scatter), and because in the present sample the galaxies with multiple components have been resolved.

TABLE 2
Principal Component Analysis

	Eigenvectors					
	HII Galaxies			Giant HII regions		
	1	2	3	1	2	3
Parameter						
$\log L(H_\beta)$	0.68	0.22	0.69	0.69	-0.18	0.70
$\log \sigma$	0.70	0.00	-0.71	0.71	-0.03	-0.71
log(O/H)	0.19	0.98	0.12	0.15	0.98	0.10
Eigenvalues	66%	33%	1%	64%	33%	3%

However, multiplicity may affect our estimates of the Malmquist bias by broadening the luminosity function of the *parent* population. I do not have a quantitative estimate of this effect but I do not think that it can explain the difference between our estimate of Ho and the "long" distance scale since about half of the HII galaxies in the Terlevich *et al.* (1988) catalogue appear to be stellar.

REFERENCES

Copetti, M.V.M., Pastoriza, M.C., Dottori, H.A., *Astr. Ap.*, **152**, 427.

Gallagher, J.S., Hunter, D.A., 1983, *Ap. J.*, **274**,, 141. [A Giraud, E., 1987, *Astr. Ap.*, **174**, 23.

Gum, C.S., and de Vaucouleurs, G., 1953, *Observatory*, **73**, 152.

Hippelein, H.H., *Astr. Ap.*, **160**, 374.

Kennicutt, R.C., 1979, *Ap. J.*, **228**, 696.

Kennicutt, R.C., 1981, *Ap. J.*, **247**, 9.

Melnick, J., 1978, *Astr. Ap.*, **70**, 157.

Melnick, J., 1977, *Ap. J.*, **213**, 15.

Melnick, J., 1979, *Ap. J.*, **228**, 112.

Melnick, J., 1987, in *Starbursts and Galaxy Evolution*, eds. T.X. Thuan, T. Montmerle, and J. Tran Thanh Van; Editions Frontières, p. 215.

Melnick, J., 1988 in, *Towards Understanding Galaxies at Large Redshifts*, eds., R.G. Kron and A. Renzini (Kluwer:Dordrecht), p.117.

Melnick, J., Moles, M., Terlevich, R., Garcia-Pelayo, J.M., 1987a, *M.N.R.A.S.*, **226**, 849 (**MMT**).

Melnick, J., Terlevich, R., Moles, M., 1987b, *ESO preprint* **539**; *M.N.R.A.S.*, in press (**MTM**).

Roy, J.R., Arsenault, R., Joncas, G., 1986, *Ap. J.*, **300**, 624.

Roy, J.R., Arsenault, R. 1986, *Ap. J.*, **302**, 579.

Sandage, A., and Tammann, G.A., 1974, *Ap. J.*, **190**, 525 (**ST**).

Sandage, A., 1962, in *Problems of Extragalactic Research, IAU symp. No. 15*, ed. G.C. McVittie (New York: McMillan), p. 359.

Sargent, L.W.L., 1970, *Ap. J.*, **160**, 405.

Sargent, L.W.L., Searle, L., 1970, *Ap. J. (Letters)*, **162**, L155.

Sérsic, J.L., 1959, *Observatory*, **79**, 54.

Sérsic, J.L., 1960, *Zs. f. Ap.*, **50**, 168.

Skillman, E.D., Melnick, J., Terlevich, R., Moles, M., 1988, *Astr. Ap.*, **196**, 31.

Smith, M.G., Weedman, D., 1972, *Ap. J.*, **172**,, 307.

Tammann, G.A., 1987, *ESO preprint* No. 497 (**TA87**).

Terlevich, R., Melnick, J., 1981, *M.N.R.A.S.*, **195**, 839; (**TM**).

Terlevich, R., Melnick, J., Masegosa, J., Moles, M., 1988, *A Spectrophotometric Catalogue of HII Galaxies*, in preparation.

de Vaucouleurs, G., 1979, *Astr. Ap.*, **79**, 274.

DISUCSSION

G. TAMMANN: It is in general quite dangerous to derive the correlation between two galaxian parameters for a sample of galaxies and to apply the resulting correlation to the same sample for deriving distances. This is because the correlation may be dominated by the specific sample selection. A well known example is the seemingly good correlation between radio luminosity and distance for 3C sources; A. Sandage (Ap.J. 178, 25, 1972, Fig. 7) has shown that this correlation is essentially due to the flux cutoff in the 3C catalogue. I suspect that your good correlation between velocity dispersion and Hβ luminosity is similarly forced upon the data by your detection limit. In that case one expects that a fainter detection limit would bring to light still many objects which lie systematically below your mean line - which automatically would decrease your value of H_o.

J. MELNICK: The functional form and the scatter in the correlations for giant HII regions and HII glaxies are essentially identical; there must be some common physical mechanism involved. For giant HII region $L{\sim}R_o^2$ while for HII galaxies $L{\sim}\sigma^5$ (but $R{\sim}\sigma^3$) so the simplest explanation is gravity. But as long as the turbulence (σ) is generated by the same stars that produce the luminosity (i.e. the young stars) the scatter in the (L,σ) relation should not depend on the selection criteria. However, if other mechanism intervene, such as the gravity of the present galaxies, for example, the conditions will suffer. Because of the similarity with giant HII regions, I do not think that our present sample is significantly biased by the gravity of the old stellar population. The fact that we have limited our study to young (W(Hβ)>30Å) HII regions also helps. I think, however, that the effect you mention does bias our estimate of the width of the luminosity function of the parent population so our estimate of the Malmquist bias may be a little too optimistic, but just a little. I'd also like to mention that the galaxies used in our study are far from the detection limit as they are mostly nearby objects.

S. G. DJORGOVSKI: You are probably right in that the relation you show is "physical" and driven by the gravity. As long as the HII regions are well-defined units, that should work. However, as soon as you run into a patchy or multiple morphology, the relation may break down. Have you looked into this?

DISUCSSION

J. MELNICK: The galaxies I have used here are close enough to be resolved from the ground. I have used the individual components to define the correlations and derive distances but I agree with you that multiplicity will limit the redshift range to which the method can be applied. This is unfortunate because HII galaxies are known out to very large redshifts ($\underline{z} \sim 0.6$).

J. MOULD: What is the r.m.s. dispersion of the final calibration realation? How many of the calibrators are HII regions and how many are galaxies?

J. MELNICK: The r.m.s. scatter in the $(L(H\beta), \sigma^5/O/H)$ relationship is 0.5 magnitudes. There are 29 HII galaxies and 14 giant HII regions but only 12 of the HII galaxies have $cz > 5000$ km s^{-1}.

PARALLAX CALIBRATION OF THE POPULATION II MAIN SEQUENCE

THOMAS E. LUTZ
Program in Astronomy, Washington State University
Pullman, WA 99164-2930 USA

ROBERT B. HANSON
Lick Observatory, University of California
Santa Cruz, CA 95064 USA

WILLIAM F. VAN ALTENA
Department of Astronomy, Yale University
New Haven, CT 06511 USA

ABSTRACT The trigonometric parallax calibration of the field subdwarf main sequence is of vital importance for determining the distance moduli of globular clusters. We have performed a multivariate analysis to assess the dependence of M_V on $(B-V)$ and [Fe/H], using 50 stars having both (a) very accurate trigonometric parallaxes and (b) spectroscopic [Fe/H] determinations. We also discuss possible effects on the calibration due to stellar surface gravity and age. The Lutz-Kelker ΔM corrections are very small for these stars.

1. Introduction

In two previous studies we have used the trigonometric parallaxes of seven field subdwarfs to calibrate the dependence of M_V on [Fe/H]. Lutz *et al.* (1987) used preliminary data from the new *Yale Parallax Catalogue* (van Altena 1988, in preparation), assuming a main-sequence slope from cluster studies, and calibrating the deviation from the main sequence as a function of [Fe/H]. Van Altena *et al.* (1988) updated the calibration with new corrections to absolute parallax, comparing the same subdwarfs with the Hyades main sequence. Many more stars, over a wide metallicity range, can now be added to the data base used by these earlier studies.

Our expanded data base for the present study consists of all main-sequence stars which are (a) listed in the *Catalogue of* [Fe/H] *Determinations* (Cayrel de Strobel *et al.* 1985) and (b) have $\sigma/\pi \leq 0.15$ in the new *Yale Parallax Catalogue*. Fifty stars, ranging in [Fe/H] from -1.83 to $+0.23$, and in $(B-V)$ from 0.51 to 1.20, are used in the present calibration.

Rather than assuming a main-sequence slope as before, we used a multivariate statistical model of the form:

$$M_V(\pi) = a + b(B-V) + c\,[\mathrm{Fe/H}] + d\log g\,,$$

to assess empirically the dependence on color, metallicity, and surface gravity.

2. Data Analysis

Preliminary results from this four-parameter model agreed reasonably well with theoretical stellar evolutionary models. However, for many stars in the [Fe/H] *Catalogue* $\log g$ is *not* derived from the spectroscopic observations, but rather is assigned on the basis of the stellar color or presumed evolutionary state, or is derived indirectly from the trigonometric absolute magnitude and stellar evolutionary theory. The lack of observational $\log g$ determinations means that this term cannot be used in our model. Instead, we enforced a color restriction to minimize the effects of stellar age uncertainty on our calibration.

Consider a grid of isochrones determined from stellar evolutionary models for a given chemical composition (we used those of VandenBerg and Bell 1985). Regardless of age, the low mass stars lie on a very narrow main sequence. Proceeding up the main sequence to higher masses, the spread in absolute magnitude due to age becomes significant. Since the ages of our calibrating stars are not known, we must restrict our sample to the narrow lower main sequence. We take as our criterion that, for a given [Fe/H], the main-sequence spread corresponding to 10^{10} years of main-sequence evolution is 0.25 mag at $(B-V)_{min}$. By using only those stars redder than $(B-V)_{min}$ we can reduce the age uncertainty in the stellar absolute magnitudes to the same level as the other observational uncertainties (parallax, [Fe/H]). $(B-V)_{min}$ ranges from 0.65 at [Fe/H] = 0.0 to 0.40 at [Fe/H] = −2.0. We note that very few real subdwarfs are removed from the calibration by the $(B-V)_{min}$ criterion; most of the stars affected are F-type stars of near-solar metallicity.

3. Results

From an unweighted least-squares solution for 50 stars, we obtain the result:

$$M_V(\pi) = 1.41 + 5.17(B-V) - 0.94[\mathrm{Fe/H}].$$

The RMS residual is 0.31 mag; the standard deviation of the whole calibration is 0.04 mag. The main sequence slope found here is somewhat shallower than the "cluster" slope (5.6) we used previously. This is because the stars used here average about 0.2 mag redder than the seven subdwarfs. The [Fe/H] coefficient is very close to the value (−0.87) found by van Altena *et al.*(1988). The residuals show no evidence of a quadratic trend with metallicity. This is contrary to theoretical expectations, which predict the slope of M_V with [Fe/H] to flatten at lower metallicities. We note that we have not used weights in our solution. This is because the trigonometric parallaxes are not the limiting factor in the calibration. Since the [Fe/H] coefficient is near unity, a typical observational error of 0.2 dex in [Fe/H] contributes as much uncertainty as the trigonometric parallax of a typical star with $\sigma/\pi \simeq 0.1$. Hence the further improvement of the subdwarf calibration will depend on better spectroscopic data as well as on astrometric observations.

REFERENCES

Lutz, T. E., Hanson, R. B., and van Altena, W. F. 1987, *Bull. Amer. Astron. Soc.* **19**, 675.

Cayrel de Strobel, G., Bentolila, C., Hauck, B., and Duquennoy, A. 1985, *Astron. Astrophys. Suppl.* **59**, 145.

van Altena, W. F., Lee, J. T., Hanson, R. B., and Lutz, T. E. 1988, Van Vleck Observatory Workshop on *The Calibration of Stellar Ages*, May 1988.

VandenBerg, D. A., and Bell, R. A. 1985, *Astrophys. J. Suppl.* **58**, 561.

ASTROMETRIC DISTANCES OF GLOBULAR CLUSTERS

KYLE M. CUDWORTH
Yerkes Observatory, Univ. of Chicago, Box 258, Williams Bay, WI 53191

RUTH PETERSON
Whipple Observatory, Smithsonian Institution, c/o Steward Observatory, Univ. of Arizona, Tucson, AZ 85721

INTRODUCTION

The distance scale of globular clusters has generally been based upon horizontal branch stars, especially RR Lyrae variables, as standard candles, or on main sequence fitting to subdwarfs. While we believe the current distance scale for globulars is probably close to correct, it is clearly useful to check it with a totally independent technique when possible, especially to provide some additional constraints on the long-elusive question of metallicity effects on the absolute magnitudes of RR Lyraes. Cudworth (1979), Lupton et al. (1985), and Cudworth and Rausher (1987) and have used Yerkes proper motions and various radial velocities to derive astrometric distances for M3, M13, M92, and M2. The results agreed well with standard candle distances, but the individual uncertainties were too large to provide definitive checks (though the agreement in the mean is probably significant).

Cross-correlation techniques are now capable of yielding radial velocities of globular cluster giants with errors < 1 km/s, compared with typical cluster velocity dispersions > 5 km/s. At 5 kpc (a very nearby globular) a velocity dispersion of 5 km/s yields a proper motion of 20 milliarcsec/cent., which is measurable in favorable cases with modern photographic astrometry.

APPLICATION TO M22

Cudworth (1986) has derived proper motions for several hundred stars in M22 using plates from Harvard, Lick, and Yerkes. We have also used the MMT to obtain radial velocities for over 100 M22 giants. Except for the faintest stars measured, the proper motion errors are much less than the intrinsic proper motion dispersions. Proper motions, of course, will reveal any radial anisotropy in the velocity dispersion, and none is seen at the radii observed.

The cluster is somewhat elongated as seen on plates (White and Shawl 1987, WS87). The best proper motion data show a larger apparent dispersion along the major axis found by WS87 than along the minor axis, as would be expected from rotation. A plot of radial velocity vs. position angle also shows a sine wave with approximately the correct phase to be rotation about the projected minor axis. For our analysis we take the best data in the annulus from 180" to 400" in radius, and then rotate the proper motion axes to the major and minor axes from WS87 and use the minor axis dispersion of 46 ± 3.8 mas/cent. A sine curve is then fit to the radial velocity vs. position angle plot for the stars in this annulus, leading to an amplitude ~ 2.5 km/sec and a dispersion about this curve of 6.13 ± 0.48 km/sec. These values then yield a distance of 2.8 ± 0.3 kpc. (This derivation, of course, does not account for any variation of rotational velocity with radius, and thus will tend to overestimate the dispersion in radial velocity and thus also overestimate the distance. For comparison, ignoring rotation entirely yields a distance of 2.4 ± 0.2 kpc.)

Previously derived distances to M22 have generally been larger than 3 kpc, based on RR Lyraes and a reddening E(B-V) < 0.35. Crocker (1988) has derived a larger reddening for M22 from her spectroscopy of several blue HB stars: 0.42 ± 0.02, which is also consistent with a reasonable color for the blue edge of the instability strip (in contrast to previous lower reddenings). We adopt this and derive M_V(HB) = +0.5 ± 0.3, within the uncertainties of most other derivations. (The derivation ignoring rotation yields +0.85 ± 0.23.)

FUTURE WORK

Although the M22 data are among the best, the large reddening makes this cluster less than ideal for calibration of the HB luminosity. We have also obtained MMT radial velocities for over 100 stars each in M4 and M15, and new proper motion reductions are in progress, but reddening is also some problem in both of these. Proper motions are currently being reduced for the metal-rich, lightly-reddened cluster 47 Tuc, for which many radial velocities are available. Significant improvements are possible and planned for the previous Yerkes proper motion studies of the lightly-reddened clusters M5 and M92. We hope to obtain distance moduli to about ± 0.2 mag. for these clusters, leading to a good calibration over a range of metallicity, totally independent of any standard candle assumptions.

ACKNOWLEDGMENTS

We thank the Harvard and Lick Observatories for the loan of plates, and the schedulers of the MMT for generous allotments of telescope time. This research was partially supported by the NSF.

REFERENCES

Crocker, D.A., 1988, (preprint).
Cudworth, K.M. 1979, A.J. 84, 1312.
Cudworth, K.M. 1986, A.J. 92, 348.
Cudworth, K.M. and Rausher, B. 1987, A.J. 93, 856.
Lupton, R., Gunn, J.E., and Griffin, R.F. 1985, in IAU Symp. 113, Dynamics of Star Clusters, ed. J. Goodman and P.Hut, p.19.
White, R.E. and Shawl, S.J. 1987, A.J. 317, 246 (WS87).

DISTANCES TO GALACTIC STAR CLUSTERS CONTAINING CEPHEIDS: NGC 7790

MARIO MATEO
Mt. Wilson and las Campanas Observatories
Pasadena, CA 91101

BARRY MADORE
Infrared Processing and Analysis Center
California Institute of Technology
Pasadena, CA 91125

INTRODUCTION

Cepheid variables in Galactic open clusters remain important in locally establishing the zero points of the Cepheid PL and PLC relations. This contributed paper briefly describes an ongoing Northern hemisphere CCD photometric survey of a number of these clusters, and presents initial results for one of the program clusters, NGC 7790 which contains three Cepheids. A complementary CCD survey of southern Cepheid clusters has been undertaken by Walker and collaborators (1987 and refs therein).

THE SURVEY AND NGC 7790

Observations for this survey of Cepheid clusters consist primarily of UBVR CCD frames obtained with the KPNO 0.9m and las Campanas 1m telescopes. Some additional deep frames were also obtained with the KPNO 4m and CFH telescopes. The program clusters include NGC 129, the CV Mon cluster, M 25, NGC 6664, and NGC 7790. Three to eight cluster fields and numerous nearby 'control' fields have been observed for each program cluster.

We focus the remainder of our discussion on NGC 7790 because it illustrates many of the problems inherent in attempting to derive precise distances to open clusters. The cluster is relatively sparse and superimposed on a rich stellar field (see figures in Sandage 1958; Pedreros, *et al.* 1984); thus, it is difficult to weed out cluster members from the more numerous non-members in the field. Also, the cluster's reddening is rather large and variable. In all, eight cluster CCD fields, and four nearby 'control' fields were observed for NGC 7790. Previous distance estimates for the cluster have ranged from $(m - M) = 11.98$ to 13.1.

RESULTS AND DISCUSSION

Figure 1 shows the UBV two-color diagram for the brighter stars in the NGC 7790 CCD fields along with the intrinsic main sequence relation. The reddening is estimated to be E(B−V)= 0.56 ± 0.04. The V-(B−V) CM diagram is shown in Figure 2 along with the theoretical ZAMS (VandenBerg and Bridges 1984), and the Pleiades ZAMS (Turner 1979). The resulting moduli (for A_V = 3.1E(B−V)), are 12.20, and 12.46, yielding an average estimate of $(m - M)_{\mathrm{NGC\ 7790}} = 12.33 \pm 0.15$.

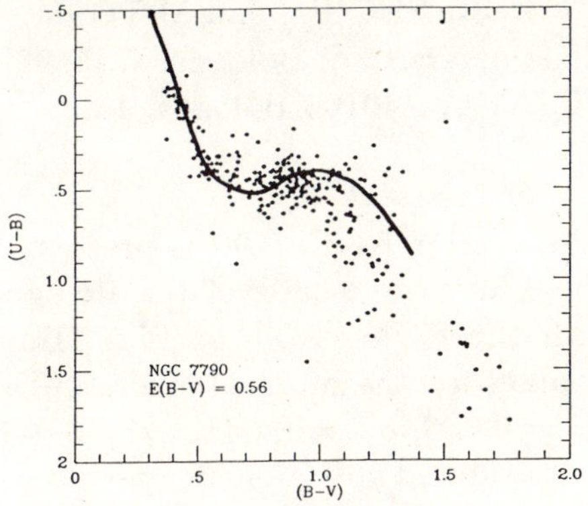

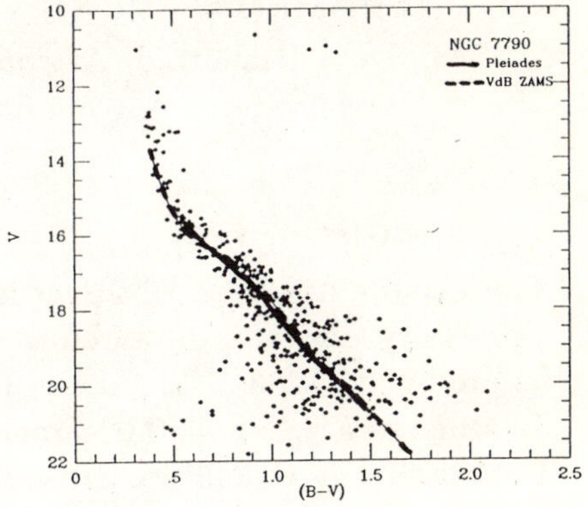

Figure 1 (*left*) - UBV two color diagram of NGC 7790. Figure 2 (*right*) BV CM diagram of NGC 7790.

Many of the practical difficulties involved in main sequence fitting are illustrated in the Figures. For example, although Figure 2 represents a preliminary attempt to remove field stars, it is apparent that field star contamination is an important problem which limits the precision with which the locus of the lower main sequence can be determined. No attempt has been made yet to reduce the scatter in Figure 1 by correcting for differential reddening, but some of the scatter is undoubtedly due to this effect. Uncertainties in reddening parameters can also introduce significant errors in the derived distance to a highly reddened object such as NGC 7790. Nevertheless, Figure 2 suggests that the uncertainty in $(m - M)_{\mathrm{NGC\ 7790}}$ is not dominated by photometric errors but by uncertainties in models and distances to calibrating clusters such as the Pleiades. Our distance modulus is consistent with the Feast and Walker (1987) scale in which $(m - M)_{\mathrm{LMC}} = 18.47 \pm 0.15$.

REFERENCES

Feast, M. W., and Walker, A. R. 1987, *Ann. Rev. Astr. Ap.*, 25, 345.

Pedreros, M., Madore, B. F., and Freedman, W. L. 1984, *Ap. J.*, 286, 563.

Sandage, A. 1958, *Ap. J.*, 128, 150.

Turner, D. 1979, *Pub. A. S. P*, 91, 642.

VandenBerg, D. A., and Bridges, T. J., 1984, *Ap. J.*, 278, 679.

Walker, A. R., 1987, *M. N. R. A. S.*, 229, 31.

CCD-PHOTOMETRY OF OPEN CLUSTERS CONTAINING CEPHEID VARIABLES: NGC 7790

G.ROMEO[1], A.BONIFAZI[2], F.FUSI PECCI[1], and M.TOSI[2]
[1] Dipartimento di Astronomia, Università di Bologna, C.P.596,
[2] Osservatorio Astronomico, C.P.596, I-40100 Bologna, Italy

A new BVRI CCD-photometry of the cluster NGC 7790 is presented. The cluster has been observed at the 1.52 m telescope of the Bologna University and the data reduced with the ROMAFOT package (Buonanno et al. 1983). The mean rms-values for the internal errors in the four bands are: $\sigma_B = 0.028$ mag, $\sigma_V = 0.016$, $\sigma_R = 0.044$, $\sigma_I = 0.041$. 149 stars, out of 343 measured, are considered to be cluster members, once the issues related to field contamination and binary systems are taken into account.

We have defined a main sequence ridge line corresponding to the blue envelope shifted toward the red by a quantity comparable to the size of the photometric error in B-V and V-I. Assuming R=3.06 at (B-V)=0.0 and fitting the observed Zero Age Main Sequence of the Pleiades (with distance modulus 5.57, as given by Walker 1985) to our ridge line, we have derived from the resulting colour-magnitude diagrams new accurate estimates of the reddening and distance modulus of the cluster: $E(B-V) = 0.54 \pm 0.04$ and $(m-M)_o = 12.65 \pm 0.15$.

The absolute magnitudes of three Cepheids corresponding to our derived values of reddening and distance modulus have been used to compute the zero-points of the period-luminosity (C1) and period - luminosity - colour relations (C2 and C3) as given by Feast and Walker (1987). These values are listed in Table 1 together with the adopted absolute magnitudes.

Table 1: Zero-points from the Cepheids in NGC 7790

Name	$< M_V >$	C2	C3	C1
CE Cas A	-3.466	-2.26	-2.42	-1.49
CE Cas B	-3.384	-2.19	-2.31	-1.57
CF Cas	-3.244	-2.07	-2.22	-1.33
adopted mean values		-2.17 ± 0.08	-2.32 ± 0.08	-1.5 ± 0.1

With the obtained parameters and assuming solar compositions, the age of NGC 7790 is $(5.0 \pm 1.5) \times 10^7$ yr using "standard" models or $(1.0 \pm 0.2) \times 10^8$ yr adopting models with overshooting. This latter estimate seems to be more in agreement with the ages obtained by applying the period-age relations to the three Cepheids member of the cluster.

REFERENCES

Buonanno, R., Buscema, G., Corsi, C.E., Ferraro, I., and Iannicola, G., 1983, Astron.Astrophys. 126, 278.

Feast, M.W. and Walker, A.R., 1987, Ann.Rev.Astron.Astrophys. 25, 345.

Kippenhahn, R. and Smith, L., 1969, Astron.Astrophys. 81, 142.

Walker, A.R., 1985, Mon.Not.R.astr.Soc. 213, 889.

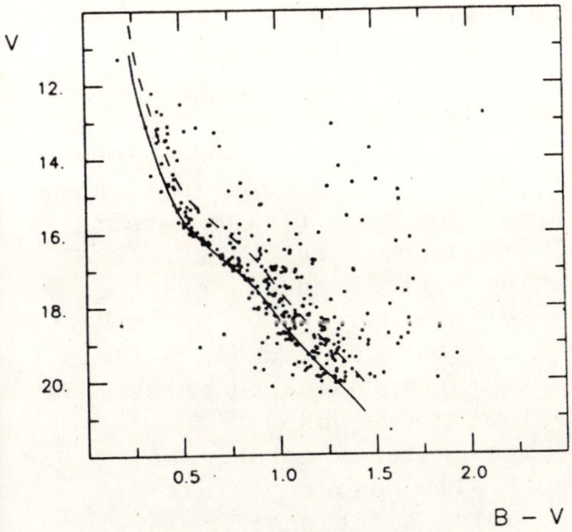

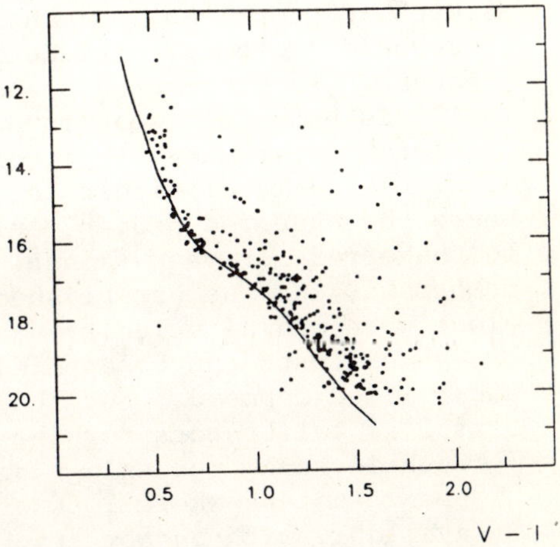

Fig.1: Colour-Magnitude Diagram (V,B-V) for all the stars measured in our CCD frames of NGC 7790. The solid line represents the observed ZAMS of the Pleiades. The dashed line shows the locus of binary systems with equal mass components, which indicates the maximum deviations from the ZAMS position due to binarism.

Fig.2: Colour-Magnitude Diagram (V,V-I) for all the stars measured in our CCD frames of NGC 7790. The solid line represents the observed ZAMS of the Pleiades. The corresponding reddening and distance modulus are $E(B-V) = 0.54 \pm 0.04$ and $(m-M)_o = 12.65 \pm 0.15$ respectively.

THE CASE FOR U CARINAE AS A LONG PERIOD CEPHEID CALIBRATOR IN A GALACTIC OB ASSOCIATION

DAVID G. TURNER
Saint Mary's University, Halifax, N.S., B3H 3C3, Canada

For a Cepheid spatially coincident with a young group of stars the following tests can be applied to establish membership in the group:

1) Cepheid luminosity comparable to that of the most luminous group members (which are typically B giants or supergiants or K and M-type supergiants);
2) Cepheid radial velocity identical to group mean;
3) Cepheid proper motion consistent with group proper motion;
4) Cepheid photometric reddening similar to reddening of nearby group members (space reddening).

Test 4 is extremely weak in practice since the space reddening of distant stars usually occurs within foreground dust clouds. Also, test 3 is difficult to use for distant groups ($d > 1$ kpc) where the observed proper motions are quite often comparable to the errors of measurement. In most instances cluster membership must be established using tests 1 and 2, and test 1 alone when radial velocity data are lacking.

Such tests can be applied to cluster Cepheids as well as association Cepheids, but the latter are considered to be less useful as calibrators for the following reasons:

1) Cepheid space reddenings are more reliably established in open clusters where lots of spatially adjacent stars are available with photometric reddening solutions;
2) Open cluster stars have a line of sight distance spread of perhaps ±5 pc, whereas the typical distance spread in OB associations may be 10 times larger. Typical uncertainties in absolute magnitude are therefore

$$\Delta M_V \simeq \pm 0.01 \text{ mag./d(kpc) for cluster Cepheids, and}$$
$$\Delta M_V \simeq \pm 0.11 \text{ mag./d(kpc) for association Cepheids,}$$

due to distance uncertainty alone.

Despite such problems association Cepheids can sometimes be established as calibrators for the distance scale. Since no long period ($P > 12^d$) classical Cepheids have been found in open clusters, association Cepheids are particularly useful for testing the high luminosity end of the PL relation.

The 39-day Cepheid U Carinae is the subject of this paper. The Cepheid lies near the Great Carina Nebula, in a field particularly rich in luminous stars. Within the overlapping 2° radius fields surrounding U Car and nearby VY Car (a 19-day Cepheid) there are 71 OB stars for which UBV photometry, slit spectral types, and Hβ photometry exist. The individual distance moduli for these stars can therefore be established to within ±0.3 in $V_o - M_V$.

Most of the stars in this field belong to the association Car OB1 at d = 2.6 kpc and $V_r = -16 \pm 3$ km/s, but a very obvious subset can also be detected at d = 1.9 kpc and $V_r = +1 \pm 2$ km/s. The distinct nature of this second association (Car OB2) is confirmed by the available radial velocity data and marked evolutionary differences. Car OB1 stars are all young O and B-type stars, whereas Car OB2 members are nearly all evolved B giants and supergiants. The optical extent of Car OB2 is at least 3 times larger than what is typically quoted in the literature (e.g. Humphreys 1978).

Car OB2 members identified here consist of three somewhat dissimilar spatial subgroups:

1) V382 Car subgroup (age $\simeq 4 \times 10^6$ yrs). V382 Car is a G0 hypergiant with a similar radial velocity (+7.3 km/s) to Car OB2 (+3.6 ± 1.7 km/s for this subgroup). It appears to be an excellent evolutionary match as well, and seems likely to be a group member with $M_{bol} \simeq -9.2$;
2) VY Car subgroup (age $\simeq 8 \times 10^6$ yrs). VY Car has a similar radial velocity (+2.0 km/s) to Car OB2 (−1.9 ± 4.1 km/s for this subgroup), but the B stars form a coeval group which is distinctly younger than the Cepheid. VY Car is unlikely to be a group member;
3) U Car subgroup (age $\simeq 17 \times 10^6$ yrs). U Car has a similar radial velocity (+1.7 km/s) to Car OB2 (+1.3 ± 1.6 km/s for this subgroup), and is an excellent match in evolutionary age as well. It seems likely to be a group member.

The spatial extent of the U Car subgroup is about ±60 pc at the distance of Car OB2, so an identical line of sight distance spread produces an uncertainty of ±0.07 mag. in the derived absolute magnitude of the Cepheid. The space reddening of U Car is established from three nearby stars (Danks 1987) to be $E_{B-V} = 0.30 \pm 0.03$ (0.34 ± 0.03 for a B0 star), while the mean distance modulus for Car OB2 is $V_o - M_V = 11.41 \pm 0.04$ s.e. The derived luminosity for U Car as a member of Car OB2 is therefore $\langle M_V \rangle = -6.22 \pm 0.13$ for a corresponding Hyades distance modulus of 3.2.

REFERENCES

Danks, T. A. 1987, M. Sc. Thesis, Saint Mary's University.
Humphreys, R. M. 1978, Ap. J. Suppl., 38, 309.

ANOMALOUS CEPHEID PERIOD-LUMINOSITY RELATIONSHIPS

James Nemec (U.B.C.), Amelia Wehlau (U.W.O.) and Clâudia Mendes de Oliveira (U.B.C.)

ABSTRACT: The P-L relationship for anomalous Cepheids (ACs) splits into two well-defined lines in the log P - M_B plane. One line corresponds to pulsation in the fundamental mode, and the other corresponds to the first-overtone. If these P-L relationships are universal, then they can be used to estimate distances to nearby dwarf galaxies. Knowledge of pulsation modes of the ACs in Draco suggests a mass range of 1.04 to 1.77 $M_\odot$.

Anomalous Cepheids have periods in the range $0^d\!.26$ to $2^d\!.37$. They are present in all the nearby dwarf galaxies, in the SMC, and in the globular cluster NGC 5466. We have recently completed a major study of the ~95 variable stars in the Ursa Minor dwarf galaxy. Four new ACs were identified, and the period of one of the known ACs (V6) was corrected. Combining these new results with previously published information on well-studied ACs in Draco, Sculptor, Leo II and NGC 5466, the log P - M_B diagram shown in Fig.1 was constructed. This diagram shows for the first time that ACs follow two distinct linear P-L relationships. The least squares equation of the upper line is M_B=-5.30 logP - 1.862, with σ=0.12 mag; we postulate that these stars are first-overtone pulsators (as has been suspected for many years). The lower regression (fundamental mode pulsators) is given by M_B=-2.433 logP - 0.422, with σ=0.10 mag.

Assuming that the P-L relationships shown in Fig.1 are universal, then they provide a way of estimating accurately the distances to other nearby systems. Fig.2 shows the ACs in the Leo I and Carina dwarf galaxies, in the SMC, and XZ Ceti, a candidate AC in our Galaxy. The systematic shift of the ACs in Leo I suggests that its distance modulus needs to be increased by 0.3 mag to $(m-M)_0$=22.0 mag. The ACs in Carina are the bright variables from Saha et al. Unlike the Leo I variables which are mostly fundamental pulsators, all the Carina variables appear to be first-overtone pulsators. The ACs in the SMC appear to be a mix of both types of pulsators. The large uncertainty in the luminosity of XZ Ceti makes it difficult to assign a pulsation mode.

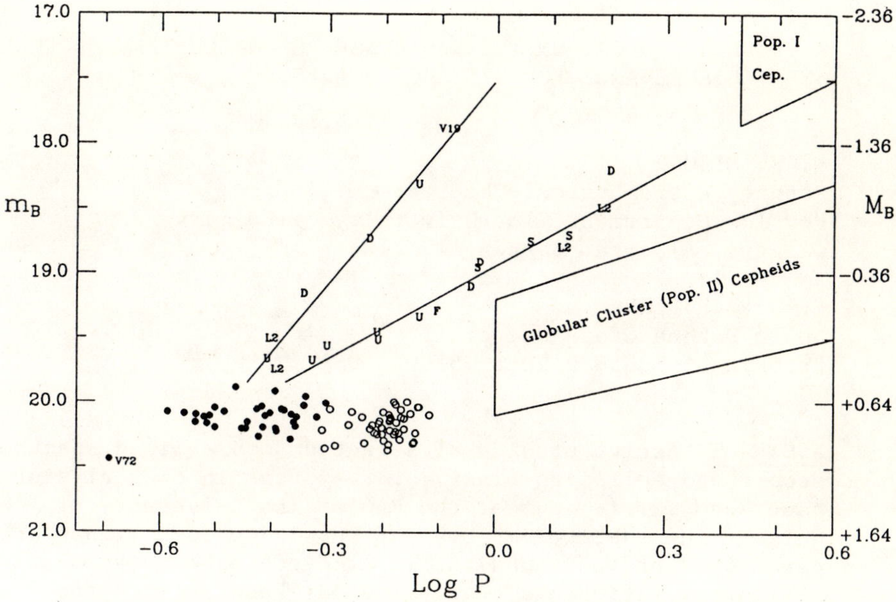

Fig.1 - Period-luminosity relationship for well-studied anomalous Cepheids. The RR Lyrae stars in the Ursa Minor dwarf galaxy are also shown.

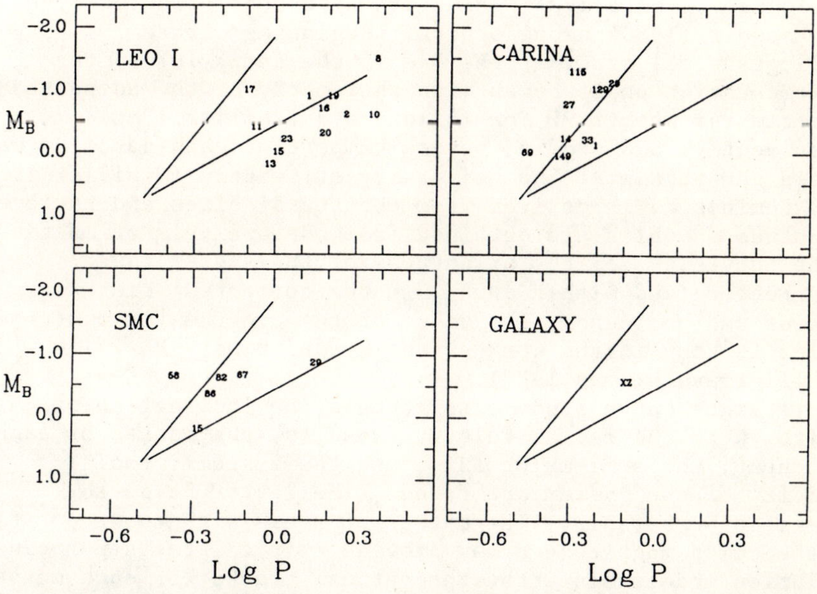

Fig.2 - P-L diagrams for the ACs in Leo I, Carina, the SMC, and XZ Ceti in the Galaxy. The lines represent the fundamental and first-overtone P-L relations from Fig.1.

A PROGRESS REPORT ON THE RAO BINARIES-IN-CLUSTERS PROGRAM AND ITS RELEVANCE FOR DISTANCE SCALES.

E. F. MILONE
Rothney Astrophysical Observatory
Physics Department, The University of Calgary
Calgary, Alberta, Canada T2N 1N4

S. J. SCHILLER
South Dakota State University
Brookings, South Dakota 57006 USA

ABSTRACT Active studies of selected double-lined spectroscopic and eclipsing binary star systems in open clusters have been carried out at the RAO of The University of Calgary in order to improve our knowledge of the absolute parameters of individual stars at discernible stages of stellar evolution and to improve our knowledge of the galactic clusters of which these systems are members.

The value of such a program has long been recognized. Catalogues of binaries-in-clusters have been produced by Kraft and Landolt (1959) among others. Beginning with a bibliographical study at the RAO in the early 1970s, the implementation phase began with photometry of the suspected Coma cluster member RW Com (Milone and Schiller 1984). Subsequently, SJS studied several important candidates as part of his PhD thesis at The University of Calgary (Schiller 1986).

Luminosity is derived from the stars' sizes and surface brightnesses which are obtained from the analysis of multicolor light curves. From the brightness of the system the uncorrected (and with R and E_{B-V}, the corrected) distance modulus can be found. The value of the program for distance scales is seen in the work on the Hyades binary HD 27130 (Schiller and Milone 1987).

A light curve study also reveals the inclination of the orbit. With the radial velocity results, the masses of each component, the semi-major axis, and the systemic radial velocity of the system are found. The latter helps to ascertain membership. The masses of stars with known positions on the color-magnitude array provide critical tests for cluster evolution models since the theoretical tracks for such masses (of the correct composition) must pass through the points on the array represented by the components. The work on DS Andromedae (Schiller and Milone 1988) illustrates the value of the technique for cluster evolution studies.

The cluster-candidate binaries are selected on the basis of likelihood of membership, suitability for photometric and radial velocity study (carried out at the RAO, Table Mountain Obs., McDonald Obs. and DAO), and availability of photometric and kinematic cluster data. The multicolor light- and radial velocity curve analyses are done with the Wilson-Devinney program on the Cyber 205 Supercomputer at The University of Calgary. Systems for which the work has been completed are: RW Com (shown to be a non-member of Me 111; Milone, Wilson, and Hrivnak 1987); HD 27130; DS And; and OX Cas (collaborative light curve study by Crinklaw and Etzel). In addition to the successful studies of cluster binaries thus far, the systems noted in Table I are expected to yield equally useful results.

TABLE I. Active Target Systems

Binary	Cluster	Magnitude	Sp	P(d)	Collaborator
SZ Cam	NGC 1502	7.0- 7.2(V)	O9V	2.70	A. J. Wesselink
QX Cas	NGC 7790	10.5-10.9(V)	B1V	6.00	
V518 Cas	Stock 2	10.8-11.1(V)	B3	3.	
DH Cep	NGC 7380	8.6 (V)	O6+O.5.5	2.11	
EM Cep	NGC 7160	7.0- 7.2(V)	B0+B1	0.81	
V444 Cyg	Berk. 86	8.3- 8.6	B1+WN5	4.21	A. Underhill
V447 Cyg	NGC 6871	13.1-14.5	A3+F5	2.20	
SS Lac	NGC 7209	10.1-10.5	B7	14.42	
UZ Lyr	NGC 6791	9.8-11.0	A0	1.89	
BS Sct	NGC 6705	11.0-12.4	A0	3.82	
EY Sct	NGC 6683	11.9-12.7	..	1.17	
AH Tau	Pleiades	11.3-11.9	G1	0.33	
TY Tau	Hyades	11.5-12.	K0	1.08	
H235	NGC 752	11. (V)	F	0.37	J. E. Penfold

This work has been supported by NSERC of Canada grants to EFM, and by a Research Corporation grant to SJS.

REFERENCES

Kraft, R. P., and Landolt, A. 1959, Astrophys. J. 129, 287.
Milone, E. F., and Schiller, S. J. 1984, Publs. Astron. Soc. Pac. 96, 791.
Milone, E. F., Wilson, R. E., and Hrivnak, B. J. 1987, Astrophys. J. 318, 325.
Schiller, S. J. 1986, "Eclipsing Binary Stars in Open Clusters", PhD Thesis, The University of Calgary. 251 pp.
Schiller, S. J., and Milone, E. F. 1987, Astron. J. 93, 1471.
Schiller, S. J., and Milone, E. F. 1988, Astron. J. 95, 1466.

IV

GALAXIES BEYOND THE LOCAL GROUP

CONSTRAINTS ON H_0 FROM GLOBULAR CLUSTERS

D.A. VANDENBERG
Department of Physics and Astronomy, University of Victoria, P.O. Box 1700, Victoria, BC V8W 2Y2, CANADA

ABSTRACT On the basis of canonical stellar evolutionary computations, the maximum age of the globular clusters is suggested to be near either 14 Gyr or 18 Gyr, depending on how [O/Fe] varies with [Fe/H] in the cluster stars. The lower estimate requires that $H_0 \leq$ 65 km/s/Mpc, for all $\Omega_0 \geq 0$, if the standard Big-Bang cosmological theory ($\Lambda = 0$) is correct — while the higher age value similarly constrains the Hubble constant to be smaller than 46 km/s/Mpc. Some reduction in the upper limit to cluster ages and a consequent increase in H_0 may be expected if helium diffusion is important in Population II stars; nevertheless, values of $H_0 > 75$ km/s/Mpc still appear to be precluded unless the cosmological constant, Λ, is non-zero.

1. INTRODUCTION

How old is the oldest globular cluster (GC)? The answer to this question is of particular interest for the reason that the globulars are the oldest objects for which reliable age determinations have been made and, indeed, they may well have been the first stellar systems to form in the universe (*e.g.,* Peebles 1984). Consequently, the GCs provide an important constraint on the cosmological timescale and hence on the Hubble constant, H_0. In particular, the ages of about 17 Gyr which have generally been found for the oldest clusters (*e.g.,* Sandage 1982, Flannery and Johnson 1982, Janes and Demarque 1983, VandenBerg 1983, Gratton 1985, Peterson 1987) have frequently been used as a supporting argument in favor of small values of H_0 — notably by Sandage and Tammann (1982, and previous papers in the series). However, in recent years, stellar evolutionary models have been constructed which have relaxed the hitherto canonical assumption of scaled-solar heavy-element abundances (*e.g.,* Rood and Crocker 1985, VandenBerg 1985) or have allowed for the effects of helium diffusion (Stringfellow *et al.* 1983). One implication of these calculations is that the inferred maximum age of the GCs should be reduced, perhaps by a large factor. This paper will review these developments and provide an updated assessment of GC ages as well as their ramifications for cosmology.

2. THE AGES OF THE GLOBULAR CLUSTERS

Figure 1 illustrates 16 Gyr isochrones from the VandenBerg and Bell (1985) grid for a helium abundance Y = 0.20 and metallicities from [m/H] = -0.5 to [m/H] = -2.3. Also shown is a (still unpublished) set of fully consistent zero-age horizontal-branch (ZAHB) loci which have been computed for the same metal contents; where the lowermost curve corresponds to the most metal-rich case, the next brightest to [m/H] = -0.8, *etc.* A star which evolves up the red-giant branch (RGB) will, after undergoing the helium flash, settle into a stable core-helium burning phase somewhere on the ZAHB which is appropriate for its chemical composition — the precise location depending on the amount of mass loss that occurred during its prior evolution. (Recall that a spread in total mass of $\sim 0.2\ M_{\odot}$ is the most reasonable explanation of the large variation in color among the horizontal-branch stars in many GCs.)

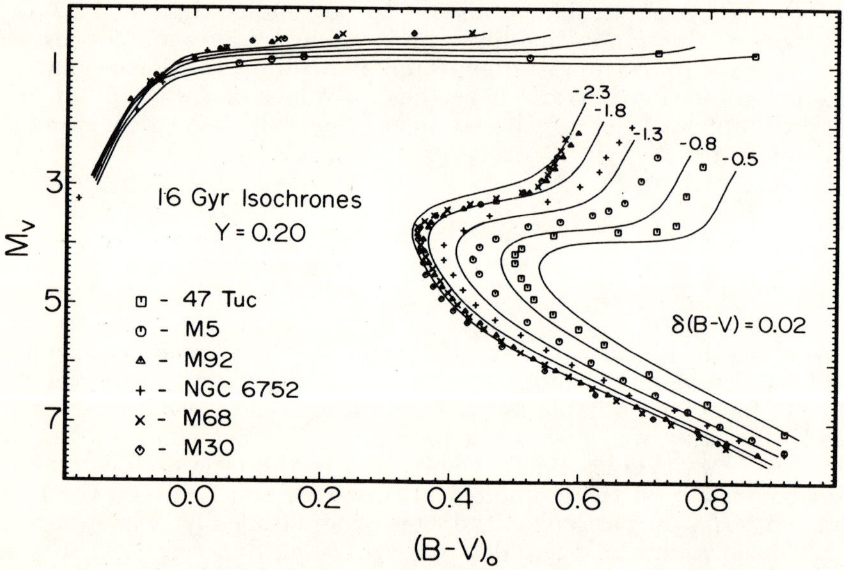

Fig. 1. - Comparison of theoretical isochrones/ZAHBs, for the indicated [m/H] values, with a number of GC fiducials.

Superposed on the theoretical curves in Fig. 1 are a number of smoothed GC fiducials which have been taken from or are based on data given in the references which are listed in the fifth column of Table I. The tabulated [Fe/H] values are essentially those adopted in the sources of the main-sequence CCD photometry — as are the E(B-V) values, although Table I reflects a preference for high reddening estimates. This admitted bias is due in large part to the fact that recent determinations of the foreground reddening to many objects tend to be appreciably larger than earlier values (*e.g.*, see Richer *et al.* 1988). However, all of the GCs in Table I have fairly low reddening, and the adopted E(B-V)'s are probably correct to within one or two hundredths of a magnitude. The distance moduli that are listed were obtained by matching the cluster ZAHBs, which can generally be defined to within 0.1 mag, to their theoretical counterparts given in Fig. 1. While this relies on the accuracy of the models, it is reassuring to note that several of the referenced observational studies determined cluster distances from main-sequence fits to the Population II subdwarf standards, and found $(m - M)_V$'s that are within 0.1 mag of those derived here.

TABLE I: GLOBULAR CLUSTER DATA

Name	[Fe/H]	E(B−V)	(m−M)	Source(s) of Photometry
47 Tuc	−0.7	0.04	13.35	Hesser et al. (1987)
M5	−1.1	0.03	14.30	Richer and Fahlman (1987) Buonanno et al. (1981) Arp (1962)
NGC 6752	−1.5	0.05	13.25	Buonanno et al. (1986) Penny and Dickens (1986) Caloi et al. (1986)
M68	−2.0	0.07	15.20	McClure et al. (1987) Harris (1975)
M92	−2.1	0.03	14.70	Stetson and Harris (1988) Sandage (1969) Sandage and Walker (1966)
M30	−2.1	0.06	14.70	Richer et al. (1988) Bolte (1987) Alcaino and Liller (1980)

In order to have consistent fits of both the horizontal-branch and the main-sequence data, the isochrone colors were adjusted redward by the indicated 0.02 mag in B-V prior to being plotted. Whether such a correction should be made and, if so, whether it is due to errors in the model temperatures or in the synthetic colors is not presently known. But certainly the resultant overall agreement of theory and observation seems excellent. Not only is the observed systematic variation of color with luminosity in complete accord with the predictions of the models, which were actually published well before most CCD photometry became available, but the cluster fiducials are located relative to the theoretical curves precisely where their [Fe/H] values suggest they should be positioned. While a closer scrutiny of cluster ages will follow, one has the initial impression that, whereas 47 Tuc appears to be close to 16 Gyr old, the most metal-poor clusters seem older by perhaps a couple of Gyr.

However, it is obviously dangerous to argue for such a slight age-metallicity relation on the basis of Fig. 1 alone. The magnitude scale is clearly rather compressed, and the vertical shifting of any of the cluster fiducials by as little as 0.15 mag would produce a change of about 2 Gyr in the inferred age. Such a shift would be generated — if the main-sequence fitting method is used to derive distances — by the choice of a reddening value that differs by only 0.03 mag or, for a fairly metal-rich cluster, a metal abundance that is different by ~ 0.2 dex (i.e., less than the typical uncertainty of current such determinations). As Iben and Renzini (1984) have argued, all of the problems associated with observed and predicted colors can be avoided if ages are obtained from a calibration of the magnitude difference between the horizontal branch and the turnoff, $\Delta V_{TO,HB}$. Moreover, as is apparent from Fig. 1, this approach minimizes the impact of abundance uncertainties since the luminosity of the horizontal branch at the color of the turnoff and the turnoff magnitude (for a given age) both become brighter as the metallicity decreases — not by exactly the same amount but near enough that an error in the assumed [Fe/H] will have only a slight effect on the derived age. Thus, for instance, the close similarity of $\Delta V_{TO,HB}$ for M30, M68, and M92 (see Fig. 1), which have comparable metal abundances, is a particularly strong argument that the ages of these three clusters are the same to within 1 Gyr.

The soft spot of the $\Delta V_{TO,HB}$ technique is that the zero point of the calibration is set by the horizontal-branch models, which necessarily requires that our canonical understanding of RGB evolution and especially the helium flash be essentially correct. Unfortunately, studies of the RR Lyrae stars do not provide unambiguous support for the evolutionary calculations. For one thing, the Baade-Wesselink work by Jones *et al.* (1988) suggests that the absolute magnitudes of these variables do not vary with metallicity — though similar analyses by Cacciari *et al.* (1988) argue otherwise. For another, Cox (1987) has asserted from his pulsational studies that the RR Lyraes must have high helium abundances (close to $Y = 0.30$). This would normally imply (see Sweigart 1987) that they have M_V's which are appreciably brighter than the 0.85 mag or so that Jones *et al.* and Cacciari *et al.* find for many of their program objects. In addition, there is the relation between the observed periods and metallicities of the RR Lyraes (Sandage *et al.* 1981) that indicates that the luminosities of these variables are more strongly correlated with [Fe/H] than the models would lead us to expect. Possible solutions to this problem have been suggested by Lee *et al.* (1987) and Caputo (1988) (see also VandenBerg and Smith 1988): the latter studies all have important implications for horizontal-branch luminosities.

Nevertheless, as already mentioned, the location of the evolutionary computations on the C-M plane is supported by the Population II field stars that have reasonably accurate parallax distances. This is shown in Figure 2 where the isochrones from Fig. 1 are compared with the usual sample of subdwarf standards, whose properties have been tabulated by van Altena *et al.* (1988), and a few subgiants, for which parallax information was provided by J.T. Lee and G.S. Da Costa (private communication). For most of the stars, the metallicities given by van Altena *et al.* have been adopted; for the rest, the indicated [Fe/H]'s represent averages of the values listed in the Cayrel de Strobel *et al.* (1985) catalogue.

It is apparent that all of the indicated field stars are located within 2σ of where the models predict they should be found — the majority within 1σ. The subdwarf with the most reliably-determined parameters

— namely Groombridge 1830 (or HD 103095) — is especially consistent with the model predictions, and indeed provides a strong argument that the colors of the isochrones should not be corrected by more than the +0.02 mag which has been assumed. A couple of other stars deserve special mention. First, HD 84937 has an observed B-V = 0.37 mag,

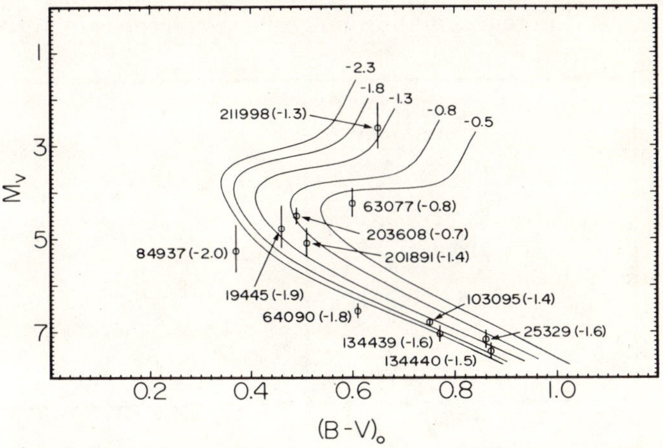

Fig. 2. - Comparison on the C-M plane of the locations of the theoretical isochrones from Fig. 1 with the properties of several nearby Population II subdwarfs and subgiants. The latter are identified by their HD numbers while the latest metallicity estimates are given within parentheses.

from which one might infer that the turnoffs of the most metal deficient GCs (and the relevant isochrones) should be at least this blue. Second, HD 63077 affords perhaps the most compelling evidence that there *are* very old stars. If this object is indeed a subgiant, as its location on Figure 2 suggests, then its measured M_V would imply an age of about 18 Gyr. Note that this is completely independent of any color uncertainty in the models. Even if future work shows that it is actually brighter by 0.8 mag (a 2σ displacement from its present position), one would still conclude that it is $\sim$ 14 Gyr old using the present isochrones.

A promising variation of the subdwarf-fitting method to derive distances to star clusters has recently been developed by VandenBerg and Poll (1988, hereafter VP88). Their procedure is most easily explained with the aid of Figure 3. The left-hand panel shows part of a zero-age main sequence (ZAMS) which VP88 have defined from the observed morphology of the Pleiades C-M diagram and which has clearly been correctly normalized to the Sun. Thus, the open circle indicates the absolute visual magnitude of a ZAMS star having the solar color (B-V = 0.64 mag) and solar abundances (Y = 0.27 and [Fe/H] = 0.0). If the solid curve were displaced in the vertical direction by an amount which is indicated by the dashed line, then the former would coincide with the dotted curve, which can be regarded as representing the location of a ZAMS for stars with the chemical composition of Gmb 1830. (Even though the latter is undoubtedly a very old star, relevant evolutionary calculations indicate that its evolutionary path would not have deviated significantly from the ZAMS locus since the 15 Gyr or so which have presumably transpired since its arrival on the main sequence.) Thus, the magnitude difference between the open circle and the filled square arises solely from the difference in abundances between the Sun and Gmb 1830.

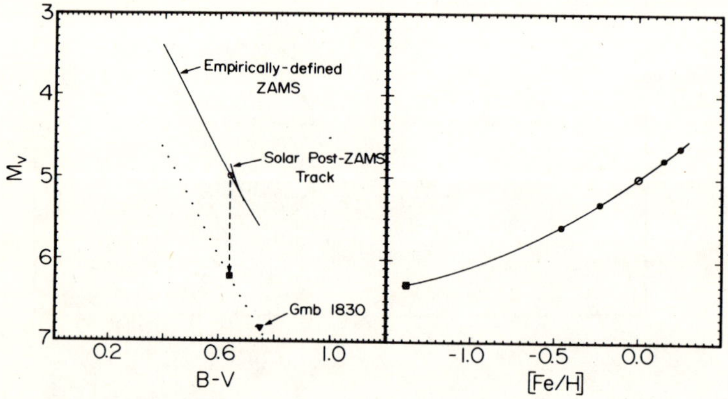

Fig. 3. - The left-hand panel indicates the magnitude difference (*dashed line*) between a solar-abundance ZAMS (*solid curve*) and one appropriate to stars with the chemical composition of Gmb 1830 (*dotted curve*). The right-hand panel plots the adopted relation between M_V and [Fe/H].

On the assumption that the subdwarf has Y = 0.24, which corresponds to the current best estimate of the primordial helium abundance (*e.g.*, Boesgaard and Steigman 1985), the effect on the position of the filled square of the small difference in Y from the solar value can be corrected for using computed ZAMS models in a differential sense. Thus, in the right-hand panel of Fig. 3, the filled square is assigned a slightly fainter luminosity than in the left-hand panel. The open circle reproduces the ZAMS model for solar abundances, while the filled circles are based on ZAMS models for Y = 0.27 and various [Fe/H]'s — where again, the latter were used only differentially.

VP88 provide analytic formulae for the solar composition ZAMS and the magnitude corrections which must be applied in order to correct that ZAMS to the position of one having different helium and/or metal abundances. When they applied the resultant semi-empirical main-sequence fitting procedure to the Hyades, they obtained a true modulus of 3.30 mag. The LMC cluster NGC 2162 — one of the two systems from which Schommer *et al.* (1984) argued for a reduced LMC distance — was also considered. VP88 found $(m - M)_V = 18.42$ if a metal abundance [Fe/H] = -0.23 is assumed or close to 18.2 if the metallicity is about one-third solar. In addition, as reproduced in Figure 4, they obtained an apparent distance modulus of 13.40 mag for the GC 47 Tuc when the Hesser *et al.* (1987) choices for Y and [Fe/H] are assumed. The Hesser *et al.* analysis of the cluster photometry is given in Figure 5 and it is apparent that they derived exactly the same distance modulus, though they did so from a match of a computed ZAHB to that observed. From the comparison of the isochrones with the main-sequence turnoff data, one infers that the cluster has an age of slightly less than 14 Gyr.

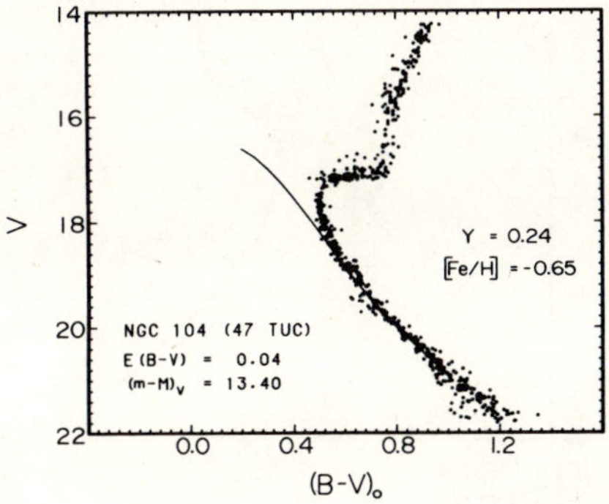

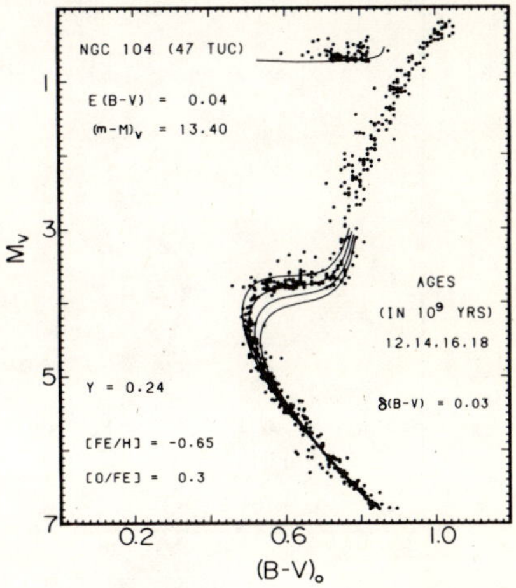

Fig. 4. - (left) Determination of the 47 Tuc distance modulus using the main-sequence fitting method developed by VP88.
Fig. 5. - (right) Hesser *et al.* 's (1987) adopted match of computed isochrones/ZAHBs to the 47 Tuc C-M diagram.

This is lower than the 16 Gyr value which was suggested from Fig. 1 simply because of differences in the assumed chemistry. Hesser *et al.* adopted Y = 0.24 because this is the value found by Caputo *et al.* (1987) from their application of the R-method to several GCs and, in addition, is the preferred value from primordial considerations. They also assumed an enhanced oxygen mix which is consistent both with the field-star relation between [O/Fe] and [Fe/H] and the results of Gratton's (1985) spectroscopic study of 47 Tuc giants. (An increase in Y makes the predicted horizontal branch brighter — resulting in a larger distance modulus and smaller age — while high oxygen tends to make the turnoff point for a given age somewhat fainter — thereby also leading to a reduction in the inferred age.) Since all of the assumptions made in the Hesser *et al.* study seem to be in complete accord with available observational constraints, one has to feel rather confident that the age of this cluster has been very well determined.

However, one cannot yet claim such confidence regarding the ages of the extremely metal-deficient GCs. As illustrated in Figure 6 for the case of M92, it is possible to argue that such clusters either are or are not significantly older than the metal-rich systems, simply by varying the assumed chemical composition in the theoretical calculations that are compared with the observed data. For instance, the left-hand panel shows that the M92 C-M diagram is well fitted by the isochrones from McClure *et al.* (1987) for Y = 0.24, [Fe/H] = -2.03, and [O/Fe] = 0.7 — leading to an inferred age of 14 Gyr (*i.e.,* essentially the same as that found for 47 Tuc). On the other hand, the right-hand panel indicates

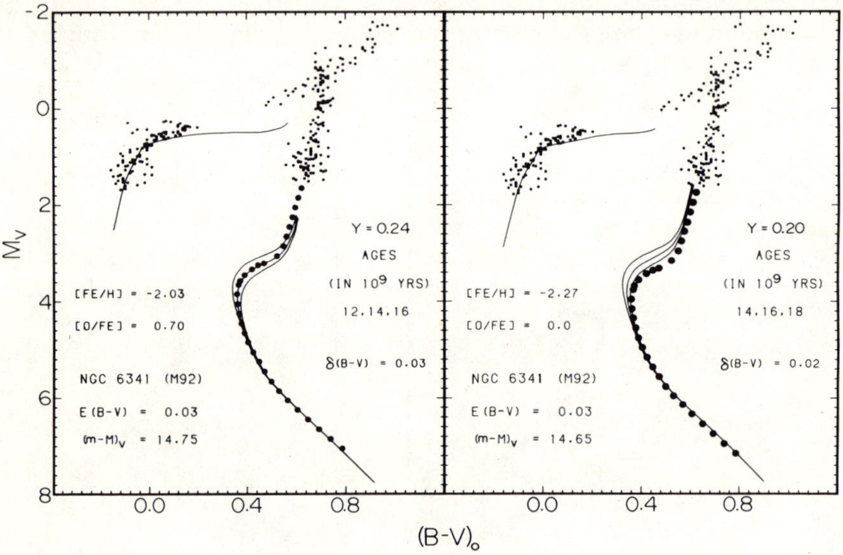

Fig. 6. - Two different matches of computed isochrones/ZAHBs, for the indicated parameters, to the M92 C-M diagram. The filled circles define Stetson and Harris' (1988) fiducial sequence; the brighter data comes from the sources listed in Table I.

from a similarly fine match of theory and observation that M92 may be $\sim$ 20 Gyr old if the VandenBerg and Bell (1985) isochrones for Y = 0.20 and [m/H] = -2.27 better represent the properties of the cluster stars. Note that since the age is basically derived from $\Delta V_{TO,HB}$, the difference in the assumed [Fe/H] values in the two sets of models has only a very small effect on the results. Moreover, had the VandenBerg-Bell computations assumed Y = 0.24, they would have predicted an age closer to 17-18 Gyr. Thus the further reduction down to 14 Gyr (in the left-hand panel) came about as a consequence of the high oxygen abundance which was assumed in the McClure *et al.* computations.

But there is a problem. Whereas [O/Fe] = 0.7 is approximately consistent with the field-star relation between the oxygen-to-iron and the iron-to-hydrogen ratios (*e.g.,* see Sneden 1985), such an oxygen overabundance does not, according to Pilachowski (1988), appear to exist in M92. Indeed, she has reported from her spectroscopic study of cluster giants that the C+N+O abundance is likely not too different from the scaled-solar value — which, if correct, would imply that field and cluster stars have had different chemical evolution histories. However, even more recent work by R. Peterson (private communication) suggests that the iron abundance in M92 may be appreciably lower than previously estimated — which, as she has noted, would tend to reconcile the GC O/Fe ratios with those in solar-neighborhood Population II stars. Such studies as those by Pilachowski and Peterson are obviously critical to our understanding of the ages of the globulars and the existence, or not, of an age-metallicity relation among them. These and further/similar efforts should be strongly encouraged and supported. In fact, until such progress is made, one really cannot say with any measure of certainty what the maximum age of the GCs is.

Still, it does appear to be difficult to obtain ages that are much less than 14 Gyr for the most metal-poor globulars. To be sure, some reduction below this value would result if $Y > 0.24$ and/or $[O/Fe] > 0.7$ were assumed in the calculations, but there is no observational basis for making such assumptions. Also, an increase in the distance scale would certainly lead to a lowering of the inferred ages, but again there is no compelling reason (*e.g.*, from the subdwarfs) that the adopted distances are significantly in error. It is also doubtful that revisions to the assumed physics in the models will have a major impact on derived ages (see the exhaustive analysis of this question by Rood 1987). This assertion is supported by the fact that the predicted turnoff magnitudes, as a function of composition and age, from the latest generation of isochrones by VandenBerg and Bell (1985) are within 0.1 mag of those computed fifteen years earlier by Iben and Rood (1970) — in spite of many improvements since then to opacity data, nuclear reaction rates, *etc.*

But one should be concerned about the possibility that the standard evolutionary calculations are missing some important physics. In particular, in a paper which has not received the attention that it deserves, Stringfellow *et al.* (1983) have argued that helium diffusion is important in old Population II stars and that the main effect of this process is to reduce deduced GC ages. For instance, we see in Figure 7 that a 12 Gyr

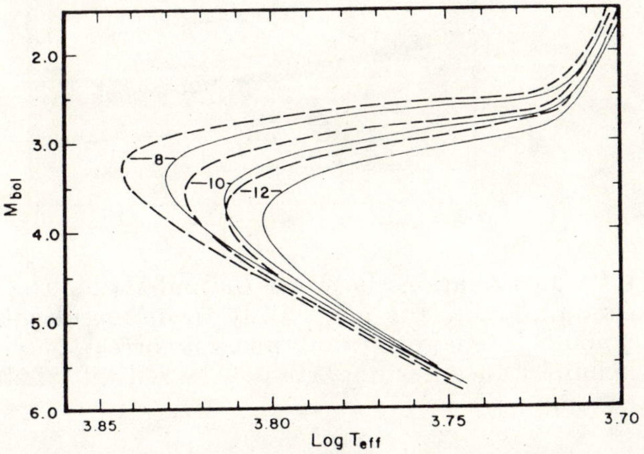

Fig. 7. - Theoretical isochrones with and without diffusion (*solid and dashed curves,respectively*) for the indicated ages. The calculations assume $Y = 0.239$ and $Z = 0.001$. (*Reproduced from the Astrophysical Journal by permission of G. Stringfellow.*)

isochrone, without diffusion, and one for an age of about 10.5 Gyr, with diffusion, are predicted to have nearly identical turnoff luminosities. That is, the ages which are determined from canonical models may be too large by about 15 percent (or more — see Stringfellow *et al.*). But, how well-determined are the diffusion coefficients and how sensitively do predicted ages depend on current uncertainties in these quantities? Will there be some differential effect of diffusion in the most metal-poor stars compared to those of higher metal abundance? What are the distinguishing characteristics of evolution with diffusion as opposed to standard computations? Although Stringfellow *et al.* comment on some of these questions, much more extensive work is needed. So, while it would be premature to abandon the canonical models for the globulars at the present time, it should be kept in mind that diffusion is a potential mechanism by which the maximum age of the GCs may be reduced.

3. RAMIFICATIONS OF GC AGES FOR H_0

Tayler (1986) has recently provided a very nice discussion of the relations between GC ages (t_{cl}), H_0, and Ω_0 (the ratio of the present density of the universe to the closure density) for various choices of the redshift of cluster formation, z_{cl}. His results are the basis of Figure 8,

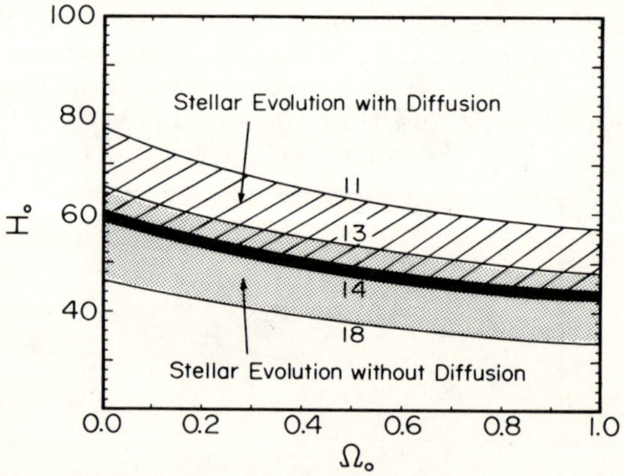

Fig. 8. - Various relations between H_0 and Ω_0 for the indicated ages. The maxiumum GC age is most likely to be found within the shaded band, if canonical stellar evolution theory is correct, or within the striped band, if helium diffusion is important. The redshift of cluster formation is assumed to be $z_{cl} = 5$.

which summarizes our current understanding of the maximum age of the globulars and the implications for standard cosmological models. Assume, for instance, that $\Omega_0 = 1$. Then, supposing that canonical stellar evolution theory is correct - in which case, $13 < t_{cl} < 18$ Gyr (the large range due mostly to abundance uncertainties) — Figure 8 indicates that the lower and upper limits to H_0 are 34 and 48 km/s/Mpc, respectively. If, instead, $\Omega_0 = 0.1$; then the Hubble constant is constrained to the range 44 – 63 km/s/Mpc. Alternatively, if our guesstimate of the effects of diffusion is reasonable, then $44 < H_0 < 57$ km/s/Mpc if $\Omega_0 = 1$ and $57 < H_0 < 74$ km/s/Mpc if $\Omega_0 = 0.1$. The heavy solid curve in Fig. 8 defines the relation between H_0 and Ω_0 for a personal best estimate of the upper limit to GC ages (about 14 Gyr). It is apparent that the inferred value of the Hubble constant in this case varies from 43 km/s/Mpc if $\Omega_0 = 1$ (*cf.* Sandage 1988) to about 60 km/s/Mpc if Ω_0 is near zero.

However, perhaps the best way to view GC ages is the following. If, by independent means, it can be shown that H_0 is sufficiently small that no difficulties are posed by high GC ages (which seem undeniable), then one might conclude that the universe can be described by standard Friedmann models. On the other hand, should future work confirm a high value for H_0 (> 75 km/s/Mpc, say) then the ages of the GCs would provide a particularly strong argument that, for instance, the cosmological constant is non-zero (*cf.* de Vaucouleurs 1983, Tayler 1986).

REFERENCES

Alcaino, G., and Liller, W. 1980, *A.J.* , **85**, 1330.

Arp, H. 1962, *Ap.J.* , **135**, 311.

Boesgaard, A., and Steigman, G. 1985, *Ann.Rev.Astr.Ap.*, **23**, 319.

Bolte, M. 1987, *Ap.J.* , **319**, 760.

Buonanno, R., Corsi, C.E., and Fusi Pecci, F. 1981, *M.N.R.A.S.* , **196**, 435.

Buonanno, R., Corsi, C.E., Iannicola, G., and Fusi Pecci, F. 1986, *Astr.Ap.* , **159**, 189.

Cacciari, C., Clementini, G., and Buser, R. 1988, *Astr.Ap.* , in press.

Caloi, V., Castellani, V., Danziger, J., Gilmozzi, R., Cannon, R.D., Hill, P.W., and Boksenberg, A. 1986, *M.N.R.A.S.* , **222**, 55.

Caputo, F. 1988, *Astr.Ap.* , **189**, 70.

Caputo, F., Martinez Roger, C., and Paez, E. 1987, *Astr.Ap.* , **183**, 228.

Cayrel de Strobel, G., Bentolila, C., Hauck, B., and Duquennoy, A. 1985, *Astr.Ap.Suppl.* , **59**, 145.

Cox, A.N. 1987, in *IAU Colloquium 95, The Second Conference on Faint Blue Stars*, ed. A.G.D. Philip, D.S. Hayes, and J.W. Liebert (L.Davis Press), p. 161.

de Vaucouleurs, G. 1983, *Ap.J.* , **268**, 468.

Flannery, B.P., and Johnson, B.C. 1982, *Ap.J.* , **263**, 166.

Gratton, R.G. 1985, *Astr.Ap.* , **147**, 169.

Gratton, R.G. 1987, *Astr.Ap.* , **177**, 177.

Harris, W.E. 1975, *Ap.J.Suppl.* , **29**, 397.

Hesser, J.E., Harris, W.E., VandenBerg, D.A., Allwright, J.W., Shott, P., and Stetson, P.B. 1987, *Pub.A.S.P.* , **99**, 739.

Iben, I., Jr., and Renzini, A. 1984, *Phys.Rep.*, **105**, 329.

Iben, I., Jr., and Rood, R.T. 1970, *Ap.J.* , **159**, 605.

Janes, K.A., and Demarque, P. 1983, *Ap.J.* , **264**, 206.
Jones, R.V., Carney, B., and Latham, D.W. 1988, *Ap.J.* , in press.
Lee, Y.-W., Demarque, P., and Zinn, R. 1987, in *IAU Colloquium 95, The Second Conference on Faint Blue Stars*, ed. A.G.D. Philip, D.S. Hayes, and J.W. Liebert (L. Davis Press), p. 137.
McClure, R.D., VandenBerg, D.A., Bell, R.A., Hesser, J.E., and Stetson, P.B. 1987, *A.J.* , **93**, 1144.
Peebles, P.J.E. 1984, *Ap.J.* , **277**, 470.
Penny, A.J., and Dickens, R.J. 1986, *M.N.R.A.S.* , **220**, 845.
Peterson, C.J. 1987, *Pub.A.S.P.* , **99**, 1153.
Pilachowski, C.A. 1988, *Ap.J.(Letters)* , **326**, L57.
Richer, H.B., and Fahlmann, G.G. 1987, *Ap.J.* , **316**, 189.
Richer, H.B., Fahlmann, G.G., and VandenBerg, D.A. 1988, *Ap.J.* , **329**, 187.
Rood, R.T. 1987, paper presented at the 13th *Texas Symposium on Relativistic Astrophysics.*
Rood, R.T., and Crocker, D.A. 1985, in *Production and Distribution of C, N, O Elements*, ed. J. Danziger, F. Matteucci, and K. Kjar (Garching bei Munchen: European Southern Observatory), p. 61.
Sandage, A. 1969, *Ap.J.* , **157**, 515.
Sandage, A. 1982, *Ap.J.* , **252**, 553.
Sandage, A. 1988, *Ap.J.* , in press.
Sandage, A., Katem, B., and Sandage, M. 1981, *Ap.J.Suppl.* , **46**, 41.
Sandage, A., and Tammann, G.A. 1982, *Ap.J.* , **256**, 339.
Sandage, A., and Walker, M.F. 1966, *Ap.J.* , **143**, 313.
Schommer, R., Olszewski, E., and Aaronson, M. 1984, *Ap.J.(Letters)* , **285**, L53.
Sneden, C. 1985, in *Production and Distribution of C, N, O Elements*, ed. J. Danziger, F. Matteucci, and K. Kjar (Garching bei Munchen: European Southern Observatory), p. 1.
Stetson, P.B., and Harris, W.E. 1988, *A.J.* , in press.
Stringfellow, G.S., Bodenheimer, P., Noerdlinger, P.D., and Arigo, R.J. 1983, *Ap.J.* , **264**, 228.
Sweigart, A.V. 1987, *Ap.J.Suppl.* , **65**, 95.
Tayler, R.J. 1986, *Quart.J.R.A.S.*, **27**, 367.
van Altena, W.F., Lee, J.T., Hanson, R.B., and Lutz, T.E. 1988, in *Calibration of Stellar Ages*, ed. A.G.D. Philip (L. Davis Press), in press.
VandenBerg, D.A. 1983, *Ap.J.Suppl.* , **51**, 29.
VandenBerg, D.A. 1985, in *Production and Distribution of C, N, O Elements*, ed. J. Danziger, F. Matteucci, and K. Kjar (Garching bei Munchen: European Southern Observatory), p. 73.
VandenBerg, D.A., and Bell, R.A. 1985, *Ap.J.Suppl.* , **58**, 561.
VandenBerg, D.A., and Poll, H.P. 1988, preprint. (VP88)
VandenBerg, D.A., and Smith, G.H. 1988, *Pub.A.S.P.* , **100**, 314.

DISCUSSION

J. GRAHAM: Are you able to say anything yet about the ages of Magellanic Cloud clusters? Is the photometry precise enough yet?

D.A. VANDENBERG: I have personally not determined the ages of Magellanic Cloud clusters from isochrone fits, but my impression is that the photometry of (especially) the oldest systems is not yet good enough to say whether they are of comparable age to, or definitely younger than, the Galactic GCs.

H. REEVES: The constancy of the Lithium abundance in Pop II stars is certainly relevant to the extent of helium diffusion. Can you comment on this?

D.A. VANDENBERG: That's a very good point. One would normally expect some depletion of the metals, as well as helium, from the stellar atmospheres if diffusion operates - though if all Pop II stars are about the same age, perhaps the star-to-star abundance variations would be too small to be detectable. I don't know enough about diffusion to be able to say. Still, you have mentioned an important observational constraint that must be accommodated by the diffusion models.

S.G. DJORGOVSKI: The Yale group claims that by fitting their isochrones, they obtain a real spread of ~ 4 Gy in globular cluster ages. You claim that all the ages are consistent with a single value. Isn't that indicative of a basic uncertainty in the isochrone fitting method?

D.A. VANDENBERG: The Yale isochrones assume solar abundance ratios of the heavy elements and, as I have described here, my models for such an element mix also lead to a significant age spread between metal-rich and metal-poor GCs. It is only when one assumes the field star relation between [O/Fe] and [Fe/H] that the ages of the globulars are predicted to be nearly the same - about 14 Gyr - independent of metallicity. So the existence, or not, of an age-metallicity relation among the GCs is more a matter of the assumed chemistry than of the isochrone-fitting procedure. At the same time, it is true that one can get very different results if one employs the main-sequence fitting method rather than the matching of predicted and observed horizontal branches to derive cluster distances (and hence ages). I believe that I have adequately discussed why the latter method is arguably the preferred one to use.

DISCUSSION

J. NEMEC: Sandage has a preprint which shows that in globular clusters the luminosity width of the red horizontal branch correlates with metal abundance. Are your theoretical models in accord with this correlation?

D.A. VANDENBERG: I have not evolved my models beyond the zero-age horizontal branch (ZAHB) but presumably Sandage's observation is saying something about the relative locations of ZAHBs and post-horizontal-branch tracks as a function of [Fe/H]. Careful comparisons of the observations with model predictions should allow some further refinement of our understanding of, say, the helium or CNO abundances in GC stars.

CEPHEIDS AND BRIGHT STARS IN NGC 3109

ALLAN R. SANDAGE
Mount Wilson and Las Campanas Observatories
813 Santa Barbara Street, Pasadena, California
91101-1292

GEORGE A. CARLSON
Citrus College
1000 West Foothill Blvd., Glendora, California 91740

ABSTRACT Cepheid variables have been found in the nearby Sm IV galaxy NGC 3109. Periods determined for 29 of the variables range from 31 to 5 days. The period-luminosity relation yields an apparent blue distance modulus of $(m-M)_{AB} = 26.4$.

The brightest supergiants have absolute magnitudes of $M_B(3)=-7.6$ for the mean of the three brightest blue stars, and $M_V(3)=-7.7$ for the corresponding mean of the three brightest red stars. These absolute magnitudes for both the red and the blue supergiants in NGC 3109 agree with the previous calibration of the brightest star luminosities as a function of the absolute magnitude of the parent galaxy.

CEPHEIDS

A series of 50 blue (103a0 + GG385) and 5 yellow (103aD + GG495) photographic plates were obtained during a 1519 day interval between 1978 and 1982 with the 2.5-m reflector at the Las Campanas Observatory. Five pairs of blue plates were blinked and 92 possible Cepheid candidates were identified. Of these, we have obtained periods and light curves for 29 confirmed Cepheids. Another 15 candidates are probably Cepheids but no periods have been determined for these.

The P-L relations at mean and maximum intensity were plotted from the Cepheid light curves. The slope and separation of the envelope lines are from the calibration used throughout this series (Sandage and Tammann 1968, Table A1), based on the combined P-L relations for Cepheids in a number of nearby Local Group galaxies. The apparent blue distance modulus was found to be 26.4 based on the zero point adopted in that reference, which averages 0.2 mag brighter in $\langle B \rangle$ and 0.1 mag brighter in B(max) than the most recent calibration by Feast and Walker (1987).

BRIGHTEST STARS

Candidates for the brightest stars were chosen by inspection of a short exposure yellow plate. In addition, a pair of yellow and blue plates were blinked to identify candidates for the reddest stars. Magnitudes for the red and blue star candidates were estimated relative to the same photometric sequence that was used to measure the comparison stars for the variables. The results are based on the mean of 3 determinations of B magnitudes, and 2 of V. We estimate the accuracy of these listings to be within about 0.15 mag in both colors.

Depending on the membership status of the brightest blue candidate, the mean absolute magnitude of the three brightest blue supergiants is $\langle M_B(3) \rangle = 7.60$. This result agrees with the calibration obtained in the last Cepheid paper of this series (Sandage and Carlson 1985).

The mean absolute magnitude of the three brightest red supergiants was found to be $\langle M_V(3) \rangle = 7.69$. There is no question as to membership since all three are variable. This result agrees with the parallel calibration of the red supergiants. The remarkably constant value of $M_V(3) = -7.8$ for the stars in parent galaxies with M_V between -14 and -18 again seems well determined. Its luminosity rise to $M_V = -9$ for the brightest parent galaxies depends on the data in M81 and M101 alone. It is of central importance to establish whether the rise is real but this can only be done either by adding more data for galaxies as luminous as M81 and M101, or by improving the data for M81 and M101.

REFERENCES

Feast, M.W., and Walker, A.R. (1987). Ann. Rev. Astr. Ap. 25, 345.

Sandage, A., and Carlson, G. (1985). Astron. J. 90, 1464.

Sandage, A., and Tammann, G.A. (1968). Astrophys. J. 151, 531.

PLANETARY NEBULAE IN THE NGC 3379 GROUP: TESTING A NEW STANDARD CANDLE

ROBIN CIARDULLO and GEORGE H. JACOBY
Kitt Peak National Observatory, National Optical Astronomy Observatories, P.O. Box 26732, Tucson, Arizona 85726–6732

HOLLAND C. FORD
Department of Astronomy, University of Michigan, Ann Arbor, MI 48109-1090

ABSTRACT The [O III] λ5007 planetary nebula luminosity function (PNLF) of three early type galaxies in the Leo Cloud are compared and used to estimate distances. The shape of the PNLF, in particular the sharp bright end cutoff, is found to be the same in each galaxy and identical to that observed in M31, implying that this shape is universal and an excellent standard candle. From maximum likelihood analyses, the most probable distances to NGC 3377, NGC 3379, and NGC 3384 are 9.7, 9.1, and 9.4 Mpc respectively, with 1σ errors of $\sim$ 12%.

There has been mounting evidence, both observational and theoretical, that the [O III] λ5007 magnitudes of the brightest planetary nebulae in external galaxies are excellent standard candles. One way to test this hypothesis is to compare the planetary nebula luminosity function (PNLF) of several different galaxies. We have done this for galaxies in the well mixed Leo Cloud. Using the Kitt Peak 4 m telescope and a 30Å wide interference filter centered on redshifted λ5007, we identified and measured 95 planetaries in the E0 galaxy NGC 3379, 102 PN in the SB0 galaxy NGC 3384, and 54 PN in the smaller, E6 galaxy NGC 3377. As displayed in Figure 1, the PNLF's shape is the same in each galaxy, and the combined luminosity function for the cluster is identical with that observed in M31 (Ciardullo *et al.* 1988).

We have analyzed a statistically complete sample of the above data using maximum likelihood statistics and assuming a universal PNLF of the form

$$N(M) \propto e^{0.307M}\left\{1 - e^{3(M^*-M)}\right\} \qquad M^* = -4.48$$

The probability contours, as a function of the true distance modulus, μ_0, and the luminosity specific PN density, α, appear in Figure 2. The most likely distances of 9.7 Mpc for NGC 3377, 9.1 Mpc for NGC 3379, and 9.4 Mpc for NGC 3384 and the tight error contours imply that all three galaxies are indeed members of a single cluster, and that the PNLF is an extremely effective standard candle. In accordance with stellar population theory (Renzini and Buzzoni 1986), the luminosity specific PN density is also observed to be approximately constant.

REFERENCES

Ciardullo, R., Jacoby, G.H., Ford, H.C., and Neill, J.D. 1988, Ap. J., submitted.

Renzini, A., and Buzzoni, A. 1986, in Spectral Evolution of Galaxies, ed. C. Chiosi and A. Renzini (Dordrecht: Reidel), p. 195.

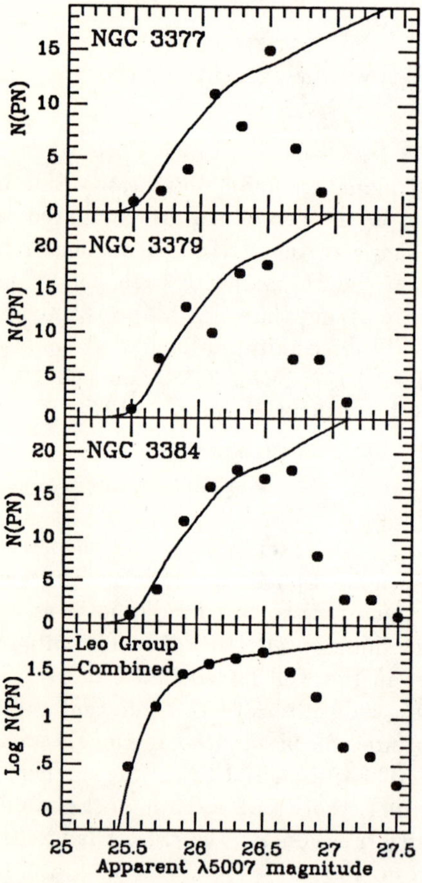

Fig 1 The PNLFs for galaxies in the Leo Cloud. The solid line is the empirical law convolved with the photometric error at the mean cluster redshift. The falloff at $m_{5007} > 26.5$ is due to incompleteness. The bottom panel plots the combined PNLF in the log.

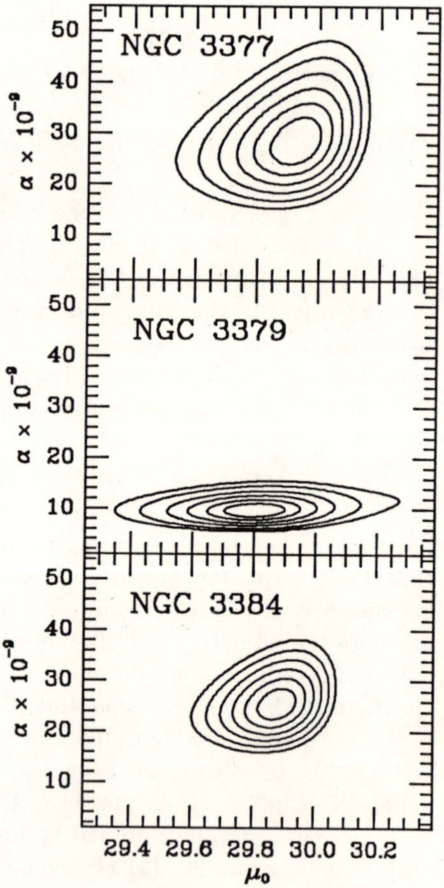

Fig 2 Confidence contours for the Leo Cloud galaxies. The absissca is the true distance modulus; the ordinate is the luminosity specific density of PN within 2.5 mag of M^*. Contours are drawn at 0.5σ intervals. An E(B-V) of 0.047 has been assumed.

CCD I–BAND PHOTOMETRY OF CEPHEIDS: DISTANCES TO THE GALAXIES M81 AND NGC 2403

WENDY L. FREEDMAN
Mount Wilson and Las Campanas Observatories
Carnegie Institution of Washington
Pasadena, California

BARRY F. MADORE
Division of Physics, Mathematics and Astronomy
California Institute of Technology
Pasadena, California

M81 and NGC 2403 are two of the closest northern hemisphere galaxies outside of the Local Group for which Cepheid variables have been detected and periods accurately determined. *I*–band CCD photometry has been obtained for the only two known Cepheids in M81, and for eight of the known Cepheids in NGC 2403. For M81, an apparent *I*–band distance modulus and formal fitting error of $(m-M)_I = 27.64 \pm 0.09$ mag are derived; while for NGC 2403 $(m-M)_I = 27.57 \pm 0.14$ mag. Taking a true distance modulus of 18.50 mag for the Large Magellanic Cloud (and $E(B-V) = 0.07$), a true modulus of 27.59 mag $\pm$ 0.31 is obtained for M81, and 27.51 mag $\pm$ 0.24 for NGC 2403 (foreground reddenings of $A_I = 0.05$ and 0.06 mag being adopted for M81 and NGC 2403, respectively, from Burstein and Heiles [1]). These moduli correspond to distances of 3.3 $\pm$ 0.5 and 3.2 $\pm$ 0.4 Mpc, respectively, but they are still upper limits since we have not yet independently determined the reddening internal to the parent galaxies.

At present, the largest factors contributing to the uncertainty in the distances to M81 and NGC 2403 are: (1) the number of Cepheids so far observed, (2) the distance to the Large Magellanic Cloud (*i.e.* the zero point of the Cepheid P-L relation) and finally (3) the internal reddening affecting each of the Cepheids in the individual program galaxies. No correction for reddening internal to each of the program galaxies has been applied in this study, but the uncertainty in the provisional true distance modulus has been increased. The total reddening can be assessed using high signal-to-noise multi-color observations, and will be dealt with in a forth-coming paper (Freedman and Madore, in preparation). Surveys for more Cepheids in M81 are underway, but are several years from completion. With phased observations of all of the known Cepheids in NGC 2403, the sample–size error for this galaxy could be further decreased by a factor of two.

The results presented here are based on observations made using prime focus CCD detectors at the Canada-France-Hawaii Telescope and the KPNO 4.0m. This work was supported in part by NSF grant AST 87-13889 to WLF. Details of the observations have been accepted for publication in the *Astrophysical Journal (Letters)*.

REFERENCES:

1. D. Burstein, C. Heiles: *Ap. J. Suppl.*, **54**, 33 (1984)

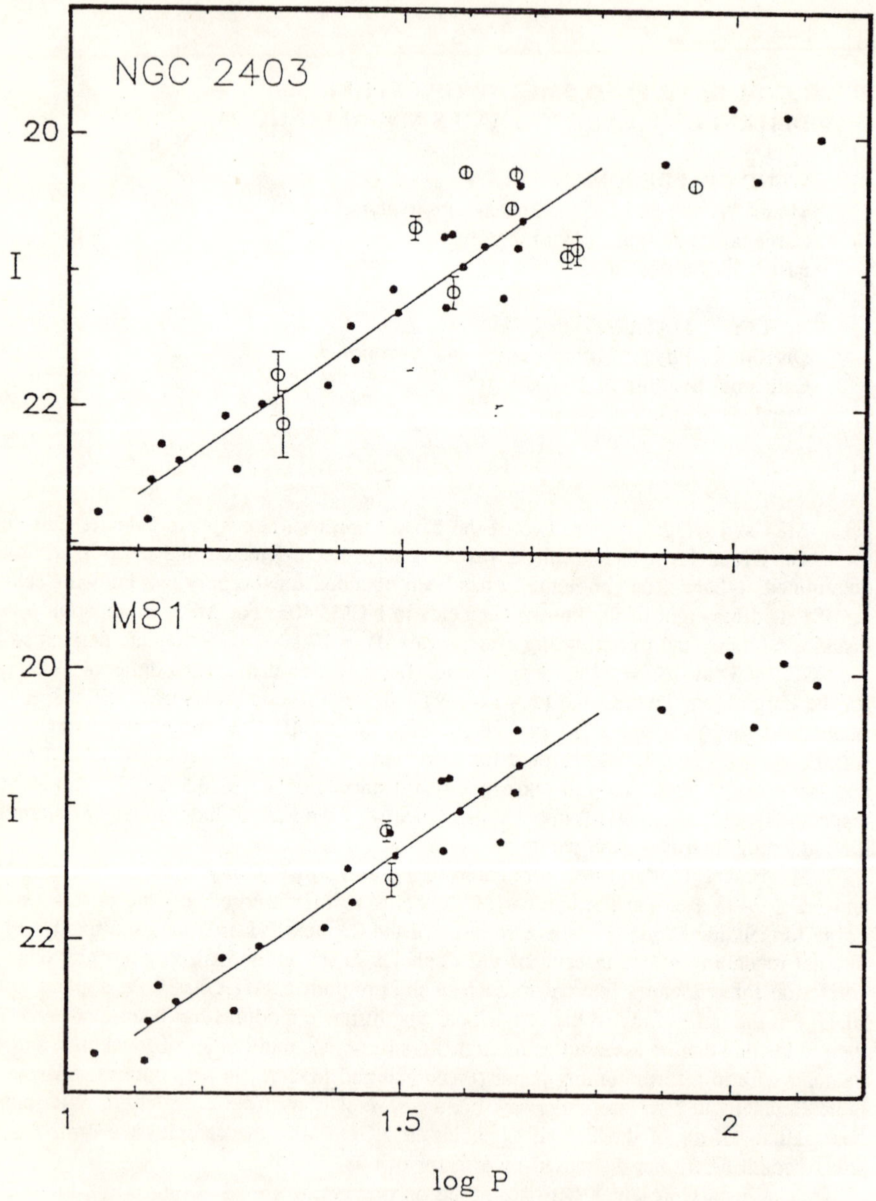

Fig. 1 – Period–Luminosity relations. *Upper Panel:* NGC 2403. *I*–band P-L relation for the averaged observations of ten Cepheids observed in NGC 2403 (open circles). *Lower Panel:* M81. Open circles mark the periods and averaged *I*–band magnitudes of for the two Cepheids (V2 and V30) with known periods in M81, Filled circles in both panels give the time-averaged photoelectric data for the Large Magellanic Cloud Cepheids, shifted in apparent modulus until minimum scatter was achieved for each galaxy.

THE DISCOVERY OF CEPHEIDS VARIABLES IN SCULPTOR GROUP GALAXIES

WENDY L. FREEDMAN
Mount Wilson and Las Campanas Observatories
Carnegie Institution of Washington
Pasadena, California

I. HOROWITZ, B. F. MADORE and J. MOULD
Division of Physics, Mathematics and Astronomy
California Institute of Technology
Pasadena, California

J. GRAHAM
Department of Terrestrial Magnetism
Carnegie Institution of Washington
Washington, D.C.

In the last three years we have been searching for Cepheids in three Sculptor Group galaxies – NGC 247, NGC 253, and NGC 7793, using the prime focus CCD at the CTIO 4m. All three of these galaxies are suitable zero-point calibrators for the infrared Tully-Fisher relation, and when we determine Cepheid distances for them we shall have doubled the current number of calibrators available for this route to the determination of the Hubble constant. In addition to working on these primary galaxies we have also, during the course of this program, obtained calibrated *BVRI* frames of known Cepheids in the galaxy NGC 300.

For NGC 247, the data have been completely reduced using DAOPHOT, and software has been developed which automatically searches for variables within prescribed color ranges and at user-defined levels of significance. A dozen variables of the correct color and magnitude for them to be Cepheids have been found after analyzing the 14 frames in one color now available for one field in NGC 247. Light curves have been obtained for these variables; however there is insufficient phase coverage at this time to determine accurate and unambiguous periods, because of aliasing problems. The program has been scheduled to continue in the fall of 1988 at CTIO, and we expect that significantly improved periods will be obtained after one more season's observing.

For NGC 7793, about 8 Cepheid candidates have been found with well-defined light curves and with consistent colors. Figure 1 shows the field in NGC 7793 being monitored, with a sampling of newly identified variables marked.

This work is being supported in part by NSF grant AST 87-13889 to WLF.

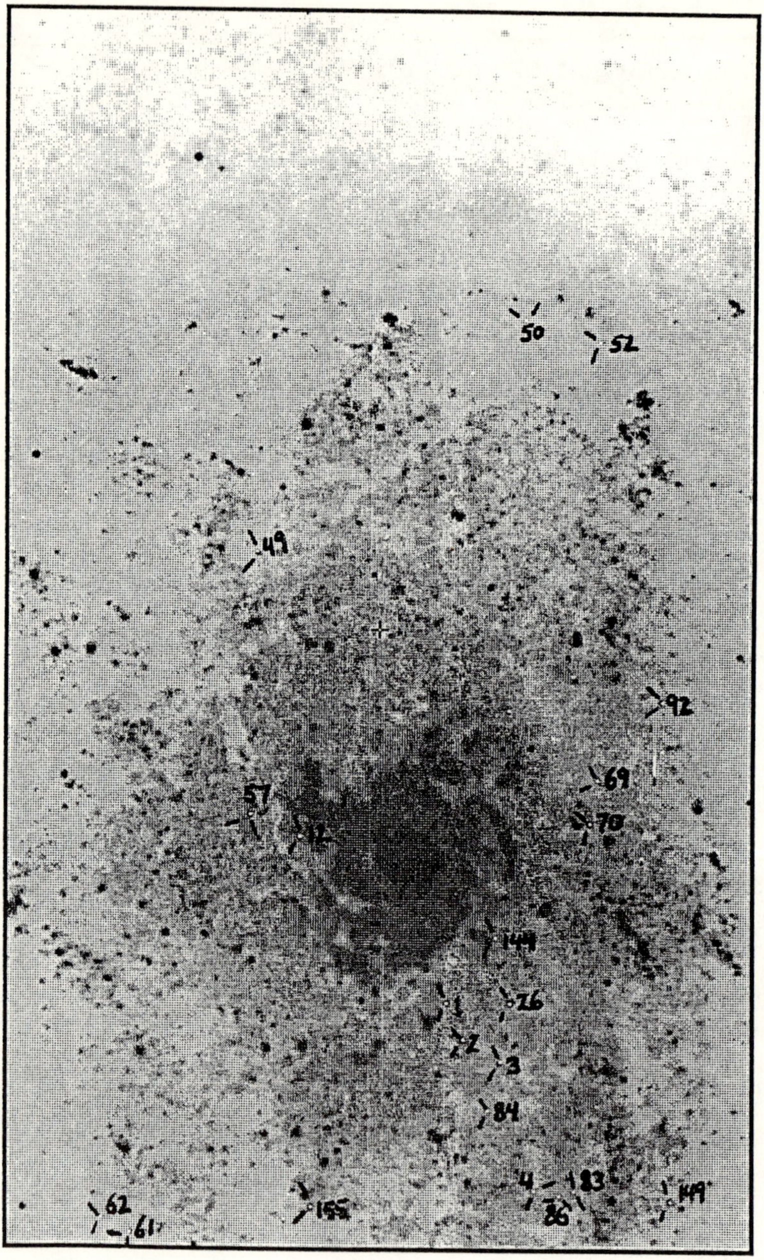

Fig. 1 *V-band* CCD exposure of the central field in NGC 7793 being monitored for Cepheid variables using the CTIO 4m. reflector. Several suspected variables are marked.

RESOLVED STELLAR POPULATIONS OF LUMINOUS STARS IN DWARF IRREGULAR GALAXIES

ROBERT RUOTSALAINEN
Department of Physics, MS-68
Eastern Washington University, Cheney, Washington 99004

Image tube exposures of the galaxies DDO 53 and GR 8 were obtained using the 2.2 m telescope at Mauna Kea Observatory. Color-magnitude (C-M) diagrams were produced from photoelectrically calibrated images of some 150 stars in DDO 53 and 65 stars in GR 8. These diagrams were compared with expected distributions based upon stellar evolutionary models of Becker (1981), and of Brunish and Truran (1982a, b). The color index U-B, more sensitive than B-V to the effective temperature for early-type stars, was used to model intervals of recent star formation in these two galaxies.

The modeled distributions depend on values assigned for intervals of star formation over ages t; and also on values for the distance modulus μ, initial mass function index α (of the form: $N \propto M^{-\alpha}$ for N number of stars of mass M), and

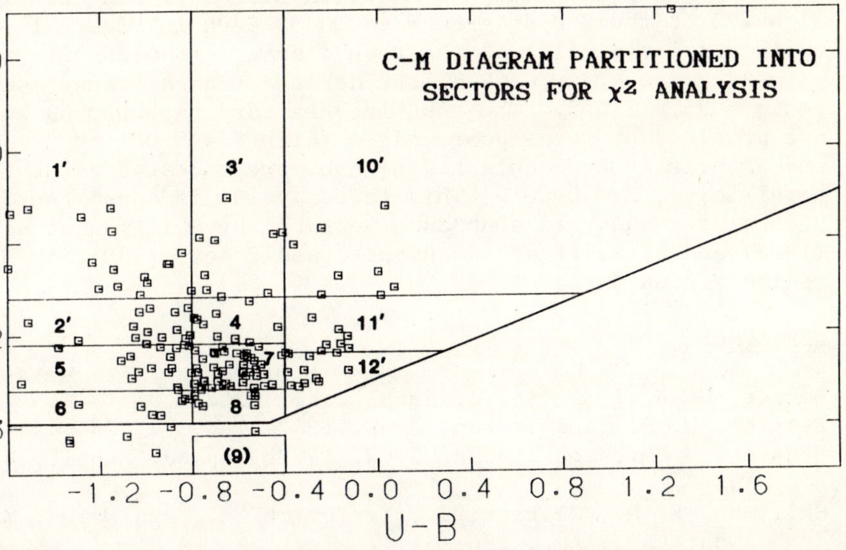

Fig. 1. Color-magnitude diagram for 152 stars in DDO 53.

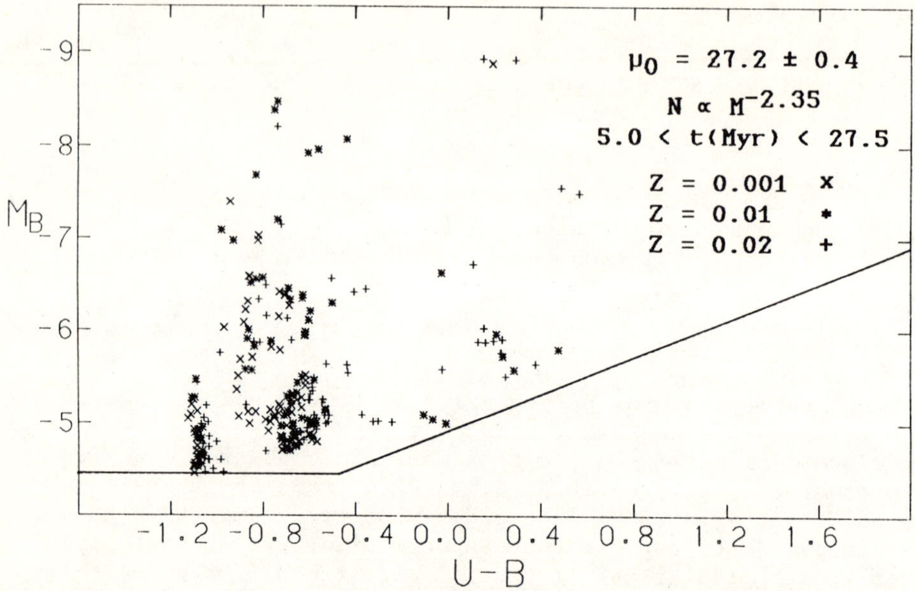

Fig. 2. Three models of 76 points each for DDO 53.

metallicity Z. Values determined for μ are especially sensitive to those points clustered about the coordinate (M_B, U-B) ≃ (-5.0, -0.7) which represent core helium-burning stars of about 10 solar masses.

The chi-square match of modeled distributions seen in Figure 2 to the C-M diagram for DDO 53 seen in Figure 1 is optimized for a value of the true distance modulus of μ_0 = 27.2 ± 0.4. For GR 8, the derived true distance modulus is μ_0 = 25.9 ± 0.5. The best matches for both DDO 53 and GR 8 are provided by metal-poor models (with Z = 0.001 or Z = 0.01), and especially by models having the lowest tested value for the metallicity, Z = 0.001. This lowest value is consistent with values of the oxygen abundance measured by Skillman et al. (1988) and by Skillman, Kennicutt, and Hodge (1989) for H II regions in GR 8 and DDO 53.

REFERENCES

Becker, S. A. 1981, Ap. J. Suppl., 45, 475.

Brunish, W. M., and Truran, J. W. 1982a, Ap. J., 256, 247.

Brunish, W. M., and Truran, J. W. 1982b, Ap. J. Suppl., 49, 447.

Skillman, E. D., Melnick, J., Terlevich, R., and Moles, M. 1988, Astr. Ap., 196, 31.

Skillman, E. D., Kennicutt, R. C., and Hodge, P. W. 1989, Ap. J., in preparation.

A NEW VALUE FOR THE HUBBLE CONSTANT FROM THE SIZE OF THE GALACTIC BULGE

D. M. TERNDRUP
Cerro Tololo Inter-American Observatory, National Optical Astronomy Observatories, Casilla 603, La Serena, Chile

ABSTRACT New surface photometry of the nuclear bulge of the Galaxy can be used to derive the relative distance between the Galactic center and the Virgo cluster. The data calibrate the velocity dispersion/size relation obtained for Virgo spirals by Dressler (1987). The surface brightness distribution of the Galactic bulge/spheroid is well represented by a de Vaucouleurs law with scale length $r_e \sim 1$ kpc, and is therefore somewhat more concentrated than previously thought. The derived Virgo distance from this method, when combined with typical values for the Galaxy's infall velocity, gives $H_0 = (50 - 60) \pm 10$ km s^{-1} Mpc^{-1}.

Dressler (1987) has presented a new method for obtaining relative distances between galaxies and the extragalactic distance scale. This "D_n-σ" method uses an observed correlation for Virgo spirals between the central velocity dispersion σ and D_n, a measure of the apparent size of the bulge. (D_n is the diameter of that aperture in which the surface brightness is $\mu_B = 19.75$ mag arcsec^{-2}.) Using M31 and M81 as calibrators, Dressler finds $H_0 = 67 \pm 10$ km s^{-1} Mpc^{-1}.

New data for our own Galaxy can also calibrate the D_n-σ relation. The steps in this calibration are: (1) measuring the central velocity dispersion σ, (2) calculating via the D_n-σ relation what apparent size D_n our Galaxy would have at the distance of Virgo, and (3) comparing the observed D_n for the Galaxy to that found in step 2, thus getting the relative distance between the Galactic center and the Virgo cluster.

There are several recent values for the velocity dispersion of K giants in the bulge. Rich (1988) found $\sigma = 104 \pm 11$ km s^{-1} from radial velocities of 88 stars. From the integrated spectrum of the bulge, Freeman et al. (1988) obtained $\sigma = 128 \pm 7$ km s^{-1}. A new astrometric analysis of more than 300 K giants in Baade's window (Jones, private communication) yields $\sigma = 122 \pm 11$ km s^{-1}. The mean of the latter two values is σ = 124 ± 9 km s^{-1}. A Virgo galaxy with this dispersion would have 16.5 ± 1.7 arcsec (Dressler 1987).

Terndrup (1988) has obtained new surface photometry in the inner bulge, which is brighter and more centrally concentrated than was previously thought. A model for the external appearance of the Milky Way based on the new photometry (upper line in the Figure) yields D_n = 15.3°. The new model has an effective radius about 1 kpc, as compared to ~ 2.5 kpc estimated by de Vaucouleurs and Pence (1978, shown as the lower line in the Figure.)

The observed D_n for the bulge is (3.3 ± 0.4) x 10^3 times that for the Galaxy if it were in Virgo, so R(Virgo) = (3.3 ± 0.4) x 10^3 R_0. With R_0 = 7.8 ± 0.6 kpc (Feast 1987), R(Virgo) = 23.4 ± 3.0 Mpc, or (m - M) = 31.8.

The infall-corrected velocity to the Virgo cluster is variously estimated to be from 1321 ± 70 km sec^{-1} (references in Dressler 1987) to 1196 ± 69 km s^{-1} (references in Sandage and Tammann 1988). For the former value, H_0 = 56 ± 10 km s^{-1} Mpc^{-1}; for the latter, H_0 = 51 ± 10 km s^{-1} Mpc^{-1}.

REFERENCES

de Vaucouleurs, G., and Pence, W.D. 1978. A.J. **83**, 1163.
Dressler, A. 1987. Ap. J. **317**, 1.
Feast, M.W. 1987. In The Galaxy, eds. G. Gilmore and P. Carswell (Dordrecht: D. Reidel), p. 1.
Freeman, K.C., de Vaucouleurs, G., de Vaucouleurs, A., and Wainscoat, R.J. 1988. Preprint.
Rich, R.M. 1988. A.J. In press.
Sandage, A., and Tammann, A. 1988. Ap. J., **328**, 1.
Terndrup, D.M. 1988. A.J. In press.

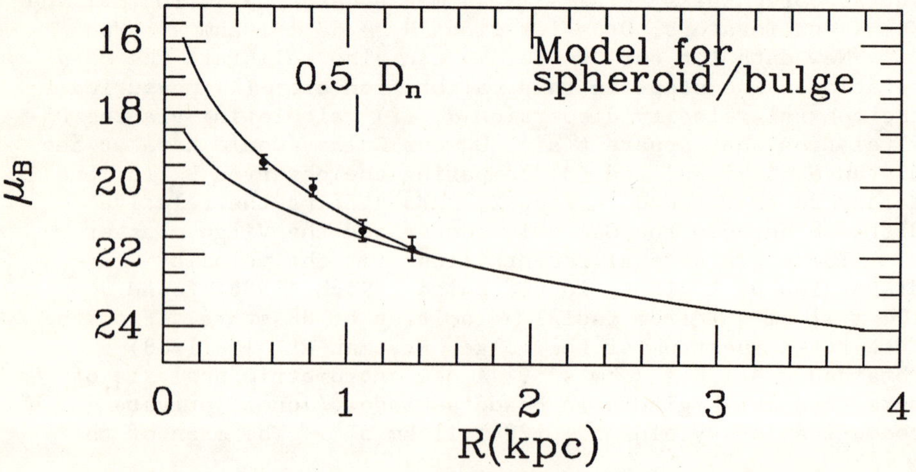

A NEW TECHNIQUE FOR DETERMINING EXTRAGALACTIC DISTANCES

JOHN L. TONRY and GERARD A. LUPPINO
MIT, 37-521, Cambridge, MA 02139

DONALD P. SCHNEIDER
Institute for Advanced Study, Princeton, NJ 08540

We describe a relatively direct technique of determining extragalactic distances that can determine accurate distances of elliptical galaxies and spiral bulges out to about 20 Mpc. The method relies on measuring the luminosity fluctuations that arise from the counting statistics of the stars contributing the flux in each pixel of a high signal-to-noise CCD image of a galaxy. The amplitude of these fluctuations is inversely proportional to the distance of the galaxy. This approach bypasses most of the successive stages of calibration required in the traditional extragalactic distance ladder; the only serious drawback to this method is that it requires an accurate knowledge of the bright end of the luminosity function.

Much of the contents of this poster summary are presented by Tonry and Schneider (1988). The basic idea behind obtaining distances from luminosity fluctuations is simple. Suppose you observe two identical galaxies with a CCD, one of which is twice as distant as the other. Ignoring the effects of seeing, the flux detected by a pixel is the flux emitted by the stars encompassed by the pixel, and this is just the number of stars times the average flux per star. In this respect these two galaxies have indistiguishable surface brightness because the more distant galaxy has 4 times as many stars per pixel, with 1/4 the flux per star. Now imagine that you measure the pixel-to-pixel fluctuations in flux. The rms fluctuation in the limit of large N is just the square root of the number of stars times a suitable average flux per star, $\overline{L}$. This rms fluctuation will vary inversely as the distance of the galaxy, since the average flux decreases as $1/d^2$, but the square root of the number goes as d.

The average luminosity here is just the second moment of the luminosity function. If a stellar population has various species with luminosity L_i and number (expectation value) n_i, the variance is just $\langle (L-\langle L\rangle)^2\rangle = \overline{L}\,\langle L\rangle$ where $\langle (L-\langle L\rangle)^2\rangle = \sum n_i L_i^2$ and $\overline{L} \equiv \sum n_i L_i^2/\langle L\rangle$. If we know how stars are distributed in the HR diagram (i.e. know the luminosity function), we can trivially calculate $\overline{L}$. Preliminary work on this indicates that $\overline{L}_V = 60 L_\odot$ or $\overline{M}_V = +0.41$

There are many sources of fluctuation in a CCD picture which can interfere with our measurement of luminosity fluctuations. Fluctuations which originate local at the CCD include photon shot noise, readout noise, flatfield shot noise, flatfield error, cosmic rays, and CCD flaws. The amplitude of these noise terms grows as the square root of time, whereas the amplitude of luminosity fluctuations increases proportional to time, and it is easy to show that the amplitudes will be equal when the detector has collected 1 electron from a source of luminosity $\overline{L}$. This requires an exposure of a few minutes on a large telescope for a galaxy at a distance of 20 Mpc.

Other sources of fluctuations include foreground stars, background galaxies, globular clusters, and dwarf companions. Globular clusters and background galaxies (at distances greater than 10 Mpc) will dominate the fluctuation signal unless they are identified and excised. This can be done in 1" seeing to a sufficiently faint cutoff if the object in question is nearer than about 20 Mpc. Loose associations of stars, HII regions, and fluctuations from patchy obscuration in the galaxy can be important, but are not a problem if we restrict ourselves to early-type systems.

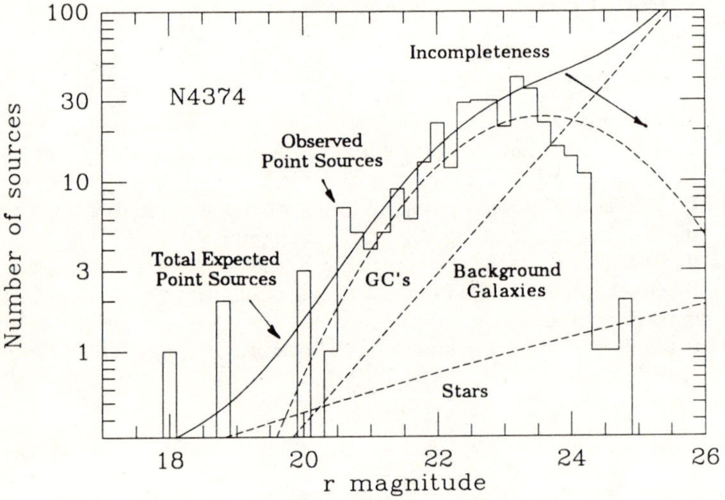

Fig. 1 The point sources found in an image of N4374. The observed sources are fitted by a model and the deficit at faint magnitudes measures the incompleteness. The fluctuation variance is corrected for residual variance from the sources that are missed.

In the absence of seeing, the variance from luminosity fluctuations is just the flux from a single source of luminosity $\overline{L}$ times the local surface brightness. We can recover this variance even when it is diminished by the effects of seeing by measuring the zero wavenumber limit of the spatial power spectrum, as shown in figure 2. This is easy to do in practice, and provides a nice separation of the noise terms local to the CCD (whose power spectrum is constant with wavenumber) and fluctuations from sources outside the atmosphere (whose power spectrum follows that of the point spread function).

We have observed a number of galaxies and measured the amplitude of their luminosity fluctuations. Preliminary distances include 0.83 Mpc for M32, 9.8 Mpc for N3379, and 17 Mpc for N4374, subject to a more definative calibration of $\overline{L}$. These observational results were obtained in mediocre seeing with a modest amount of telescope time, and demonstrate that the *apparent* $\overline{m}$ can be easily and accurately measured to distances of at least 20 Mpc. The foremost observational difficulty is identifying and removing faint galaxies and globular clusters from the image, so good seeing is critical. With subarcsecond images it may be possible to extend this method to about 40 Mpc.

Measurement of the apparent magnitude $\overline{m}$ cannot yield a distance until $\overline{M}$ is known, and $\overline{M}$ depends on the details of the stellar population. Thus we must seek to determine $\overline{M}$ from observables such as color, line strengths, and detailed spectra.

$\overline{M}$ brightens by about 35% when metallicity decreases by a factor of 2, and distance depends only on the square root of $\overline{M}$, so even a crude measure of $\overline{M}$ can give us useful distances. Of course, an absolute calibration of $\overline{M}$ from theoretical stellar populations would give us H_0, subject only to calibration of ZAMS. Even with the current large uncertainties, the observation of N4374 implies that $H_0 = 70 \pm 20$ km/s/Mpc.

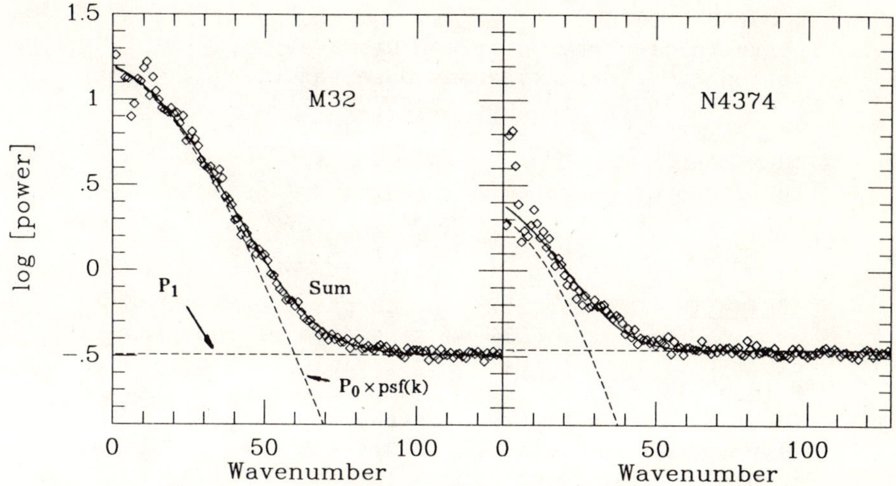

Fig. 2 The fluctuation signal for M32 and N4374. The variance is fitted by a constant P_0 times the psf power spectrum plus another constant P_1: $P(k) = P_0 \times psf(k) + P_1$.

If we know a galaxy's distance, $\overline{M}$ gives us a second moment of the luminosity function (average flux is the first moment). Measurement of $\overline{M}$ in different colors is a very useful diagnostic of stellar populations, when individual stars cannot be seen. For example, the blue fluctuation amplitude relative to the red will tell us the luminosity of the brightest blue stars relative to the red stars. Another use of these fluctuations is to distinguish populations in S0 bulges and disks, or to measure population gradients.

Other work is in progress to observe a sample of Virgo elliptical galaxies. This will allow us to calibrate empirically $\overline{M}$ as a function of color and line strength, and in terms of Virgo's distance. We can then measure *accurate* relative distances and peculiar velocities in the local supercluster and distances to other clusters.

These relative distances can be more accurate than 10% which is considerably more accurate than luminosity-velocity calibrations of spiral or elliptical galaxies. They are also more reliable because stellar populations surely are less sensitive to systematic environmental effects than galaxy structure. Thus we can hope to accurately measure the Virgo infall and any deviations from spherical symmetry imposed by a Great Attractor.

REFERENCES

Tonry, J.L., Schneider, D.P. 1988, *A.J.*, **96**, 000.

THE NGC 1399 GLOBULAR CLUSTER SYSTEM

DOUGLAS GEISLER
Cerro Tololo Inter-American Observatory,
National Optical Astronomy Observatories
P.O. Box 26732, Tucson, AZ 85726.

JUAN CARLOS FORTE
Instituto de Astronomía y Física del Espacio
Casilla de Correo 67, Suc. 28, Buenos Aires, Argentina.

ABSTRACT Deep Washington CCD photometry has been obtained with the CTIO 4m telescope of the globular cluster system of the central elliptical galaxy in the Fornax cluster, NGC 1399. The resulting luminosity function apparently reaches beyond the peak in the cluster distribution. Preliminary data indicate a distance modulus of 30.4±0.3. The metal abundance distribution of some 250 clusters is also derived.

INTRODUCTION

Extragalactic globular cluster systems offer unique opportunities to investigate the oldest stellar components of galaxies. They are also valuable as independent distance indicators. The globular cluster system of NGC 1399, the central elliptical galaxy in the Fornax cluster, is a particularly interesting and accessible one. The recent study by Hanes and Harris (1986) found a very high specific frequency of ~17, possibly higher than that for M87. It is also relatively close and therefore provides a very large sample of nearby clusters. Since it lies in the Local Supercluster anticenter direction, an independent distance determination is important for investigating the peculiar motion of the Local Group.

RESULTS

Washington integrated photometry of NGC 1399 globular clusters was obtained with the CTIO 4m and RCA PFCCD during 3 nights in December. Approximately 80% of the star-like images in the central 3'x5' frame are globular clusters in NGC 1399. This frame contains ~500 such clusters.

After standard removal of instrumental signature, the frames were further processed by removing the background galaxy light using various IRAF routines. The resulting images were then reduced using Stetson's DAOPHOT routines. The derived M-T_1 versus T_1 color-magnitude diagram shows a strong concentration of globulars between M-T_1 = 0.5 - 1.0. A luminosity function was determined for all objects falling in the above color range. Note that no effort has been made yet (but will be shortly) to correct for the background star or galaxy contribution, incompleteness or bad photometry. Nevertheless, the data suggest that the turnover in the globular cluster luminosity function has been reached, at $T_1 \sim 22.8 \pm 0.2$. This is more than a mag. brighter than the limiting mag. We can then equate this number to the peak absolute magnitude M_V of the globular cluster systems with known distances. Harris (1987) shows that this value equals -7.2 and is nearly idependent of parent galaxy luminosity. Converting T_1 into V and assuming a reddening of 0.03, we find $(m-M)_o \sim 30.4 \pm 0.3$. This number is quite preliminary and the error, if not the value, will be improved. Nevertheless, the result agrees closely with the distance derived by Caldwell and Bothun (1987) for the Fornax cluster. It is likely that the Fornax system is ~0.5 mag closer than Virgo.

Finally, the available colors can be used to investigate the metallicity distribution of the globular clusters. Harris and Canterna (1977) have shown that one can form an integrated, reddening-free color index, Q_{CMT}, from Washington photometry that is quite sensitive to metal abundance, as corroborated by Hesser and Shawl (1985). This index is tightly correlated with Zinn's (1985) metallicities for Galactic globular clusters. The NGC 1399 metallicities are derived from the C-M: M-T_1 color-color plot for the globulars brighter than 22.5. The clusters are found to have a metallicity distribution rather similar to that of M87, with a reasonable number of clusters exceeding the solar metallicity.

In the future, we will refine this analysis and add the observations of three surrounding galaxy fields and a comparison field. This will greatly improve the luminosity function derivation and allow abundances to be derived for a total of some 400 globular clusters.

REFERENCES

Caldwell, N., and Bothun, G.D. 1987, A.J., 94, 1126.
Hanes, D.A., and Harris, W.E. 1986, Ap. J., 309, 564.
Harris, W.E. 1987, Pub. A.S.P., 99, 1031.
Harris, H.C., and Canterna, R. 1977, A.J., 82, 798.
Hesser, J.E., and Shawl, S.J. 1985, Pub. A.S.P., 97, 465.
Zinn, R. 1985, Ap. J., 293, 424.

V

THE VIRGO CLUSTER

NOVAE AND SUPERNOVAE AS DISTANCE INDICATORS

SIDNEY VAN DEN BERGH
Dominion Astrophysical Observatory, National Research Council of Canada, 5071 W. Saanich Road, Victoria, B.C. V8X 4M6

ABSTRACT As calibrators of the extra-Galactic distance scale novae have a number of advantages over Cepheids. More observations are, however, required to determine the intrinsic dispersion in the maximum magnitude versus rate-of-decline relation for novae. Supernovae of type Ia are found to have $M_B(\max) = -18.45 \pm 0.2 + 5 \log h$ with an observed dispersion of 0.8 mag. Supernovae of type II appear to exhibit a relationship between $M_B(\max)$ and the intrinsic luminosity of their parent galaxy. For galaxies brighter than $M_B^{oi} \simeq -14.5 + 5 \log h$ the dispersion about this relationship is 0.9 mag. It is not yet known how much observational errors and uncertainties in the adopted absorption values contribute to the luminosity dispersions for SNIa and SNII quoted above. At maximum SNIb are fainter in blue light than SNIa. Too few SNIb are presently known to determine either their $\langle M(\max)\rangle$ or the dispersion about this value.

VAE AS DISTANCE INDICATORS

e following are reasons for believing that novae might be re suitable than Cepheids as calibrators of the extra-lactic distance scale:

1. At maximum light novae are ~3 mag brighter than pheids (see Fig. 1) and can therefore be observed out to gnificantly greater distances.

2. Novae occur in photometrically smooth regions such the nuclear bulges of spirals and in elliptical galaxies, ereas young long-period Cepheids are strongly concentrated spiral arms where crowding problems are severe.

3. Since dust and gas are concentrated in spiral arms sorption and reddening problems are severe for Cepheids. is contrasts with the situation for novae which can be served in low-absorption areas such as the nuclear bulges

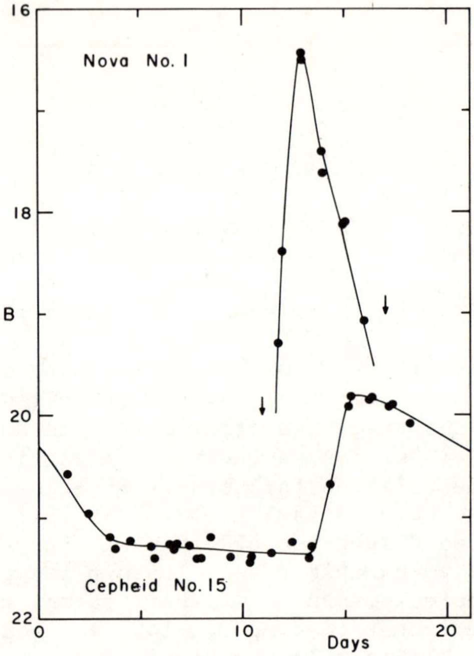

Fig. 1 Comparison between the lightcurve of Baade and Swope (1963) for Cepheid No. 15 (P = 21.26 days) and Arp (1956) nova No. 1 in M31.

of spirals and in elliptical and SO galaxies.

A number of problems do, however, remain with the use of novae as distance calibrators:

1. Observations of the maximum magnitude versus rate-of-decline (MMRD) relationship for novae in M31 (see Fig. 2) suggest (van den Bergh and Pritchet 1986) that a small subclass (comprising ~8% of all novae) is about a magnitude brighter than normal novae, which exhibit a dispersion of ~0.3 mag about the adopted M31 MMRD relation.

2. The MMRD relation for fast novae in M31 is not well defined by extant observations. This is due to the fact that few fast novae have so far been discovered in M31 and because the exact time of maximum may have been missed for some of them.

3. A systematic difference of 0.2 mag remains between the M31 MMRD relations derived from the photographic observations of Arp (1956) on the one hand and Rosino (1964, 1973) on the other. Part of this problem is, no doubt, due to the difficulty of doing photographic photometry of objects that are superposed on a bright background.

The most promising method to resolve these problems would appear to be the systematic study of novae with CCD

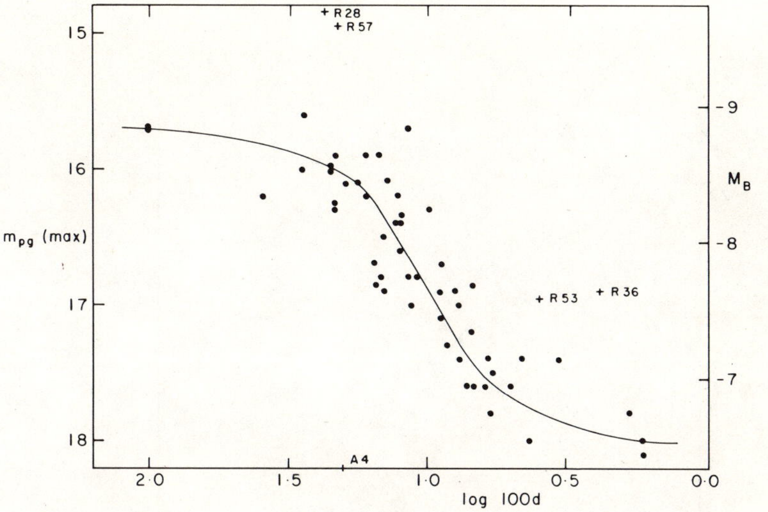

Fig. 2 Maximum magnitude versus rate-of-decline relationship for novae in M31 from photographic observations by Arp (1956) and Rosino (1964, 1973). In the figure d is the number of days a nova required to decline by 2.0 mag from maximum light.

detectors on a 1-m class telescope in objects such as the nuclear bulge of M81 and NGC5128. For accurate absolute calibration of the MMRD relation for novae we shall, however, have to continue to rely on the novae in M31.

Recently Pritchet and van den Bergh (1987) have used the MMRD relation for 7 novae in 4 bright Virgo elliptical galaxies (see Fig. 3) to derive a value for the distance modulus of the Virgo cluster. The main difficulty with this approach turned out to be that neither the maximum magnitudes nor the rates of decline of individual novae could be determined accurately enough from 14 nights of observation spread out over a 6-week period. Almost daily coverage of the lightcurves of novae will be required to overcome this problem. This places an almost unprecedented demand on scarce observing time on a 4-m class telescope at a site with excellent seeing and weather. In future it should be possible to employ a focal reducer, which would increase the area that can be observed with a single CCD exposure, thus increasing the number of novae that could be discovered in single exposures of Virgo ellipticals.

With 4-m class telescopes at sites that enjoy good seeing extra-bright novae, such as R28, R36, R53, and R57 in M31, could be observed as far away as the Coma cluster. It would therefore be very important to obtain CCD observations

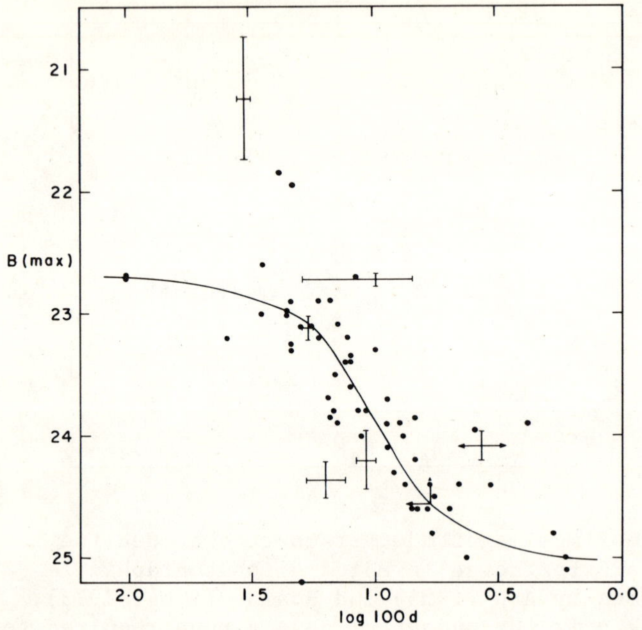

Fig. 3 Crosses show measurements of the maximum magnitudes and the rates of decline for 7 novae in the Virgo cluster. Also shown is the MMRD relation for M31 shifted by $\Delta B = 7.0$ magnitudes. The large scatter in the Virgo observations is mainly due to the difficulty in determining M(max), and the rate of decline d, from observations separated in time by a few days. The brightest Virgo nova might belong to the class of super-bright novae; 4 of which have also been observed in M31.

of such objects in M81 and NGC5128 to see if these extra bright novae define a well-behaved MMRD relation. It will also be interesting to see if Hα observations (Ford, this conference) can be used to establish a well-defined MMRD relation that can be used out to even larger distances.

LUMINOSITIES OF SUPERNOVAE

For some time now it has been realized (e.g. Panagia 1985) that there are at least 3 spectroscopically distinguishable types of supernovae viz. SNIa, SNIb and SNII. Supernovae of type Ia are believed to result from the deflagration of a CO white dwarf (Iben and Tutukov 1984), which has been pushed over the Chandrasekhar limit by mass transfer from a close companion. Supernovae of types Ib and II are thought to result from core collapse in massive main sequence stars. Van den Bergh (1988a) has presented arguments suggest that the progenitors of SNII have main sequence masses in the range $8 \lesssim M \lesssim 16\ M_\odot$, whereas the progenitors of SNIb probably have $M \gtrsim 16\ M_\odot$. Note, however, that the transition mass between SNIb and SNII might be sensitive to metallicity.

Luminosities of SNIa

A recent listing of supernovae which, on the basis of their spectra, have been assigned to the SNIa class is given by Branch, Drucker and Jeffrey (1988). The values of B(max), or in some cases m_{pg}(max), for those objects that occurred before January 1 1984, were taken from Barbon *et al.* 1984. The corresponding absolute magnitudes M_B(max) of those supernovae that occurred in Shapley-Ames galaxies were derived from the relation

$$M_B(\max) = M_B^{oi} + B(\max) - B_T , \qquad (1)$$

in which B_T and M_B^{oi} for each supernova parent galaxy was taken from Sandage and Tammann (1981). For 13 SNIa that occurred in Shapley-ames galaxies it is found that

$$\langle M_B \rangle = -18.45 \pm 0.23 + 5 \log h, \qquad (2)$$

in which h is the Hubble parameter in units of 100 km s^{-1} Mpc. The dispersion of individual supernovae about the adopted mean is $\sigma = 0.8$ mag. It is presently not known which fraction of the dispersion in the luminosity of SNIa is due to observational errors in B(max) and to the assumption that each individual supernova suffers the same absorption as its parent galaxy. The observation that SNIa are not all identical spectroscopically (Branch 1987, Branch, Drucker and Jeffreys 1988) or photometrically (Younger and van den Bergh 1985, Phillips *et al.* 1987, Frogel *et al.* 1987) suggests that at least some of the observed dispersion in the luminosities of SNIa is probably intrinsic.

Supernovae of Type Ib

A listing of supernovae, which on the basis of their spectra, have been assigned to type Ib has recently been given by Ensman and Woosley (1988). Magnitudes at maximum for three of these objects that occurred in Shapley-Ames galaxies are provided by Barbon *et al.* (1984): M_B(max) = −16.9 + 5 log h for 1962L, M_B(max) = −16.5 + 5 log h for SN 1964L and M_B(max) = −17.7: + 5 log h for the possible SNIb 1954A. These data are consistent with Panagia's (1985) conclusion that SNIb at maximum are up to 1.5 mag fainter in blue light than are SNIa. More observations will, however, be required to determine the intrinsic dispersion within the class of SNIb. Ensman and Woosley (1988) have recently suggested that some supernovae with very massive progenitors might suffer extra extinction due to dust formation in their recently ejected envelopes. As a result some SNIb might appear quite red and subluminous at maximum light.

Supernova of Type II

Fig. 4 shows a plot of M(max) versus M_B^{oi} for those SNII for which M(max) is given by Barbon *et al.* (1984). The distance moduli, absorption values, and parent galaxy luminosities were taken from Sandage and Tammann (1981). The figure appears to show that SNII in luminous galaxies are brighter than those that occur in fainter galaxies. Excluding SN1987A one obtains $r = 0.45 \pm 0.15$ for the correlation coefficient between M(max) and the absorption-free absolute magnitude of the parent galaxy M_B^{oi}. It might be argued that this is a pseudo-correlation that results from the fact that faint supernovae are difficult to observe in luminous galaxies. Two arguments miliate against this suggestion: (1) SNI (see Fig. 5) do not show a correlation between M(max) and M_B^{oi} and (2) Fig. 5 shows no bright supernovae in faint galaxies; even though such objects would be particularly easy to discover. It is therefore concluded that the observed correlation between M(max) and M_B^{oi} for SNII is probably real.

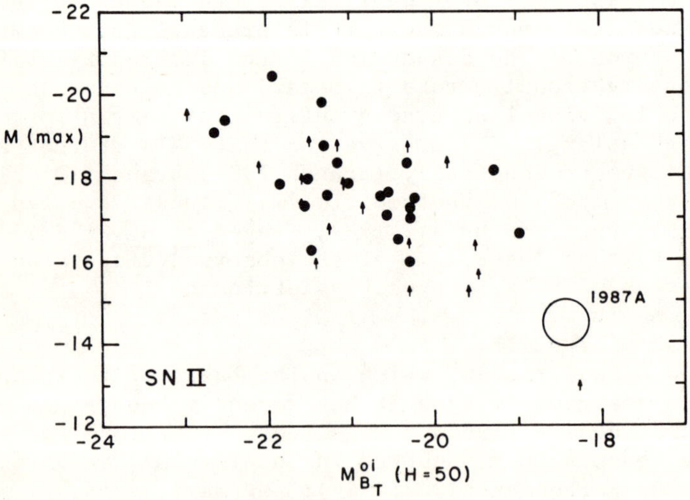

Fig. 4 Maximum magnitude of SNII versus parent galaxy luminosity. H = 50 km s^{-1} Mpc^{-1} assumed for all objects except SN1987A. Arrows indicate lower limits.

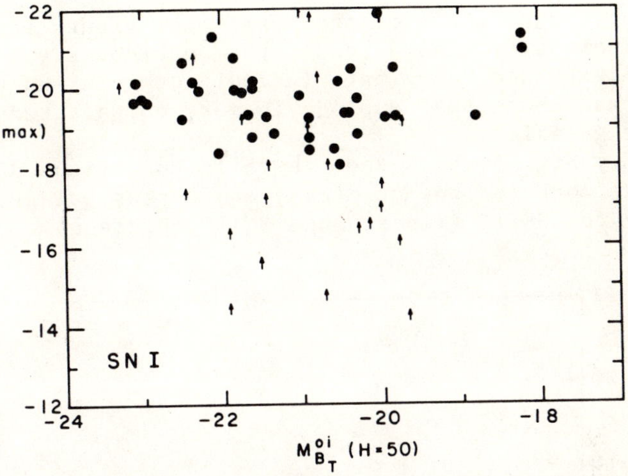

Fig. 5 Maximum magnitude of SNI versus parent galaxy luminosity. Note the absence of a correlation with parent galaxy luminosity.

Assuming that the M_B^{oi} values are error-free (and excluding SN1987A) one obtains the following relation between M(max) of SNII and the luminosities of their parent galaxies:

$$M(max) = \underset{\pm 0.2}{-16.2} + 5 \log h + \underset{\pm 0.2}{0.6} (M_B^{oi} + 19.5 - 5 \log h) \quad (3)$$

A more detailed discussion of this result will be given in van den Bergh (1988b). The observed dispersion of M(max) about this regression line is 0.9 mag. Uncertainties in the adopted absorption values and observational errors in M(max) undoubtedly contribute to the observed scatter. More photoelectrically determined light and color curves will be required to determine what the intrinsic dispersion is in the relationship between M(max) and parent galaxy luminosity.

The most plausible explanation for the observed dependence of M(max) on M_B^{oi} is that M(max) of SNII actually correlates with the metallicity of the supernova progenitor. It has been argued by Chevalier (1976), Trimble (1982) and Shklovskii (1984) that supernovae in compact progenitors will be less luminous than those that explode in giants with more extended photospheres. The reason for this is that the extra energy required to lift a compact stellar envelope will be derived from the radiation field of the exploding supernova. It has been known for many years (e.g. van den Bergh 1975, Faber 1977) that luminous galaxies are metal-rich compared to faint galaxies. One would therefore expect the photospheres of evolved giants in luminous (metal-rich) galaxies to be more extended than those of the metal-poor giants in galaxies of lower luminosity. As a result the supernovae exploding inside the extended red-giants in luminous galaxies should be brighter than those that explode in more compact red giants in fainter galaxies.

Finally Fig. 6 shows that there is no obvious difference between the correlations of M(max) and parent galaxy luminosity for SNIIL (linear) and SNIIP (plateau).

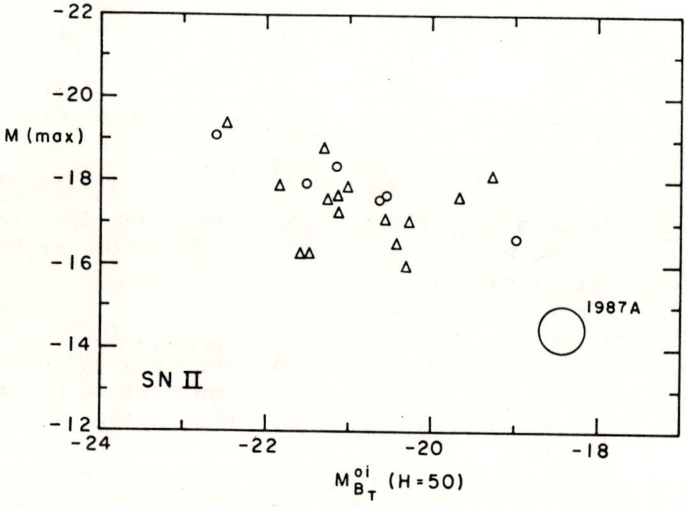

Fig. 6 Maximum magnitude of SNII versus parent galaxy luminosity for SNIIL (linear-circles) and SNIIP (plateau-triangles). No significant difference is seen between SNIIL and SNIIP.

REFERENCES

Arp, H.C. 1956, Astrophys. J., 61, 15.
Baade, W., and Swope, H.H. 1963, Astron. J., 68, 435.
Barbon, R., Cappellaro, E., Ciatti, F., Turatto, M., and Kowal, C.T. 1984, Astron. Astrophys. Suppl., 58, 735.
Branch, D., 1987, Astrophys. J. (Letters), 316, L81.
Branch, D., Drucker, W., and Jeffrey, D. 1988, preprint.
Chevalier, R.A. 1976, Astrophys. J., 208, 826.
Ensman, L.M., and Woosley, S.E. 1988, Astrophys. J., in press.
Faber, S.M. 1977, in The Evolution of Galaxies and Stellar Populations, ed. B.M. Tinsley and R.B. Larson (New Haven: Yale University Press), p. 157.
Frogel, J.A., Brook, G., Kawara, K., Laney, D., Phillips, M.M., Terndrup, D., Vrba, F., and Whitford, A.E. 1987, Astrophys. J. (Letters), 315, L129.

Iben, I., and Tutukov, A.V. 1984, Astrophys. J. Suppl., 54, 335.
Panagia, N. 1985, in Supernovae as Distance Indicators, ed. N. Bartel (Berlin: Springer), p. 14.
Phillips, M.M., et al. 1987, Pub. Astron. Soc. Pac., 99, 592.
Pritchet, C.J., and van den Bergh, S. 1987, Astrophys. J., 318, 507.
Rosino, L. 1964, Ann. Astrophys. 27, 498.
Rosino, L. 1973, Astron. Astrophys. Suppl., 9, 347.
Sandage, A., and Tammann, G.A. 1981, A Revised Shapley-Ames Catalog of Bright Galaxies (Washington: Carnegie Institution).
Shklovskii, I.S. 1984, Soviet Astron. Letters, 10, 302.
Trimble, V. 1982, Rev. Mod. Phys., 54, 1183.
van den Bergh, S. 1975, Ann. Rev. Astron. Astrophys., 13, 217.
van den Bergh, S., and Pritchet, C.J. 1986, Pub. Astron. Soc. Pac., 98, 110.
van den Bergh, S. 1988a, Astrophys. J., 327, 156.
van den Bergh, S. 1988b, Astron. J., in press.
Younger, P.F., and van den Bergh, S. 1985, Astron. Astrophys. Suppl., 61, 365.

DISCUSSION

A. MIKALOW: How many supernovae do we have CCD data on at this time?

S. VAN DEN BERGH: Very few.

J. GUNN: In the Virgo nova project, did you take into account the fact that if you miss the real maximum, the modulus is always biased in the same direction?

S. VAN DEN BERGH: Yes! Computer simulations show that this changes the difference in distance moduli of Virgo and M31 from $\Delta = 7.0 \pm 0.4$ to $\Delta = 6.8 \pm 0.4$.

D. LYNDEN-BELL: The radial dependence of metallicity within any galaxy should also show up as a dependence of supernova magnitude on position in the galaxy if your explanation is correct.

S. VAN DEN BERGH: I have searched for this effect and have not found it. Possibly the higher luminosity of metal-rich supernovae near the centers of galaxies is compensated for by high absorption in metal-rich dusty interstellar clouds.

J. COHEN: We need amateur organizations to monitor supernova light curves as the AAVSO does for novae.

Also an enormous amount of large telescope time has gone into searching for extragalactic Cepheids; the amount of time required to search for novae in Virgo is not that large in comparison. The problem is the need to have the frames properly spaced (i.e. consecutive nights).

GLOBULAR CLUSTER SYSTEMS AS DISTANCE INDICATORS

William E. Harris
Department of Physics, McMaster University

ABSTRACT: New CCD photometry of the globular cluster systems around three Virgo Cluster ellipticals (NGC 4365, 4472, 4649) has now been obtained to a limiting magnitude of $B = 26.2$. These data clearly show the luminosity function turnover (the peak frequency of the number of globular clusters versus luminosity) in these systems for the first time. In all three of these 'normal' ellipticals *and* in M87, the peak location occurs uniformly at $B_0 = 24.7 \pm 0.25$, with the same gaussian LF shape $\sigma \simeq 1.4$ mag. Matching directly to the LF for the Milky Way globular cluster system then yields a distance of (21.9 ± 2.2) Mpc for Virgo. The potential for applying the LF method to more distant systems and large E galaxies in particular is discussed.

I. INTRODUCTION

With the advent of CCD imaging, globular clusters can now be detected rather easily in the halos of large galaxies out to $\sim$ 100 Mpc. In this review, I will concentrate on just one aspect of globular cluster systems, namely their potential use as standard candles. After a brief review of the basic method and its history, I will summarize our present state of knowledge (particularly with regard to the many recent results that have been obtained for the keystone Virgo galaxies) and then suggest where the globular cluster distance calibrations may be leading in the near future. Broader reviews of the many aspects of globular cluster systems in general can be found in Harris and Racine (1979) and Harris (1988). An extremely comprehensive review which places these extragalactic systems within the context of current globular cluster research is given by Hesser (1988).

Globular clusters are very luminous stellar objects, and for galaxies more distant than $\sim$1 Mpc they provide virtually the only chance we have to calibrate the distance scale by a pure Population II route, which is based ultimately on the RR Lyrae stars. The basic idea of the method is extremely simple: in the Local Group galaxies that have well defined halo cluster populations (the Milky Way, M31, and the dwarf ellipticals), we see that globular clusters follow a distribution in absolute magnitude $n(M_B)$ that is clearly single-peaked and, to first order, can be quite well described by a simple gaussian curve with only two free parameters: the *centroid*, which I will denote as $M_B{}^0$, and the *standard deviation* σ; see Figure 1. In more distant galaxies, we attempt to measure the cluster distribution in apparent magnitude, $n(B)$, and then match up the peak point B_0 and shape of the distribution with the intrinsic one established in the Local Group. The required magnitude shift then gives the distance modulus directly.

From a purely practical viewpoint, this method has certain very strong advantages when put against other 'standard candles' of comparable range, such as supergiants, Cepheids, novae, supernovae, or disk rotation curves:
(a) Globular clusters are old halo objects, and thus can be selected to be almost totally free of the difficulties with reddening, inclination-angle corrections, metallicity variations and so forth that affect all Population I indicators;
(b) They are nonvariable stellar objects, and thus do not require long time-series of observations to determine periods, peak luminosities, etc.;
(c) Since the globular cluster luminosity function (GCLF) represents a pure Population II method, it is especially useful in early-type galaxies where clusters are seen in largest numbers, and thus can be a potentially powerful check on the numerous methods that apply principally to late-type galaxies;
(d) As will be seen below, the potential range of applicability of the method may be anywhere from 1 to ~50 Mpc *in a single step from the Milky Way*, even with ground based photometry.

No standard candle, however, is without its characteristic disadvantages too. The validity of the GCLF method rests on the operating assumption that globular clusters have similar luminosities everywhere, i.e. that they have a 'universal' LF. At some level of precision, this broad-brush assumption must be false. The whole issue of their usefulness then relies on just what that level of precision is. A more specific (and practical) form of the assumption is therefore that GCLFs will be useful standard candles if we learn how the parameters of the LF depend on galaxy size and type.

II. MOTIVATION FOR THE METHOD

The basic outline of the method may sound attractive, but what reasons do we have to believe that the key assumption is valid? On observational grounds, the pieces of evidence that have been assembled tend to fall into three different areas, and are at least partly circumstantial:
(a) The range of globular cluster colors and deduced metallicity distributions has repeatedly been found to be almost identical in an impressive variety of galaxies, on the basis of broadband photometric indices and low-dispersion spectral indices. This conclusion applies to all the Local Group members (van den Bergh 1969; Hanes 1977c; Harris and Racine 1979; Elson and Walterbos 1988; da Costa and Mould 1988), M87 and the other Virgo ellipticals (Hanes 1971; Ables et al. 1974; Strom et al. 1981; Forte et al. 1981; Cohen 1988; Racine et al. 1978; Hanes and Brodie 1986; Mould et al. 1987; Huchra and Brodie 1987), and the peculiar elliptical NGC 5128 (H.Harris et al. 1984, 1988). The only pieces of evidence that run counter to this trend (and that are undergoing reinvestigation!) are the $H\beta$ and CN anomalies in the M31 cluster spectra (Burstein et al. 1984), and the suggestion that clusters exist in NGC 5128 that are quite a bit redder than anything in the Milky Way system (H.Harris et al. 1988). However, these second-order anomalies do not seem to have affected the mean *luminosities* of the clusters in any obvious way.

(b) The *shape* of the GCLF (essentially, its intrinsic dispersion σ) is a distance-independent parameter whose similarity from one galaxy to another can be tested directly. For all galaxies with sufficient LF information, the dispersion consistently appears to be in the range ~1.2–1.5 mag (e.g. Hanes 1977b; Harris and Racine 1979; van den Bergh 1985; H.Harris et al. 1984; van den Bergh et al. 1985; Hanes and Whittaker 1987; Harris 1988; and see section IV below). Minor differences at the level of ± 0.2 in σ have been quoted from one galaxy to another (see the references cited), but it is entirely possible that these are the expected fluctuations from size-of-sample considerations. More will be said about this later.

(c) The turnover or mean cluster luminosity $M_B{}^0$ — perhaps the most critical parameter of the argument — has been shown to vary from galaxy to galaxy by no more than $\sim \pm 0.5$ mag over a factor of $\gtrsim 4000$ in parent galaxy luminosity, and over Hubble types that include dwarf ellipticals, large ellipticals, and the spheroids of large spirals (Harris 1987a). This conclusion relies only on the distance scale within the Local Group (which is assumed to be known to well within half a magnitude!), plus the very conservative assumption that the value of H_0 is in the range 80 ± 40 km/s/Mpc. To put it another way, if we use GCLFs as distance indicators with the very simplest assumptions, then we already deduce H_0 values that are fully within the range currently under debate. The whole purpose of pursuing this method is then to find out whether it can be sharpened from the rough $\sim \pm 0.5$ mag level of precision, to something approaching ± 0.2 mag.

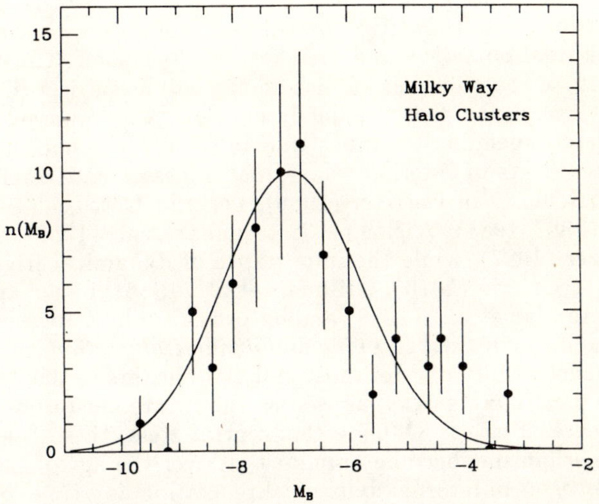

Fig. 1. The luminosity distribution for the globular clusters in the Milky Way halo. The number of clusters per 0.4-magnitude bin is plotted against absolute blue magnitude M_B. This sample includes only clusters with galactocentric distances larger than 4 kpc and $E_{B-V} < 0.6$ (see section V of the text).

Are there other reasons on theoretical grounds to believe in a universal LF, or some version of it? This question falls rather naturally into two parts: (1) Were globular clusters *formed* similarly in all galaxies? and (2) Has their *dynamical evolution* since the time of formation permitted them to have similar luminosities now, from one galaxy to another? Question (1) is an extremely challenging one, and may eventually require a far more sophisticated understanding of the physical conditions in protogalaxies than we now have. Fall and Rees (1985, 1988) postulate a sequence of events in which protoclusters form at a characteristic mass $\mathcal{M}_{cl}$ because the gas is at low metallicity and condenses out in pressure balance with the hotter, less dense halo gas; the dependence of $\mathcal{M}_{cl}$ on the galaxy mass then nearly cancels out its opposite dependence on the halo gas temperature. By contrast, Larson (1988) visualizes clusters forming as cores within much bigger ($\sim 10^8$–10^9 $M_{\odot}$) gas-rich starburst-like complexes, which could be assembled in arbitrary numbers to build larger galaxies. In either of these schemes, it is possible to argue that the characteristics of the globular clusters at formation can be almost independent of galaxy size. Of course, the argument becomes much easier if the clusters are *pregalactic* in origin (Peebles and Dicke 1968; Fall and Rees 1988; Rosenblatt et al. 1988). All of these models still contain arbitrary assumptions, and need to be made far more quantitative.

Question (2) – modeling the effect of dynamical evolution – has seen increasing activity recently, and has essentially developed from the classic Fall and Rees (1977) argument that the clusters we now see are survivors of several disruptive or limiting dynamic processes. The cluster has an overall *radius* set by the *tidal field* of the galaxy; and internal relaxation of the cluster stars, binary formation, and the drive toward core collapse cause *evaporation of stars* past the tidal boundary and shrinkage of the cluster. Clusters on highly eccentric orbits or those in a major disk galaxy will be subject to *tidal shocking* by passages through the disk or galactic nucleus. And *dynamical friction* will eliminate clusters spending too much time in the galactic nucleus region itself. Models of the effects of internal relaxation processes and tidal shocking are described particularly in Ostriker (1988), Lee and Ostriker (1987), Chernoff and Shapiro (1987, 1988), Wielen (1988), Aguilar et al. (1988), and the recent book of Spitzer (1987); while the importance of dynamical friction has been evaluated by Tremaine (1976), Grillmair et al. (1986), Lauer and Kormendy (1986), and Aguilar et al. (1988) among others. These investigations show that the twin effects of *tidal shocking and dynamical friction* from the galactic bulge should successfully disrupt most globular clusters in the inner ~1–2 kpc region of a typical large galaxy, as well as those rare clusters with extremely high orbital eccentricities. Outside this central zone, these "environmental" destruction mechanisms become unimportant over the age of the galaxy, and the slower processes of internal dynamical relaxation and core collapse (which drive cluster mass loss and eventual dissolution by coupling with the tidal field) will be dominant. Some very basic uncertainties in the model assumptions remain (for example, how steep is the initial mass function of star formation in globular clusters? Does a protocluster form already filling up its tidally bounded region, and what is its initial density profile? What distribution of orbital eccentricities is appropriate for the system of globular clusters?). These problems need to be answered by a better theory of formation (back to question

1!) Nevertheless, the available analyses strongly suggest that typical globular clusters well *outside* the nuclear region (i.e. with galactocentric distances $r_{gc} \gtrsim 3$–4 kpc) have had no difficulty surviving to the present time. That is, the array of destructive processes affecting the more fortunate clusters in the middle-to-outer halo is much less severe, and in addition much less sensitive to the details of the environment (i.e., the size and type of the parent galaxy). Their properties as seen at the present epoch are then determined primarily by their initial state (this is particularly evident in the synthesis of Aguilar et al.)

In summary, the theory of cluster formation and evolution does not yet seem to present any serious barrier to the pursuit of globular clusters as standard candles. In fact, there is visible promise that further developments in theoretical modeling might be able to tell us in much greater detail exactly why these old-halo clusters resemble each other so closely over an amazing range of galactic environments, and that their use as standard candles may eventually be placed on a firm theoretical foundation. The single most important warning to take from current models is that we should avoid using clusters close to galactic centers, where dynamical modifications to the GCLF are the strongest. Interestingly, in practice, the photometry of globular clusters in other large galaxies imposes almost exactly the same radial limit ($r_{gc} \lesssim 4$ kpc; see below), since the innermost clusters are prohibitively difficult to measure against the bright background of their parent galaxy's nuclear bulge. Thus we find an encouraging convergence of theoretical and observational limiting factors when we put the method into effect.

III. SOME HISTORY

The first globular cluster system outside the Local Group to be used as a distance indicator was the outstandingly populous one in M87, at the center of the Virgo System. At first, the assumption was made that the brightest visible ones (at the tip of the LF) were similar to the brightest clusters in the Milky Way or M31 (Baum 1955; Racine 1968; Sandage 1961, 1968; van den Bergh 1969), or that the brightest clusters could be matched after a correction for the cluster population size (de Vaucouleurs 1970; Hodge 1974). Eventually, these preliminary efforts were abandoned when it became clear that the luminosity level at the top of the LF was a statistical result of sample size, and that the brightest clusters alone are not usable in a consistent way unless one is comparing strictly similar populations, or unless the full shape of the LF is already known to much fainter levels. However, the brightest-cluster method might well experience something of a comeback when we press outward to extremely distant galaxies, particularly the supergiant ellipticals with the largest cluster populations (Harris 1983). Since the bright end of the GCLF does decrease quite steeply toward increasing luminosity, its level may still prove to be useful in a *rough* (± 0.5 mag) sense for such galaxies. A recent example of this is the discovery of globular clusters around the giant Coma galaxies at $\sim$ 100 Mpc (Harris 1987b; Thompson and Valdes 1987), where a self-consistent distance relative to Virgo was obtained by matching just the visible top $\sim$ 1.5 mag of the globular cluster LF to that of M87.

The modern approach of matching the LF over its entire observed range, and of eventually using the turnover point as the primary fiducial "marker", is first fully realized in the work of Hanes (1977a,b, 1979). Hanes obtained photographic photometry of the GCLFs in several of the largest Virgo galaxies to a photometric limit equivalent to $B_{lim} = 22.5$. By combining the five largest observed cluster populations (in the ellipticals NGC 4374, 4406, 4472, 4486, 4649), he obtained a mean Virgo GCLF which was then matched to the Milky Way LF to obtain $(m-M)_B$(Virgo) ~ 30.5. The match was done in two ways: first, by using a gaussian interpolation curve to predict the magnitude level of the turnover point in Virgo; and second, by directly matching the integrated LFs and adjusting the assumed distance for a best fit. More photographic data for the clusters around 4 Virgo ellipticals to $B_{lim} \sim 23$, and with similar conclusions, were presented by Strom et al. (1981) and Forte et al. (1981). Unfortunately, the technological limitations of the day meant (as will be seen below) that these photographic surveys fell 2 magnitudes or more short of the actual turnover position; this, coupled with inadequate knowledge of the correct intrinsic dispersion σ of the $\phi(M_B)$ distribution, meant that tight constraints on the LF fit could not be made. Other interesting attempts to use photographic starcount data for rough GCLF distance estimates can be found in Strom et al. (1977), Harris and Racine (1979), van den Bergh and Harris (1982), Harris et al. (1983, 1984), de Vaucouleurs (1983), and Hanes and Harris (1986); all of these now have only academic interest in today's framework of CCD photometry.

The Virgo System holds almost the same importance for the extragalactic distance scale as does the Hyades for stellar astronomy. For the GCLF method in particular, it offers the chance to compare globular clusters in a wide variety of galaxies all at essentially the same distance, and thus to make several crucial consistency tests of the method. In the early 1980's, with the advent of CCD imaging techniques and their far fainter limits, the search could be mounted in earnest to find the true shape of the GCLF in the Virgo galaxies, and especially the magnitude level of the turnover.

The first new indications that a turnover might have been reached, or at least clearly approached, appeared in the LF for M87 by van den Bergh et al. (1985) and Grillmair et al. (1986), from CCD photometry with B_{lim} =25.4 obtained at the CFH telescope. (These authors also published data for the much closer Leo system NGC 3379 which clearly do pass the turnover, but the cluster population in this galaxy is small enough and the radial velocity corrections for Leo are uncertain enough that its applications to the distance scale are minor; see Pritchet et al. 1985.) Again by fitting a gaussian interpolation curve to their observed LF, they derived B_0(M87) $\simeq 25.0$ and then $(m-M)_0$=31.4 by matching to the combined M31/Milky Way globular cluster sample. Still more recently, Cohen (1988) has published GCLFs in the Gunn g magnitude for larger samples of clusters around three Virgo E's (NGC 4406, 4472, and M87) to a similar photometric limit (B(equivalent) ~ 25). At least to the magnitude level for which the photometric image detection is virtually complete, Cohen's data give very much the same results for the deduced turnover magnitude and LF shape as does the van den Bergh et al. sample (see Figure 2).

These first CCD surveys demonstrate beyond serious doubt that the LFs in the large Virgo ellipticals clearly flatten out past $B \sim 24$. A nagging worry is that the actual identification of a turnover relies on just the very last magnitude bin or two for which the counts are seen to drop, but for which completeness and background corrections are of course the largest. This point has been driven home in an independent analysis of the M87 LF data by Hanes and Whittaker (1987), who show that the deduced turnover magnitude B_0 depends significantly on the assumed dispersion σ of the gaussian interpolation curve. Through simulations of GCLFs, Hanes and Whittaker also demonstrate that a *systematically correct* value for the turnover level *can* be deduced even from data that fall surprisingly far short of the turnover, as long as the correct σ is used (this point will be revisited in section VII below).

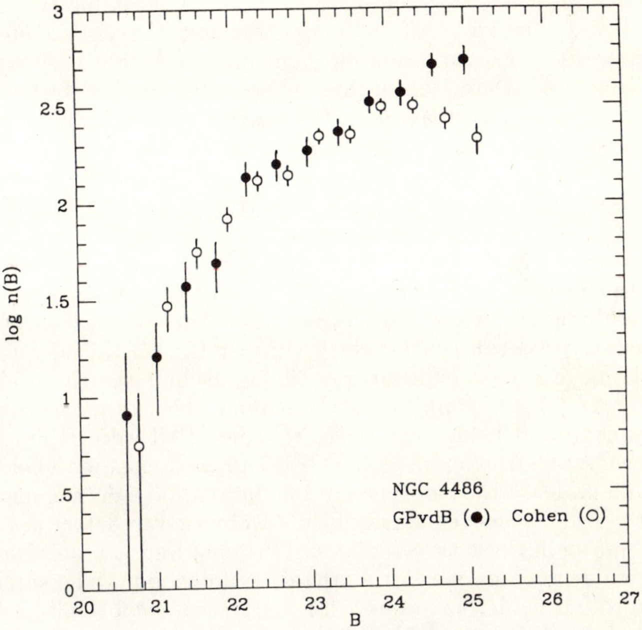

Fig. 2. The GCLF for the supergiant M87 system, as measured with CCD photometry by Grillmair et al. (1986) and Cohen (1988, transformed from g to B). Here $n(B)$ is the number of globular clusters per 0.4-mag bin; the two datasets have been normalized vertically for the same total population brighter than $B = 24$. For $B > 24$, the systematic deviation of Cohen's data may be due to problems with completeness corrections and/or background subtraction in the very crowded M87 field; see the papers cited for further discussion. (*NB:* The GPvdB data are plotted differently here than in Fig. 4 below; the bin boundaries are shifted by 0.2 mag in the later figure.)

Underlying all these concerns about the M87/Local Group GCLF fit is, of course, the still more basic worry that we have been matching the globular cluster populations in two rather different galaxy types (the disk galaxies of the Local Group versus a large Virgo elliptical). Furthermore, M87 is not just a giant elliptical, but an *anomalous* one: its globular cluster population is 2–3 times more numerous than the average Virgo elliptical (see Harris 1988 for a review of the specific frequencies of cluster systems). Though the existing photometry (Hanes 1977a, Cohen 1988) shows that the major Virgo E's have rather similar GCLFs at the bright end, they do not demonstrate whether or not the two key parameters of the LF (B_0, σ) are sensibly identical among all these galaxies

Searches for radial variations in the LF *within* one galaxy (which would be the visible result of the dynamical effects summarized above) have been carried out by Lauer and Kormendy (1986), Grillmair et al. (1986), and Cohen (1988). All found no significant differences in LF shape with galactocentric position within these large ellipticals. Thus at least for the clusters on the brighter (more massive) half of the distribution, the visible effects of any formation processes or destruction mechanisms that might vary with radius have not yet been seen. Detection of these effects may require deeper (and far more time-consuming) CCD imaging in the central regions of these galaxies.

IV. A NEW SURVEY OF GCLF'S IN THREE VIRGO ELLIPTICALS

In the spring of 1986, an opportunity arose to obtain new photometry around some of the Virgo galaxies that would – it was hoped – reach deep enough to show unambiguously where the long-sought 'turnover' point of the LF really was. Parts of two consecutive dark runs at the CFHT were used to obtain a long string of repeat exposures on fields around the three ellipticals NGC 4365, 4472, and 4649. Simultaneously, comparable exposures were obtained for 'blank' background fields near these galaxies. (All were B magnitude images with the RCA1 CCD camera at the CFHT prime focus, with field size 2.1×3.5 arcmin and scale 0.41 arcsec/pixel.) The individual time sequence exposures – one hour per night on each galaxy field – were used to detect and measure light curves of novae in these three galaxies (Pritchet and van den Bergh 1987b; see also the review by van den Bergh at this symposium). The same images, plus the background fields, have now also been used to obtain the GCLFs in the same galaxies (W.Harris et al. 1988).

The three galaxies just mentioned were deliberately chosen to be among the most 'normal' of the large Virgo ellipticals, to avoid any anomalies connected with specific frequencies, special locations, Virgo membership uncertainties, or other individual peculiarities. Thus their measured GCLFs should provide a strong test of two specific questions: (a) are the GCLFs in *large, normal elliptical galaxies* similar to one another?, and (b) are they similar in turn to the GCLF in an *anomalous giant elliptical* (M87)? Though both questions are only special cases of the the broader hypothesis that GCLFs are 'universal', they are especially important ones: the use of globular clusters as standard candles well beyond the Virgo distance will rely almost entirely on such galaxies.

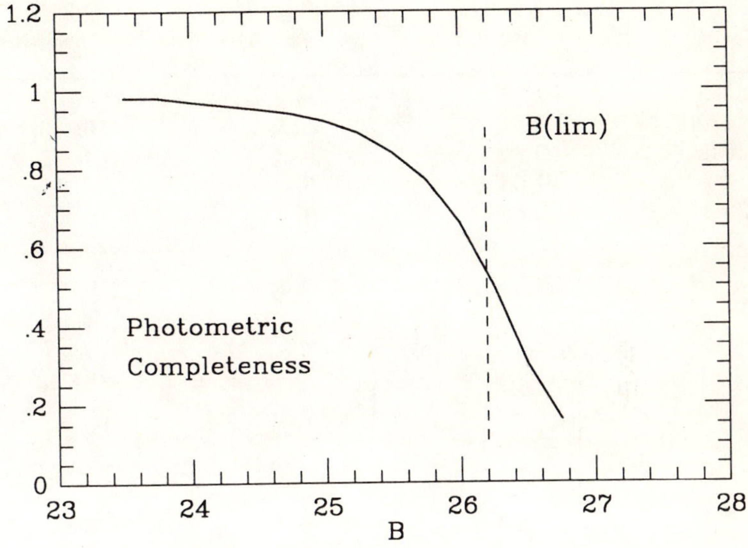

Fig. 3. The mean completeness of detection $f(B)$, for the new CFHT photometry in the Virgo elliptical fields (Harris et al. 1988). The curve shown is the average over the six program fields, although the individual frames all had similar limits (see Table I). This function is determined by repeated artificial-star experiments on each field, through the usual DAOPHOT procedures (Stetson 1987). The typical uncertainty in f at any B is near ± 0.05. The practical limit of the photometry is set at $B \simeq 26.2$, where $f \sim 0.5$.

Only a brief summary of these new observations will be given here; full descriptions are given by Pritchet and van den Bergh (1987b) and W.Harris et al. (1988). Our photometry of the cluster populations in the three Virgo fields was done first by 'stacking' (shifting-and-adding) the many nights' worth of CCD frames to construct one coadded image of each field equivalent to a single exposure of many hours' duration. Each composite picture was median-filtered to remove the bulk of the background light gradient from the central galaxy itself, thus permitting photometry of stellar images to be carried inward to ~25 arcsec of the galaxy center on a globally flattened sky background (cf., for example, Harris 1987b or Lauer and Kormendy 1986 for similar applications). All subsequent photometry of the frames was completed through Stetson's (1987) DAOPHOT code with the normal three-pass FIND, PSF, and NSTAR operations. The characteristics of the final pictures (two fields in N4472, one each in N4365 and 4649, and two adjacent background fields) are summarized in Table I.

TABLE I New CFHT Observations for Virgo

FIELD	FWHM (arcsec)	B_{lim} (f=0.5)	AREA (arcmin^2)
N4365	1.03	26.2	5.75
N4472(E,W)	0.91	26.1	11.90(E+W)
N4649	0.86	26.3	6.37
B1	1.03	26.4	7.08
B2	0.84	26.3	7.18

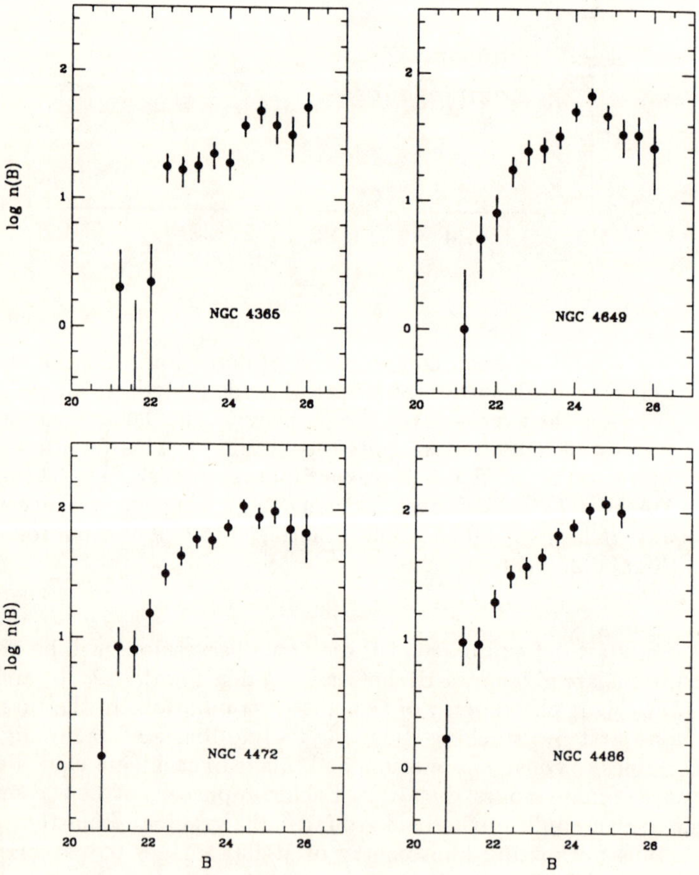

Fig. 4. CFHT CCD luminosity functions for the globular clusters around four giant Virgo ellipticals. As before, $n(B)$ is the number of clusters per 0.4-mag interval, after correction for incompleteness and subtraction of background.

During the data reduction, considerable attention was paid to deriving the detection completeness of the photometry as a function of magnitude, $f(B)$, since this is crucial to deriving the observed LF. For every program field, numerous groups of artificial stars were generated through DAOPHOT's ADDSTAR function and the simulated pictures re-reduced in the usual way (see, e.g., Stetson 1987; Hesser et al. 1987). A representative $f(B)$ curve is shown in Figure 3. The limiting magnitudes quoted in Table I correspond to completeness fractions (for starlike images) of 50%; since f itself becomes increasingly uncertain below the 50% level, our final LF data are terminated at $B = 26.2$ in what follows.

Finally, image classification was used to find and eliminate nonstellar images from the sample. Because globular clusters at the Virgo distance are completely starlike objects at any level of ground-based seeing, eliminating nonstellar images on an individual basis by a set of automatic, unbiased criteria then considerably increases the S/N of the sample (e.g. Thompson and Valdes 1987). To do this, we calculated two radial image moments, an 'isophotal' magnitude, and a roundness parameter for every image (similar to those used by Tyson and Jarvis 1979 and Kron 1980), and employed plots of these to classify image types. For $B \lesssim 25$, the effect of this procedure was to reduce the background (of *starlike* images) to less than 20% of the globular cluster population in each of the Virgo galaxy fields. The background exceeds the cluster population only in the last half-magnitude of the photometry, where the image-classification parameters become very uncertain.

The luminosity functions for four Virgo ellipticals (the three with new data as just described, plus the M87 LF from Grillmair et al. 1986) are displayed in Figure 4. The data points represent the remaining population of globular clusters around each galaxy, after correction for photometric completeness and background; the region closer than r = 40 arcsec (or r $\simeq$ 4 kpc) from the center of each galaxy was excluded from the final LF data. The new data reach a magnitude deeper than any previous material, and for N4472 and N4649 especially the conclusion seems clear that the 'turnover' has been passed. The conclusion is less clear for N4365, which has the smallest cluster population and poorest statistics of the four; close inspection of the frame suggests that the 'high' faintest bin may be due to a local enhancement of background objects, but without superior data nothing more definite can be said.

Figure 5 shows all four distributions, now plotted on the same graph and normalized to the same total population (the curves all cross near the turnover). No differences are evident within the internal error bars of each curve; note particularly the close agreement between M87 and N4472, the two largest and most statistically significant samples. Comparison with Cohen's (1988) data for N4472 also supports this conclusion. Given that the actual distances of the individual galaxies may differ by up to $\sim \pm 0.2$ mag because of the line-of-sight depth of the central Virgo Cluster ($\pm 10\%$, corresponding to the 'six degree circle' which conventionally defines the Virgo core; de Vaucouleurs and de Vaucouleurs 1973), our conclusion from Figure 5 is that there is no evidence at all for any fundamental LF differences among these galaxies.

TABLE II GCLF Parameters for Virgo Ellipticals

	B_0	σ	$B_0(\sigma \equiv 1.4)$
N4365	25.31 ± 0.42	1.50 ± 0.22	25.14 ± 0.18
N4406	(25.19 ± 1.16	1.13 ± 0.61)	(25.2:)
N4472	24.79 ± 0.22	1.50 ± 0.13	24.64 ± 0.11
N4486	25.29 ± 0.42	1.60 ± 0.17	24.99 ± 0.12
N4649	24.46 ± 0.17	1.26 ± 0.12	24.61 ± 0.13
Mean	24.70 ± 0.16	1.42 ± 0.13	24.74 ± 0.08

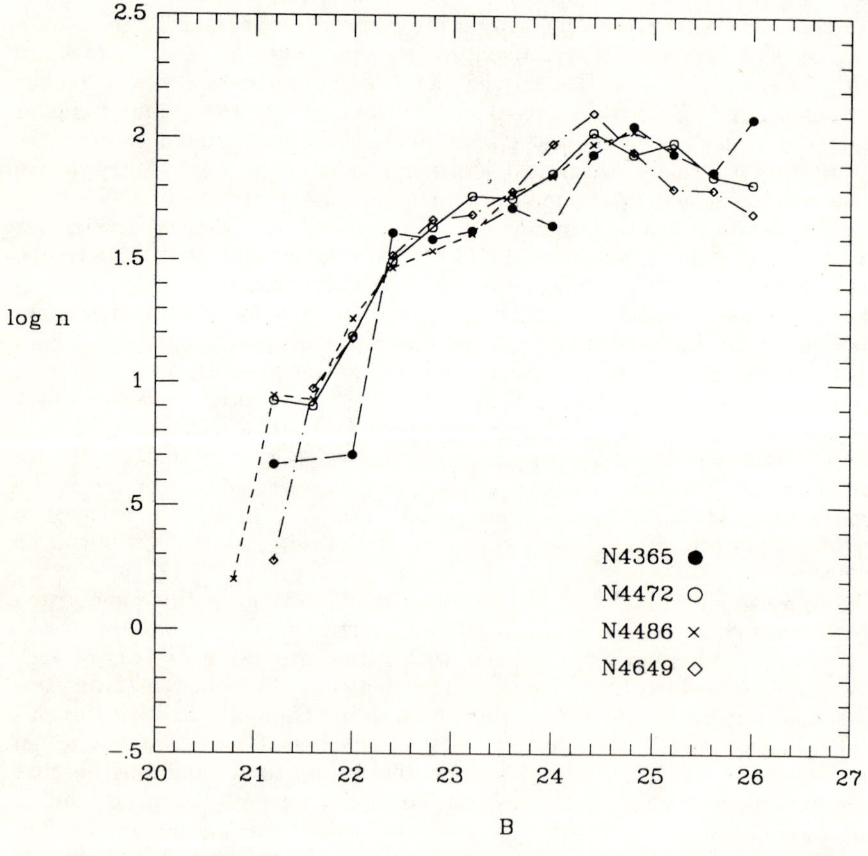

Fig. 5. The same GCLFs plotted in Fig. 4, now superimposed and normalized to the same amplitude at the 'turnover' position ($B = 24.7$).

A more quantitative comparison is summarized in Table II, which contains the best-fit parameters of gaussian interpolation curves fitted to each LF. Specifically, by nonlinear weighted least squares we fit to the data a function of the form $n(B) = A \exp[-(B - B_0)^2/2\sigma^2]$, where B_0 (the apparent magnitude of the turnover), σ (the intrinsic dispersion of the LF), and the scale factor A are all left as free parameters of the fit. In Table II, Cohen's (1988) LF for N4406 is also included for completeness; even though it is quite sparse, it does show a strong peak at the same level as the other four. The average values of B_0 and σ are given at the end of the table (note that these are weighted means, since the individual values vary quite a bit in precision). The final column of Table II shows what we obtain for the turnover level if we perform restricted 2-parameter curve fits, where now the dispersion σ is required to take on its mean value of 1.4 and only B_0 and A are allowed to float. Compare the internal errors of fit for the two B_0 columns especially.

These new results support the inference made by van den Bergh et al. and by Cohen, that the level of the turnover in Virgo is just short of $B = 25$ and that the intrinsic dispersion of the LF is near 1.4 magnitudes. (It is worth noting that, *at the turnover level*, the new data described here have a photometric completeness fraction of $\sim$ 95% (cf. Figure 3), the internal random uncertainty of the photometry is $e(B) \simeq 0.06$ mag, and the background comprises typically one-third of the total measured sample of images.) Within the errors of estimation, there are no significant trends in B_0 or σ with galaxy size, total population, or location. For the 3–parameter fit, the galaxy-to-galaxy differences in B_0 have a (weighted) scatter of ± 0.31 mag s.d.; for the restricted 2–parameter fit, the scatter is ± 0.24 mag. Subtracting out the scatter of $\simeq \pm 0.2$ mag expected from individual differences in distance (see above), we arrive at the implication that *the intrinsic scatter in* $M_B{}^0$ *among large elliptical galaxies is at the level of* ± 0.2 *magnitudes, and the LF dispersion* σ *is not sensibly different from one galaxy to another.* This statement by itself is already remarkable, since it specifically includes the anomalous M87 cluster system with the other, quite normal, ellipticals. When we add this new LF information to the strong similarities in cluster color indices (Cohen 1988) and in their spatial distributions around the parent galaxies (Harris 1986), *the globular clusters in M87 are indistinguishable from those in the other Virgo ellipticals in all known respects except for sheer numbers.* This will be an interesting challenge for formation models of galactic halos to reproduce.

V. A GLOBULAR CLUSTER DISTANCE FOR THE VIRGO SYSTEM

We would now like to ask whether the new Virgo data can be successfully matched to the "calibrating" Local Group GCLF. Since the method must assume that $M_B{}^0$ and σ are intrinsically similar in both systems, we have now (thankfully!) lost virtually any freedom to adjust the matchup in arbitrary ways. By hypothesis, we are required to take our calibrating function $n(M_B)$, move it faintward by an amount $(B_0 - M_B{}^0)$ (having now *directly observed* B_0), and then scale it to normalize the total populations in the two samples. The two LFs must then match satisfactorily over the entire observed ranges of both, if the full hypothesis is correct.

Our starting point for the calibrating function $n(M_B)$ is the Milky Way globular cluster system (Figure 1). The necessary data for the cluster integrated V magnitudes, $B-V$ colors, reddenings, and distances are taken from the catalogs of Webbink (1985), Peterson (1986), and Peterson and Reed (1987), augmented by many other recent references for individual clusters. In calculating the integrated M_B luminosities for the clusters, I have assumed $M_V(RR) = 0.6$ for the absolute magnitude level of the horizontal branch, $R_0 = 8$ kpc for the Milky Way distance scale, and (for the reasons outlined previously) have rejected clusters with galactocentric distances less than 4 kpc (a few additional objects with $E_{B-V} > 0.6$ and very uncertain data were also dropped from the list). This selection procedure essentially treats the Milky Way sample as if it were seen from the outside, making it more strictly comparable with the Virgo material. The resulting sample contains 72 clusters brighter than M_B = -4. Least-squares fits of gaussian curves to this distribution in the same manner as described above give $M_B{}^0$ = -7.0 ± 0.15, regardless of the assumed σ (since there are about as many faint clusters as bright ones, any symmetric interpolation curve will yield very much the same peak point). [NB: Although the inner-halo clusters have been deliberately rejected, it is interesting to note that including them would have no effect on the LF peak position: recent suggestions that the LFs of the inner- and outer-halo samples were systematically different (Zinn 1985, Hesser and Shawl 1985) have now disappeared with improvements in the fundamental cluster data (Armandroff 1988).]

TABLE III GCLF Parameters for Local Group Galaxies

	# Clusters	$M_B{}^0$	σ
Milky Way	72	-7.00 ± 0.15	1.2 ± 0.2:
4 dE's combined*	21	-6.40 ± 0.70	1.3 ± 0.7
M31*	104	-6.85 ± 0.11	1.0 ± 0.2:

* $(m-M)_V = 24.6$ assumed (Pritchet and van den Bergh 1987a)

In principle, any other globular cluster systems within the Local Group galaxies might also be put in to the calibrating function as long as we can assume that the Local Group distance scale is far more secure than the Virgo scale. Table III summarizes the parameters for:
(a) the Milky Way sample as just described;
(b) the clusters in the four dwarf ellipticals Fornax, NGC 147, 185, and 205, with data from Harris and Racine 1979 and Webbink 1985;
(c) all known clusters in the *outer halo* of M31, i.e. ones lying outside the M31 disk. These objects should not be subject to systematic errors due to internal reddening corrections, which can become quite large for clusters seen in projection on the disk. Data are from a merged catalog of existing M31 photometric surveys compiled by Harris and Allwright 1988; see also Battistini et al. 1987 and Elson and Walterbos 1988.

The late-type galaxies in the Local Group (M33, LMC, SMC) are not included because the old globular clusters are evidently quite rare in galaxies of type Sc or later, and they become extremely difficult to separate unambiguously from other types of clusters in these galaxies.

Which of these galaxies should go into the calibration function? Whether or not the few dE clusters are added has no important effect. However, a decision on the M31 system is much more critical, since adding them would more than double the calibration sample size. The available literature on the M31 system (see especially Crampton et al. 1985; van den Bergh 1985; Battistini et al. 1987; Elson and Walterbos 1988) gives two reasons *not* to include them that I find rather compelling. *First* is the quality of data: the fainter M31 globular cluster candidates are subject to dismayingly large internal photometric errors, since most of them have been measured only by Schmidt photographic photometry; and (even worse) they contain a significant but not clearly known degree of contamination by other types of objects (primarily background elliptical galaxies and old open clusters). These problems begin very near the luminosity level of the turnover. Thus it is not yet possible to establish the GCLF for M31 from a 'clean' database extending clearly past $M_B{}^0$; this is a frustrating situation that richly deserves attention. *Second* is the distance of M31 itself, which is not generally settled at less than ±0.2 mag (see, for example, Pritchet and van den Bergh 1987a or Sandage and Tammann 1988 for good indications of the current level of agreement). Thus the statistical gain in sample size for $n(M_B)$ would be more than offset by the loss of external accuracy in $M_B{}^0$.

For the present time, I have chosen to define the calibrating function as *the GCLF for the Milky Way alone.* Our fit to the Virgo $n(B)$ distribution, then, relies on a single jump outward from the Galaxy. The apparent distance modulus is $(m - M)_B = (24.7 \pm 0.16) - (-7.0 \pm 0.15) = 31.7 \pm 0.22$, and the complete matchup of the two LFs is shown graphically in Figure 6. The functional fit is entirely satisfactory over the ~4 magnitude range of overlap, either to the eye or to the usual statistical tests. If correct, this comparison would suggest that the brightest clusters in the giant ellipticals are a bit more than 1 magnitude more luminous than the brightest seen in the Milky Way (ω Cen). On grounds of self-consistency, this does not seem implausible, since the composite Virgo data contain ~30 times as many clusters as the calibrating distribution.

To complete the distance calculation and a formal derivation of H_0, I adopt A_B(Virgo) $\simeq 0.0 \pm 0.1$, and a mean Virgo radial velocity corrected for infall of $V = 1336 \pm 54$ km/s; these parameters are chosen to make the result strictly comparable with the nova method of Pritchet and van den Bergh (1987b). These then give d(Virgo) $= 21.9 \pm 2.2$ Mpc, and $H_0 = 61 \pm 7$ km/s/Mpc. There are several second-order assumptions in this calculation which can be adjusted quite plausibly: for example, if the Virgocentric infall were as low as ~200 km/s (Sandage and Tammann 1988), then we would obtain $H_0 \simeq 55$. Conversely, if A_B(Virgo) $\simeq 0.1$ and $M_V(RR)$ is actually ~0.7 or fainter (see below), then we would obtain $H_0 \simeq 70$. In any case, it should be stressed that this calculation *bypasses any uncertainties in the M31/Local Group distance scale.*

VI. THE ERROR BUDGET: UNCERTAINTIES

The true (external) uncertainty in d(Virgo) is, as usual, determined by the operating assumptions. At this point, there are just two important stages where it seems most likely that systematic errors could enter the method:
(a) The luminosities of the Galactic globular clusters are calculated with a given RR Lyrae distance scale (adopted here to be $M_V(RR) = 0.6$).
(b) The step to Virgo is taken assuming that the GCLF turnover luminosity for the Milky Way is strictly similar to those in the large ellipticals.

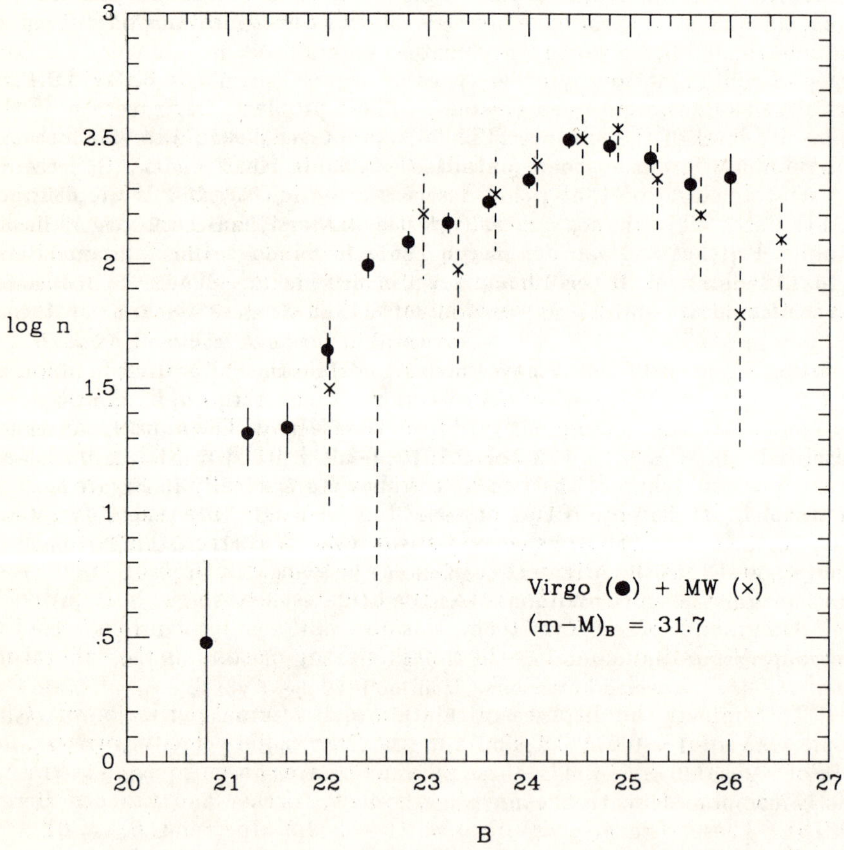

Fig. 6. The GCLF for the Milky Way (from Fig. 1) matched to the composite Virgo GCLF (from Fig. 4, with all 4 galaxies combined). The adopted distance modulus is $(m - M)_B = 31.7$, and the Milky Way LF has been shifted vertically by $\Delta \log n(B) \simeq +1.5$ to superimpose the curves at the turnover.

Assumption (a) – the RR Lyrae luminosity scale – continues to be a matter of active debate, but current observations have progressed to the point that major surprises (i.e. errors at the half-magnitude level) may no longer be lurking. The value I have adopted here is essentially the mean of the horizontal-branch luminosities deduced from recent color-magnitude studies for nearby globular clusters, from either isochrone fitting or subdwarf parallaxes (see Hesser 1988 for an excellent review of the many recent papers in this area). These studies currently give $M_V(HB) = 0.6 \pm 0.2$, only weakly dependent on cluster metallicity. By contrast, recent determinations of absolute magnitudes of the field RR Lyraes from statistical parallax (Strugnell et al. 1986; Barnes and Hawley 1986) give $M_V \simeq 0.7$, while the Baade-Wesselink method (Jones et al. 1987) now gives $M_V \simeq 0.9$. Notice that *random* cluster-to-cluster errors of ± 0.2 mag that might be generated by, e.g., incorrect reddening estimates or a variation of $M_V(HB)$ with metallicity, have no important effect on the $n(M_B)$ function since its intrinsic dispersion σ is far larger; but any adjustment in the adopted RR Lyrae luminosity will translate directly into a *systematic* shift of the $n(M_B)$ curve and on outward from there. If indeed $M_V(RR)$ is as faint as 0.9, then d(Virgo) would become 19 ± 2 Mpc.

Assumption (b) is the one major item that still needs direct testing at the ± 0.2 mag level, and which would make the GCLF a fully mature distance determination technique. The analysis described above demonstrates that the GCLFs in normal, large disk galaxies (M31 and the Milky Way) have a similar zero point to ± 0.2 mag, and the GCLFs in the Virgo ellipticals appear similar to one another to the same precision. I want to stress that it is not necessary to prove that the GCLFs in these two types of galaxies are *identical* in luminosity level; instead, what we ultimately need to know is the value ΔM_B of the "transfer correction" between the two types of LFs. With the numerous consistency tests that we now have on hand, my guess is that this correction factor will prove to be well within the ± 0.5 mag range that has already been established from elementary arguments (section II).

The deduced Virgo distance scale here clearly favors a 'low' H_0, and is larger than any previous globular-cluster-based distance determination. The difference between this new estimate and the slightly smaller ones derived by van den Bergh et al. and by Cohen is due primarily to changes in the Milky Way calibrating function and $M_B{}^0$. This, in turn, has been the result of many revisions and improvements in the fundamental data for individual globular clusters in the past few years. On the other hand, the difference between the present conclusion and the $\gtrsim 1$ mag lower values published in the last decade by Hanes (1977b, 1979) or Harris and Racine (1979) are due almost entirely to the adopted σ for the fitting function and to the large extrapolation that was necessary at that time to predict the turnover level. One lesson that has been learned from the new, much deeper Virgo photometry is that the 'correct' σ is much closer to 1.4 than it is to the $\sim$1.1–1.2 range that had usually been used before 1980. It has now become clear that the Local Group GCLFs cannot be relied upon to *define* the intrinsic dispersion of the hypothesized 'universal LF', simply because they are based on too small a number of objects. By example, a little LF simulation work shows that the addition of even one or two more clusters in the outermost bins of the Milky Way distribution will immediately raise the formally deduced σ from 1.2 (Table III) to $\sim$1.4. This vagueness inherent in the estimated dispersion of a sparse sample is the explanation for the apparent paradox noted by van den Bergh et al. (1985) and Hanes and Whittaker (1987): that the Virgo and Local Group GCLFs have formally different σ's (Tables II,III) and yet fit perfectly well when matched at their peak points (Figure 6).

By contrast, the centroid of the distribution, $M_B{}^0$, is a far more stable quantity. Its uncertainty (± 0.15) is exactly what would be expected for $\sim$75 objects distributed with an intrinsic dispersion of $\sim$1.4 magnitudes. This simple and well understood feature of the GCLF then makes it possible to predict the ultimate precision that this distance determination method is capable of. If the apparent magnitude of the turnover B_0 for a typical distant system can be determined at best to ± 0.15 mag (see the last column of Table II), then distance moduli ($B - M_B{}^0$) for single galaxies will have an irreducible uncertainty of ± 0.2 mag, irrespective of external calibration errors. The calculated internal error budget of ± 0.22 mag for the Virgo ellipticals (section V) is then already nearly at this fundamental limit.

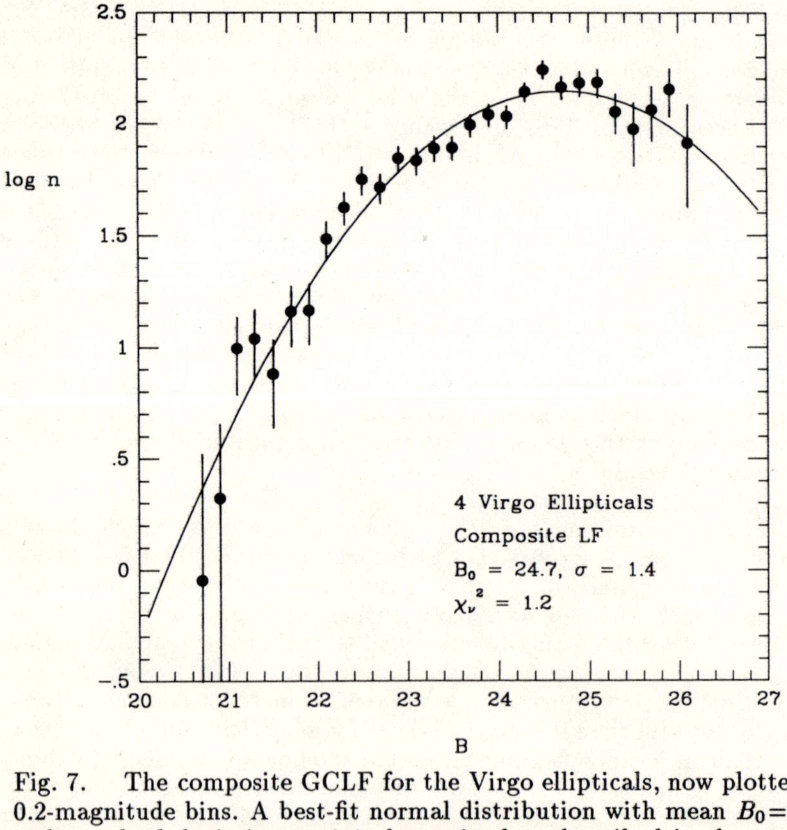

Fig. 7. The composite GCLF for the Virgo ellipticals, now plotted in 0.2-magnitude bins. A best-fit normal distribution with mean B_0=24.7 and standard deviation σ=1.4, determined as described in the text, is drawn through the points.

VII. FUTURE PROSPECTS: HOW MUCH FARTHER?

Figure 7 shows again the composite LF for the four large Virgo ellipticals NGC 4365, 4472, 4486, and 4649. The simple gaussian function drawn through the data continues to be a remarkably accurate first-order description of the LF over more than a factor of 100 in cluster luminosity. However, at the second-order level it may not be premature to suggest the existence of fine structure in the distribution (an apparent 'plateau' around $B = 23$, also visible in the individual plots of Figure 4). A Kolmogorov test, which takes into account the ordering of the points as well as the total χ^2, shows that the deviation of the composite data from the best-fit gaussian is significant at the 90% level – enough to be tantalizing but not convincing. Larger samples will be needed to pin this feature down further.

The last question I want to address in this review is to ask how much farther out into Hubble space we might eventually push this remarkably promising GCLF method. Although the distance for Virgo itself needs to be refined by settling the 'transfer' question between the Milky Way and the large ellipticals, we are already in an excellent position to measure the GCLFs in other, more distant, giant ellipticals and to establish their distances *relative to the Virgo E's.* In this way we will essentially be using the composite Virgo LF (Figure 7) as the fiducial marker, and comparing strictly similar types of galaxies. The M87/'normal'-gE comparison tells us that it makes no difference whether the target galaxy has a 'normal' or anomalous 'supergiant' cluster system, so we are free to use the biggest and most populous systems that we can find.

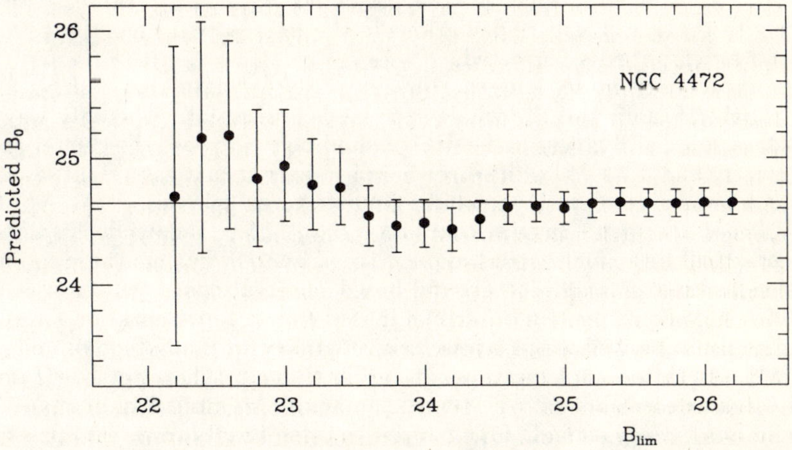

Fig. 8. A simulation of the GCLF method for more distant galaxies. Each plotted point represents the predicted turnover magnitude B_0 that would have been derived from the NGC 4472 data, if the sample had reached only to the level B_{lim}. In each case, a gaussian interpolation curve with σ=1.4 is fitted to the remaining data to estimate B_0. The 'true' B_0=24.64 for this galaxy (Table II) is the horizontal line.

A remarkable and prophetic clue to just how far our reach may extend, can be found in Hanes' (1977b) work more than a decade ago. In his analysis of the Virgo data – which at that time stopped 2 magnitudes short of the peak, as we now realize with the full benefit of hindsight – Hanes showed explicitly that the deduced turnover level would depend strongly on the assumed dispersion σ. The result could be expressed as an equation, which converted to the presently used B system is $B_0(\text{Virgo}) \simeq 18.90 + (4.20 \pm 0.13)\sigma$. Entering $\sigma = 1.4$ yields $B_0 = 24.8 \pm 0.2$. The correct answer was already available 11 years ago, if only we had known how to extract it!

I have attempted to simulate this same process of estimating B_0 in galaxies that are progressively further away by using the NGC 4472 sample from the new CFHT data. (Hanes and Whittaker 1987 carry out very much the same thing with GCLFs simulated directly from gaussian functions.) Starting with the faintest bin, we take the data points in the NGC 4472 LF and remove them one by one: after each truncation, we fit a gaussian curve with $\sigma = 1.4$ to the remaining points and estimate B_0 in the same manner as was done for Table II above. This procedure approximately mimics how accurate our distance estimates will become as we look at more and more distant galaxies and see progressively fewer of their globular clusters. Figure 8 shows the results, for LFs all the way from the original sample ($B_{lim} = 26.2$) down to one that was truncated 2.5 magnitudes short of the actual turnover ($B_{lim} = 22.2$). Several inferences may be drawn from this figure:

- If the observations do reach to the turnover or anywhere beyond, then the predicted curve-fit B_0 converges strongly to the 'true' value.
- For data that fall short of the turnover by $\simeq$ one magnitude or less, the predicted B_0 remains close to the true value, with $\lesssim 0.4$ mag internal uncertainties. The obvious 'wave' in the predicted B_0 as B_{lim} decreases is due entirely to the fine-structure deviations in the LF that were mentioned above, and can be empirically corrected.
- The uncertainty in B_0 increases dramatically for data that fall $\gtrsim 1.5$ mag short of the turnoff, and the whole exercise begins to fail. The only way that such data could still be used effectively would be to increase the cluster sample size greatly and beat down the error bars by sheer statistical weight, as Hanes (1977a,b) was able to do by combining many Virgo galaxies.

These arguments suggest that $(B_0 - B_{lim}) \lesssim 1.0$ magnitude is an acceptable practical limit for the technique. To estimate what can be done under existing techniques and with ground-based observations, I will assume $B_{lim} = 26.0$ (or its equivalent in other bandpasses) for a representative CCD photometric limit, as well as $\lesssim 1$ arcsec image quality so that effective nonstellar image classification-and-rejection can be performed. Then we have $B_0(\text{max}) \simeq 27.0$, which corresponds to systems 2.3 magnitudes more distant than Virgo. This limit is already enough, for example, to bring the Hydra/Centaurus galaxies within range at redshifts 3000–4000 km/s. If we learn to reach even deeper photometric limits routinely, or can employ even better seeing conditions, then these limits will be pushed still further. The extensions to the regime of HST are obvious.

On the basis of existing methodology, then, I suggest that it should be possible to use the GCLF method effectively for giant ellipticals up to 60 Mpc, i.e., about three times farther away than the Virgo System. If the distance moduli of candidate galaxies at this range can be estimated relative to Virgo with ±0.2-0.3 mag precision (see Fig.8), then distances with ±15% internal errors will result, with stronger leverage on the far-field value of H_0 than Virgo can provide. Going still farther (cf. section III above), it will be possible just to *resolve* the brightest clusters in giant elliptical galaxies out to $\gtrsim$100 Mpc; although in such cases the full GCLF fitting method cannot be used, a new version of the brightest-cluster method may still prove to be effective at a rougher level. This distance range is one which can be attacked directly, among primary Population I indicators, only by supernovae and the Tully-Fisher method. The obvious first (and easiest) targets should be the supergiant E's at the centers of large clusters, since these are the only known locations of the anomalously populous globular cluster systems.

The outlook for this subject is far brighter than it was even 5 years ago. From the theoretical side, increasingly sophisticated modeling of the dynamical changes in clusters produced by the environments of their parent galaxies have already taught us much about their probable evolution. A quantitative theory for their formation, however, is still a clear need. On the observational side, the primary agents of change have been the CCD imaging and photometric reduction techniques, coupled with the excellent seeing that it is now possible to achieve on outstanding sites such as Mauna Kea. The first forays into this wider landscape have already begun: from CTIO and the CFHT, deep luminosity function observations have been obtained during the past few observing seasons for several additional giant E galaxies, with the data currently in the reduction stages (Hanes, Bridges, H.Harris, and W.Harris 1988). This material includes the central supergiants in the Fornax system (NGC 1399), Hydra (NGC 3311), Centaurus (NGC 4696, 4709), the Coma supercluster (NGC 3842, 4874), and several other objects in smaller groups. The especially interesting and relatively nearby systems in NGC 5128 (H.Harris, Hesser, G.Harris, and Geisler 1988) and NGC 4594 (Hanes and Bridges 1988) are also undergoing new investigation in a variety of ways. It is not hard to predict that globular clusters are going to make an increasingly strong contribution to the extragalactic distance scale.

ACKNOWLEDGEMENTS

In both the observational and theoretical arenas, this subject is experiencing the beginning of an exciting new epoch. It is a pleasure to express sincerest thanks to the staff and time allocation committees of the CFHT and Cerro Tololo, where most of the observations discussed here have been carried out. Financial support from the Natural Sciences and Engineering Research Council of Canada, through operating grants to the author, is gratefully acknowledged.

REFERENCES

Ables, H.D., Newell, E.B., and O'Neil, E.J. 1974, *Pub.Astron.Soc.Pacific*, **86**, 311.

Aguilar, L., Hut, P., and Ostriker, J.P. 1988, preprint.

Armandroff, T.E. 1988, Ph.D. thesis, Yale University.

Barnes, T.G., and Hawley, S.L. 1986, *Ap.J.Letters*, **307**, L9.

Battistini, P., Bonoli, F., Braccesi, A., Federici, L., Fusi Pecci, F., Marano, B., and Borngen, F. 1987, *Astron.Astrophys.Suppl.*, **67**, 447.

Baum, W. 1955, *Pub.Astron.Soc.Pacific*, **67**, 328.

Burstein, D., Faber, S.M., Gaskell, C.M., and Krumm, N. 1984, *Ap.J.*, **287**, 586.

Chernoff, D.F., and Shapiro, S.L. 1987, *Ap.J.*, **322**, 113.

Chernoff, D.F., and Shapiro, S.L. 1988, in *Globular Cluster Systems in Galaxies (IAU Symposium No.126)*, eds. J.E. Grindlay and A.G.D. Philip, (Dordrecht: Reidel), p.283.

Cohen, J.G. 1988, *Astron.J.*, **95**, 682.

Crampton, D., Cowley, A.P., Schade, D., and Chayer, P. 1985, *Ap.J.*, **288**, 494.

da Costa, G.S., and Mould, J.R. 1988, *Ap.J.*, in press.

de Vaucouleurs, G. 1970, *Ap.J.*, **159**, 435.

de Vaucouleurs, G. 1983, *Ap.J.*, **268**, 468.

de Vaucouleurs, G., and de Vaucouleurs, A. 1973, *Astron.Astrophys.*, **28**, 109.

Elson, R.A.W., and Walterbos, R.A.M. 1988, *Ap.J.*, in press.

Fall, S.M., and Rees, M.J. 1977, *M.N.R.A.S*, **181**, 37p.

Fall, S.M., and Rees, M.J. 1985, *Ap.J.*, **298**, 18.

Fall, S.M., and Rees, M.J. 1988, in *Globular Cluster Systems in Galaxies (IAU Symposium No.126)*, eds. J.E.Grindlay and A.G.D.Philip, (Dordrecht: Reidel), p.323.

Forte, J.C., Strom, S.E., and Strom, K.M. 1981, *Ap.J.Letters*, **245**, L9.

Grillmair, C.J., Pritchet, C.J., and van den Bergh, S. 1986, *Astron.J.*, **91**, 1328.

Hanes, D.A. 1971, M.Sc. thesis, University of Toronto.

Hanes, D.A. 1977a, *Mem.R.A.S.*, **84**, 45.

Hanes, D.A. 1977b, *M.N.R.A.S.*, **180**, 309.

Hanes, D.A. 1977c, *M.N.R.A.S.*, **179**, 331.

Hanes, D.A. 1979, *M.N.R.A.S.*, **188**, 901.

Hanes, D.A., and Brodie, J. 1986, *Ap.J.*, **300**, 279.

Hanes, D.A., and Harris, W.E. 1986, *Ap.J.*, **304**, p.599.

Hanes, D.A., and Whittaker, D.G. 1987, *Astron.J.*, **94**, 906.

Harris, H.C., Harris, G.L.H., Hesser, J.E., and MacGillivray, H.T. 1984, *Ap.J.*, **287**, 185.

Harris, H.C., Harris, G.L.H., and Hesser, J.E. 1988, in *Globular Cluster Systems in Galaxies (IAU Symposium No.126)*, eds. J.E. Grindlay and A.G.D. Philip, (Dordrecht: Reidel), p.205.

Harris, W.E. 1983, *Pub.Astron.Soc.Pacific*, **95**, 406.

Harris, W.E. 1986, *Astron.J.*, **91**, 822.

Harris, W.E. 1987a, *Pub.Astron.Soc.Pacific*, **99**, 1031.

Harris, W.E. 1987b, *Ap.J.Letters*, **315**, L29.

Harris, W.E. 1988, in *Globular Cluster Systems in Galaxies (IAU Symposium No.126)*, eds. J.E.Grindlay and A.G.D.Philip, (Dordrecht: Reidel), p.237.

Harris, W.E., and Allwright, J.W.B. 1988, in preparation.

Harris, W.E., Allwright, J.W.B., Pritchet, C.J., and van den Bergh, S. 1988, in preparation.

Harris, W.E., Harris, H.C., and Harris, G.L.H. 1984, *Astron.J.*, **89**, 216.

Harris, W.E., and Racine, R. 1979, *Ann.Rev.Astron.Astrophys.*, **17**, 241.

Harris, W.E., Smith, M.G., and Myra, E.S. 1983, *Ap.J.*, **272**, 456.

Hesser, J.E. 1988, A.S.P. Conference Proceedings Series, *Progress and Opportunities in Southern Hemisphere Optical Astronomy*, eds. V.M.Blanco and M.M.Philips, in press.

Hesser, J.E., Harris, W.E., VandenBerg, D.A., Allwright, J.W.B., Shott, P., and Stetson, P.B. 1987, *Pub.Astron.Soc.Pacific*, **99**, 739.

Hesser, J.E., and Shawl, S.J. 1985, *Pub.Astron.Soc.Pacific*, **97**, 465.

Hodge, P.W. 1974, *Pub.Astron.Soc.Pacific*, **86**, 289.

Huchra, J., and Brodie, J. 1987, *Astron.J.*, **93**, 779.

Jones, R.V., Carney, B.W., Latham, D.W., and Kurucz, R.L. 1987, *Ap.J.*, **314**, 605.

Kron, R. 1980, *Ap.J.Suppl.*, **43**, 305.

Larson, R.B. 1988, in *Globular Cluster Systems in Galaxies (IAU Symposium No.126)*, eds. J.E.Grindlay and A.G.D.Philip, (Dordrecht: Reidel), p.311.

Lauer, T.R., and Kormendy, J. 1986, *Ap.J.Letters*, **303**, L1.

Lee, H.M., and Ostriker, J.P. 1987, *Ap.J.*, **322**, 123.

Mould, J.R., Oke, J.B., and Nemec, J.M. 1987, *Astron.J.*, **93**, 53.

Ostriker, J.P. 1988, in *Globular Cluster Systems in Galaxies (IAU Symposium No.126)*, eds. J.E.Grindlay and A.G.D.Philip, (Dordrecht: Reidel), p.271.

Peebles, P.J.E., and Dicke, R.H. 1968, *Ap.J.*, **154**, 891.

Peterson, C.J. 1986, *Pub.Astron.Soc.Pacific*, **98**, 1258.

Peterson, C.J., and Reed, B.C. 1987, *Pub.Astron.Soc.Pacific*, **99**, 20.

Pritchet, C.J., and van den Bergh, S. 1985, *Astron.J.*, **90**, 2027.

Pritchet, C.J., and van den Bergh, S. 1987a, *Ap.J.*, **316**, 517.

Pritchet, C.J., and van den Bergh, S. 1987b, *Ap.J.*, **318**, 507.

Racine, R. 1968, *J.R.Astron.Soc.Canada*, **62**, 367.

Racine, R., Oke, J.B., and Searle, L. 1978, *Ap.J.*, **223**, 82.

Rosenblatt, E.I., Faber, S.M., and Blumenthal, G.R. 1988, *Ap.J.*, **330**, 191.

Sandage, A. 1961, in *Problems of Extragalactic Research (IAU Symposium No.15*, ed. G.C.McVittie, (New York: Macmillan Co.), p.359.

Sandage, A. 1968, *Ap.J.Letters*, **152**, L149.

Sandage, A., and Tammann, G.A. 1988, *Ap.J.*, **328**, 1.

Spitzer, L. 1987, *Dynamical Evolution of Globular Clusters* (Princeton: Princeton University Press).

Stetson, P.B. 1987, *Pub.Astron.Soc.Pacific*, **99**, 191.

Strom, K.M., Strom, S.E., Jensen, E.B., Moller, J., Thompson, L.A., and Thuan, T.X. 1977, *Ap.J.*, **212**, 335.

Strom, S.E., Forte, J.C., Harris, W.E., Strom, K.M., Wells, D.C., and Smith, M.G. 1981, *Ap.J.*, **245**, 416.

Strugnell,P., Reid, N., and Murray, C.A. 1986, *M.N.R.A.S.*, **220**, 413.
Thompson, L., and Valdes, F. 1987, *Ap.J.Letters*, **315**, L35.
Tremaine, S. 1976, *Ap.J.*, **203**, 345.
Tyson, J.A., and Jarvis, J.F. 1979, *Ap.J.Letters*, **230**, L153.
van den Bergh, S. 1969, *Ap.J.Suppl.*, **19**, 145.
van den Bergh, S. 1985, *Ap.J.*, **297**, 361.
van den Bergh, S., and Harris, W.E. 1982, *Astron.J.*, **87**, 494.
van den Bergh, S., Pritchet, C., and Grillmair, C. 1985, *Astron.J.*, **90**, 595.
Webbink, R.F. 1985, in *Dynamics of Star Clusters (IAU Symposium No.113*, eds. J.Goodman and P.Hut, (Dordrecht: D.Reidel), p.541.
Wielen, R. 1988, in *Globular Cluster Systems in Galaxies (IAU Symposium No.126)*, eds. J.E. Grindlay and A.G.D. Philip, (Dordrecht: Reidel), p.393.
Zinn, R.J. 1985, *Ap.J.*, **293**, 424.

DISCUSSION

D. TERNDRUP: Your observations are in B. Is the contrast between G.C.'s and galaxies different at other wavelengths?

W. HARRIS: Very different. In general the underlying parent galaxy is a lot redder than its globular clusters, so to maximize the contrast of the clusters over the background light you need to stay well to the blue side of the spectrum. The best signal-to-noise is in either B or V.

J. MOULD: How does the luminosity function of your Virgo sample behave as a function of radius in the galaxy?

W. HARRIS: Dividing up the samples by radius (3 to 8 kpc, and 8 to 14) shows no detectable LF differences all the way down to B=26. I don't know what is happening inside the 3 kpc boundary of our data; finding that out will need a somewhat different observing strategy than we used here.

B. TULLY: The turnover in the luminosity function is only one magnitude above your cutoff. To be really convincing that there is a universal turnover with small dispersion it would be nice to work several systems in a group such as the one in Leo that is a magnitude or so closer.

W. HARRIS: Pritchet and van den Bergh have published a LF for NGC 3379 (A.J. **90**, 2027, 1985), and I have some new photometry (unreduced as yet!) for NGC 3377. For these there should be no difficulty to get far past the turnover. As far as the Virgo material is concerned, the turnover is actually 1.5 magnitudes brighter than the photometric limit, and at the turnover the completeness is 95% and the background extremely low.

C. PRITCHET: What fraction of all the images at B_o are rejected by the image classifier, and how many "background" images are left after this procedure?

W. HARRIS: For magnitudes brighter than $B \simeq 25$, I found that about 2/3 or a bit more of all the background field images could be clearly classified as nonstellar. In that range, this meant that about 10% to 20% of the remaining images (starlike) were background objects in the Virgo galaxy fields.

S. DJORGOVSKI: If you assume a universal LF, can you apply it to the Coma cluster data by Thompson and Valdes?

W. HARRIS: The Thompson-Valdes data (and mine) just show the top 1.5 magnitudes of the LF, so a full LF fit is not really feasible. A match of just the brightest clusters to those in M87 gives the same distance ratio Coma/Virgo as the radial velocity ratio, so the method is self consistent.

C. PRITCHET: One big difference in the limiting magnitude of the M87 data, and the Virgo ellipticals that you discuss here, is crowding. Crowding is much more severe in the M87 field. The implication of this is that, if one moves out to 60 Mpc, crowding of globulars will be much more severe, so that it may be difficult to achieve the goal of B(lim) = 26 for such observations.

W. HARRIS: That's a good point, and it emphasizes the importance of obtaining imaging at the best possible scale and especially in the best possible seeing. I can also mention the work published last year by Thompson and Valdes, and by me, which shows that it's possible, under subarc-second conditions, seeing to do photometry fainter than B=26 for the clusters even in the Coma galaxies.

ON INFALL INTO THE VIRGO CLUSTER

John P. Huchra
Harvard-Smithsonian Center for Astrophysics
60 Garden Street, Cambridge, MA 02138 USA

ABSTRACT The history of the study of our motion with respect to the Local Supercluster is reviewed from the original suggestions of de Vaucouleurs, and Silk and Peebles through today. Direct measurements of the infall pattern using velocity independent galaxy distances suggest an infall component at the Local Group of ~ 250 km/s. This is consistent with the velocity expected if the average mass-to-light ratio of galaxies is similar to that dynamically measured in galaxy clusters, a value that gives an Ω of only ~ 0.3. The infall pattern is also consistent with galaxy density maps produced by IRAS surveys, although it may be only marginally consistent with the "Great Attractor." On smaller scales, there is strong evidence that most of the spiral galaxies near the cluster core are still falling in — that the spiral galaxies in the core of the cluster do not form a dynamically relaxed system. Using the best distance estimates to the Virgo cluster, the local value of the Hubble constant is ~ 85 km $s^{-1}Mpc^{-1}$.

I. INTRODUCTION

The study of infall into the Virgo Cluster has essentially been the study of systematic deviations from the nearby smooth expansion of the universe — otherwise known as the Hubble Flow. The first searches for a velocity field were actually done in the early 1950's by V. Rubin (1951) and K. Ogorodnikov (1952), with inconclusive results. This early work, as well as many current studies, were attempts to measure the galaxy's motion with respect to to some global, large-scale or cosmic reference frame. In the early 1950's, the existence of what we now call the Local Supercluster was not generally accepted; the clustered distribution of bright galaxies was thought to be due probably to galactic extinction.

Although the existence of a large concentration of galaxies in a small region of the sky was noticed by Shapley in the early 1930's (Shapley and Ames 1932; Shapley 1934), the notion of the Local Supercluster (LSC) with the Virgo Cluster at its core was really introduced in the 1950's by Gerard de Vaucouleurs (1956), who called it the "Local Supergalaxy." De Vaucouleurs (1958, 1972 and references therein) was the first to recognize that the LSC was a real, 3-dimensional dynamical entity whose mass distribution could affect the motions of galaxies around it, and the first to suggest the existence of both an infall motion and a possible rotation. Figure 1 is a plot taken from the first Reference Catalogue of Bright Galaxies (de Vaucouleurs and de Vaucouleurs 1964) showing the positions in de Vaucouleurs' supergalactic coordinates.

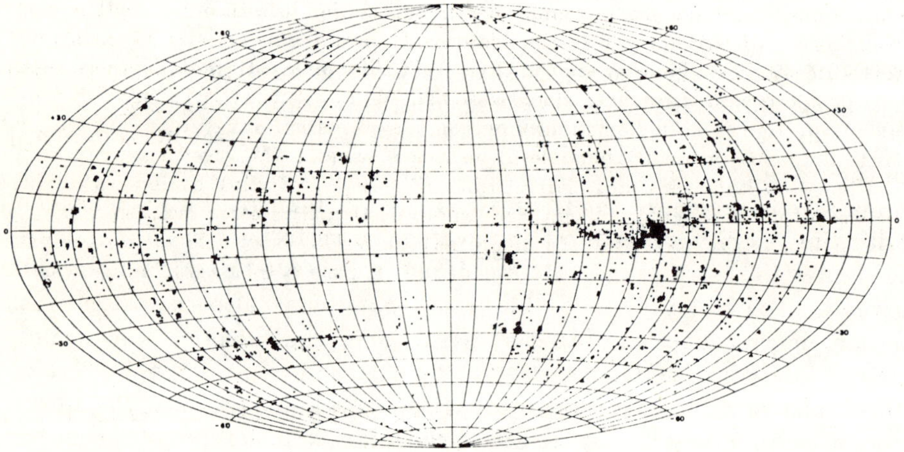

Figure 1. The distribution of 2599 galaxies in supergalactic coordinates. Virgo easily stands out. (From the *Reference Catalogue of Bright Galaxies*).

Over the ensuing decades, experimental work on non-Hubble velocity fields proceeded along these above two lines: searches for global motions and searches for motions associated with the LSC. Stewart and Sciama (1967) used the distances of nearby groups of galaxies to derive the rotation of the galaxy about the supercluster (and got the rather large value of 340 km s^{-1}) and to place a crude upper limit on the infall (less than about 200 km s^{-1}). De Vaucouleurs and coworkers (eg. de Vaucouleurs 1976; de Vaucouleurs and Peters 1968, 1981; de Vaucouleurs *et al.* 1981), in his quest for the global value of the Hubble Constant, used distances to both individual galaxies and galaxy groups to study the motion of the Local Group (LG). Sandage, Tammann and Hardy (1972) conducted a search for velocity perturbations using the magnitudes of first ranked cluster galaxies. Rubin *et al.* (1976) searched for a motion of the LG with respect to a distant ($V_r \sim 5000$ km s^{-1}) sample of Sc I galaxies. And in the 1970's, a large number of groups were searching for an anisotropy in the microwave background radiation that might signal LG motion with respect to the most distant frame measureable (Weiss 1980, gives an excellent review of the early work).

Another important milestone in the study of Virgo infall came in the mid 1970's when Silk (1974, 1977) and Peebles (1976) calculated simple models of the distortion of the local Hubble Flow by the LSC. They pointed out that the infall velocity depends on the mass overdensity of the LSC. This meant that a measurement of the infall could be used to estimate Ω, the ratio of the mean mass density of the universe to the closure density. This was predicted to be an extremely interesting measurement because it had then recently been suggested (Ostriker, Peebles and Yahil 1974) that the mass-to-light ratio of systems of galaxies seemed to increased with scale. Clusters appeared more massive than groups, which were more massive than binaries, etc. It was expected that a measurement of the mass and mass-to-light ratio of the LSC, on a scale of $\sim$ 15 to 20 Mpc, would exceed the closure value.

The next major development that stimulated work on the LSC was the great improvement in the measurements of the microwave background (μwB) dipole that took place in the late 1970's. The dipole anisotropy of the μwB is most easily interpreted as a motion of the sun or the LG with respect to a cosmic reference frame. In an impressive series of papers starting in 1979, the uncertainties in the direction and amplitude of our motion with respect to the μwB dropped to less than 3° and less than 30 km s^{-1}(Cheng *et al.* 1979; Smoot and Lubin 1979; Fabbri *et al.* 1980; Boughn, Cheng and Wilkinson 1981; Gorenstein and Smoot 1981; Fixsen, Cheng and Wilkinson 1983; Lubin, Epstein and Smoot 1983; Lubin *et al.* 1985). (As an aside, it was extremely interesting to see the competition between a few competing groups lead to better and better results!)

With the IAU standard correction to the centroid of the Local Group, (300 sin(l)cos(b)), the current results give the μwB vector as 600$\pm$30 km s^{-1}towards $\alpha = 10.^h5\pm0.^h2$ and $-26°\pm3°$. This is an extremely large velocity in the context of the previously assumed quiet Hubble flow (Sandage and Hardy 1973). And, by definition, the first place the optical observers searched for the culprit causing this motion was the LSC.

Over this same period of time, several new, and sometimes accidental, attempts were being made to fit the Virgo infall to the galaxy field. These were based on the application of two new techniques for measuring distances to galaxies. Marc Aaronson, Jeremy Mould and I had been working on an improvement to the distance estimator proposed by Tully and Fisher (1977). They had found a surprisingly good relation between the luminosity or diameter of a galaxy and the rate at which it was rotating (as measured by 21-cm neutral hydrogen profiles). However a great argument arose as to how the technique should be applied in the presence of large absorption corrections to edge on galaxies, different useage of local calibrators, the use of photographic or ancient photoelectric photometry, etc. (Sandage and Tammann 1976b).

We realized that many of the problems could be either minimized or eliminated by performing accurate photometry in the infrared. The H band at 1.6μ was chosen to minimize sky brightness and to sit on the peak of the K giant energy distribution, a direct tracer of the old main sequence which should be a good measure of a galaxy's mass. Unfortunately, (or fortunately depending on your point of view!) our first results for the Hubble constant, based only on distances to the Virgo Cluster and the Ursa Major Cloud, were midway between the values advocated by Sandage and Tammann and by Tully and Fisher (Aaronson, Huchra and Mould 1979). To rectify this problem, and because it was believed that there might be a problem with internal motions in the LSC or in the near field (eg. Rubin *et al.* 1976), we measured distances to 4 clusters well outside the LSC (Aaronson *et al.* 1980). These clusters all gave a much ***higher*** value for the Hubble constant than Virgo and Ursa Major, and the simplest explanation for this effect was a motion in the direction of Virgo of 480 ± 75 km s^{-1}.

Paul Schechter, who had just started working with us and Brent Tully on the infall problem, was also interested in the velocity dispersion – luminosity relation for elliptical galaxies (the Faber-Jackson relation). While this relation does not have the tightness (small scatter) of the infrared version of the Tully-Fisher relation (IRTF), it can be used in places where there are no spirals and also as a check on the IRTF. Schechter developed elegant solutions for the infall based on Silk's (1974, 1977) formulation and applied them to a small sample of 32 elliptical galaxies, the majority of which were outside the LSC. He derived an infall velocity of 190 ± 130 km s^{-1}.

Here it is useful to point out the differences between near and far-field measures of the motion. Almost all studies, including the μwB measurements, rely on the establishment of a reference frame. In the above galaxy studies, the reference frames are galaxies or galaxy clusters at moderate redshifts (~ 5000 km s^{-1}), but nonetheless well outside the bounds of the LSC. These studies are "far-field" w.r.t the LSC and may measure a lot more than just Virgo infall. In contrast, "near-field" studies of the LSC generally depend on Virgo itself as the reference frame and are dominated by galaxies inside the LSC. Examples of early near field studies are those of Hoffman, Olsen and Salpeter (1980), who studied the mass distribution of Virgo by examining the cluster's velocity dispersion as a function of radius, $\sigma(r)$, and the study of Yahil, Sandage and Tammann (1980) who searched for LG motions by studying directional differences in the luminosity funtcion for galaxies in the Revised Shapley-Ames Catalog (Sandage and Tammann 1981, hereafter RSA), a catalog dominated by the LSC.

Tonry and Davis (1981) increased the sample of elliptical galaxy velocity dispersions and also applied the Faber-Jackson relation to S0 galaxies. For a total sample of 373 elliptical and S0 galaxies, again mostly *outside* the LSC, they derived a motion towards Virgo of $\sim 450 \pm 60$ km s^{-1}. This paper is also noteworthy because it contains the first pictorial description of what has become the standard dipole flow field. This can be seen in Figure 6 below, and has also even graced the front cover of the CfA phone book. In the cylindrical coordinate system with major axis running between the LG and Virgo, lines of constant radial velocity versus distance are distorted by a dipole that looks like an apple centered on Virgo. If there were no flow but the Hubble flow (if Virgo was massless), lines of constant velocity would simply be circles centered on the LG. For any non-zero infall, at large distances the lines of constant velocity are centered on a point offset from the LG by the amplitude of the infall at the LG. An important insight that is easily seen in this plot is the existence of the degenerate or multi-valued region inside the turnaround radius where objects at several different distances can have the same apparent velocity. The size and shape of this region depends on the amplitude of the flow (essentially the mass and mass distribution inside the supercluster core) and extreme care must be taken fitting data to flow models in this region. Estimates of the actual projected turnaround radius of Virgo range between 20° and 30.°

The first, and so far the only good attempt to directly fit the near-field was made by Aaronson *et al.* (1982, hereafter AHMST). AHMST used Tully and Fisher's sample of edge on spiral galaxies with velocites ≤ 3000 km s^{-1}, and thus mostly *inside* the LSC. This work will be described further below.

The most recent contribution to the Virgo velocity problem is the review by Tammann and Sandage (1985). Tammann and Sandage first derive their own estimate of the infall using distances to 9 groups and clusters of galaxies, including Virgo itself, most of which are on the far side of the supercluster. Their distance estimates for the groups are heavily weighted by derivations based on galaxy color-magnitude relations, which have large scatter, and on the corrected first ranked galaxy magnitude, also with large scatter and with corrections that have been derived assuming there *was* no Virgo flow. One of the points in their fit is Virgo itself, which should be the origin in a flow model. Examination of the residual plot for the groups shows that their result of 220 km $s^{-1}$is actually based on only two points, one of which is near the triple valued region. Their estimate should therefore be given relatively low weight. They then compute a grand average value of the velocity towards Virgo from "selected" measurements. Unfortunately their final value is heavily dependent on several estimates based on essentially the same technique applied to the *same* data set, and which are thus not statistically independent.

The AHMST study of the near-field has not yet been surpassed. The Infra-Red Tully-Fisher relation is still one of the best galaxy distance indicators, and it is also very difficult to assemble samples of galaxies with *well measured distances* larger than that of AHMST inside the LSC. Workers who have concentrated on determining the Hubble Constant are now trying to measure accurate distances to objects well beyond the LSC, where it is hoped that the effects of velocity flows will be minimal. Work on the velocity field, since the LSC cannot explain the μwB motion, is now also concentrated on larger scales (see below).

A much more detailed summary of velocity field measurements up until 1982 can be found in the review article of Davis and Peebles (1983) who were especially interested in determinations of Ω based on measures of the infall; further new contributions have been collected in the symposium volume edited by Madore and Tully (1986). The Davis and Peebles article suffers because they believed that the best parameters for the infall into Virgo should be derived from the far field measurements. This was because they were in best agreement with the μwB velocity amplitude, and because they gave the largest values of Ω, in line with the theoretical prejudice for Ω =1. The superiority of the far-field measurements is a premise that is almost certainly wrong in light of the detection of large scale flows caused by more distant, massive superclusters (see the next section).

Theoretical models of infall into superclusters and clusters have also advanced since the early 1980's. Improvements include full non-linear calculations for spherical distributions (Yahil 1985; Shaya 1986), matches to N-body simulations (Bushouse *et al.* 1985; Villumsen and Davis 1986), and extensions to non-spherical models (eg. Hoffman 1986) — which are extremely important because the LSC is quite flat with an axial ratio of better than 2 to 1. Other "improvements" include calculations of the effects of tidal torques caused by other nearby superclusters, and will be discussed below. An interesting review of the general theoretical problem of infall patterns in redshift space has been given by Kaiser (1987).

II. RELATED TOPICS

There are several old and new issues that, while not directly concerned with the infall pattern about the LSC, significantly impact its interpretation both in terms of the value of the Hubble Constant, the use of LSC dynamics to derive a mass-density for the universe, and as consistency checks on any infall parameters derived.

The first of these is the great Virgo Velocity Debate. This started in the late 1960's and early 1970's with Sandage and Tammann (eg. Tammann 1972) and de Vaucouleurs (eg. de Vaucouleurs and de Vaucouleurs 1973; Sandage and Tammann 1976a) arguing about the actual velocity of the cluster. Based on the meager velocity data available at that time, the elliptical galaxies seemed to form a system with a much lower mean radial velocity than the spirals. Fortunately, this problem with types has now essentially gone away (Huchra 1985). However, the proper Virgo Cluster velocity is still not agreed upon.

For example, some camps argue that M87 is at the center of the cluster potential well, and its velocity should be used as the cluster velocity, especially for deriving H_0 from distances determined to elliptical galaxies. I have derived a mean cluster velocity from the complete cluster velocity histogram, shown in figure 2, with both known foreground and background members removed (Huchra 1985); this velocity is about 140 km s^{-1}lower than M87's. Sandage and Tammann (1988) prefer removal of a larger fraction of the high velocity tail of the histogram, due to contamination by the probably background Virgo W group, which yields a velocity 55 km s^{-1}lower than mine. All choices are somewhat dangerous, but the last two are probably both reasonable given the magnitude of the uncertainty in velocity, $\sim$ 50 km s^{-1}, just due to the small number statistics and the large cluster velocity dispersion. On top of those problems, there is also the question of what correction should be used to the centroid of the LG — a lecture in itself. A better value of the cluster heliocentric velocity, V_h, awaits both more galaxy velocities and relative distances for suspected foreground and background galaxies.

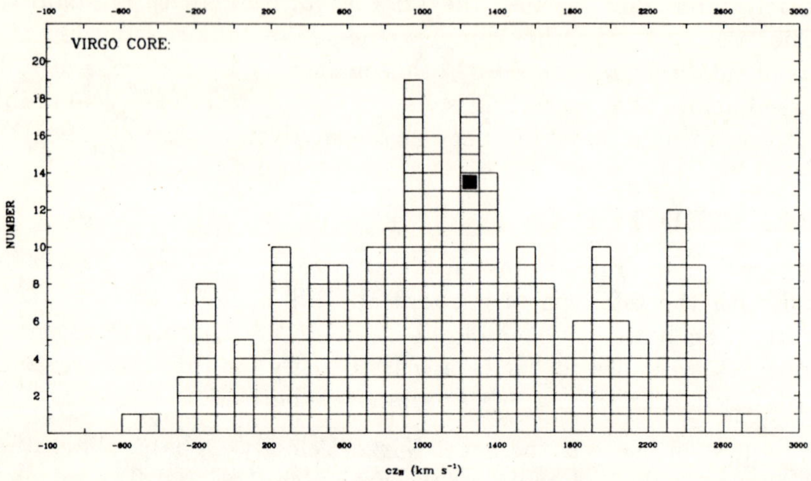

Figure 2. The velocity histogram for the core of the Virgo Cluster. The bin containing M87 has one box blacked in. (From Huchra 1985).

Another interesting problem is the relation of the Virgo Flow to motions on larger scales. The microwave background motion, now known excruciatingly well (Wilkinson 1984), is not towards Virgo. Neither is the optical dipole direction calculated from the light vector of nearby galaxies (Davis and Huchra 1982). Neither is the IRAS dipole (Yahil, Walker and Rowan-Robinson 1986; Meiksin and Davis 1986). Both the optical and IRAS dipoles lie close to each other and roughly half-way between Virgo the μwB direction, although neither are well determined in the face of fluctuations caused by nearby galaxies — for example M31 and the Magellanic Clouds completely dominate the optical and IRAS dipole unless they're removed — and both can be moved closer to the μwB (Lahav, Rowan-Robinson and Lynden-Bell 1988) by an appropriate choice of sample.

It should be noted that the interpretation of both the optical and IRAS dipoles in terms of mass concentrations and Ω is difficult due to problems with the samples. The optical surveys are extremely non-uniform, there is no single survey that is deep enough that also covers the whole sky, and galactic extinction effects dominate at low latitude. The IRAS samples, selected at 60μ, are dominated by starburst galaxies and AGN and do not contain any elliptical galaxies; the 60μ luminosity of a galaxy is likely to be a poor indication of its mass, and some of the most massive and most clustered galaxies (thus most likely to induce velocity perturbations) are not included.

The probable solution to the mismatch between the Virgo direction and the dipoles was suggested by Shaya and Tully (Shaya 1984; Tully and Shaya 1984), Tammann and Sandage (1985) and later Lilje, Yahil and Jones (1986). Basically, they estimated the effects of other nearby Superclusters, most notably Hydra-Centaurus, on the velocity field. A great supercluster at the distance and direction of Hydra-Centaurus could easily conribute tidal velocities of a few hundred km s^{-1}at the LG, and might not perturb the infall motion very much.

The notion of an additional, large, nearby, flow-perturbing mass recently has been given great support (as well as great publicity) due to the possible detection of a "Great Attractor," by the Seven Samurai (Burstein *et al.* 1986; Dressler *et al.* 1987, hereafter SS). Their results are based on an improved version of the Faber-Jackson relation for field elliptical galaxies known as the $D_n - \sigma$ relation. These authors have found evidence for a large scale flow towards an object, which is not the Hydra-Centaurus cluster, but lies in its general direction. Our overall motion determined relative to moderately distant galaxies would then be reasonably consistent with the μwB motion.

At the moment, I believe this result should be taken as very suggestive, but not yet definitive. In particular, the "Great Attractor" is in crude agreement but detailed disagreement with Infrared-Tully-Fisher studies of distant clusters (Aaronson *et al.* 1986), and with new IRTF studies of southern clusters (Aaronson *et al.* 1988). For example, we still cannot be sure that Hydra-Cen isn't the Great Attractor.

In fact, the most recent solution from the SS (Burstein *et al.* 1988) for the Great Attractor model only includes an infall motion of 80 km s^{-1}towards Virgo, a solution in very poor agreement with much more detailed studies of the interior of the LSC. While such a small Virgo infall solution does not carry much weight (since most of the SS sample lies outside the interior of the LSC), it does indicate the difficulty of accurately determining velocity fields in the presence of complicated galaxy distributions and noisy data.

In contrast to the above result, Aaronson *et al.* (1988) fit both the northern and southern cluster IRTF data to the "linear bi-infall" model of Han and Mould (1988). In this model, the infall pattern is produced by two masses, one at Virgo and the other at the Great Attractor. When the total velocity is constrained to fit the μwB amplitude, they find that the infall component towards Virgo is ~ 250 km s^{-1}, in very good agreement with the near-field studies.

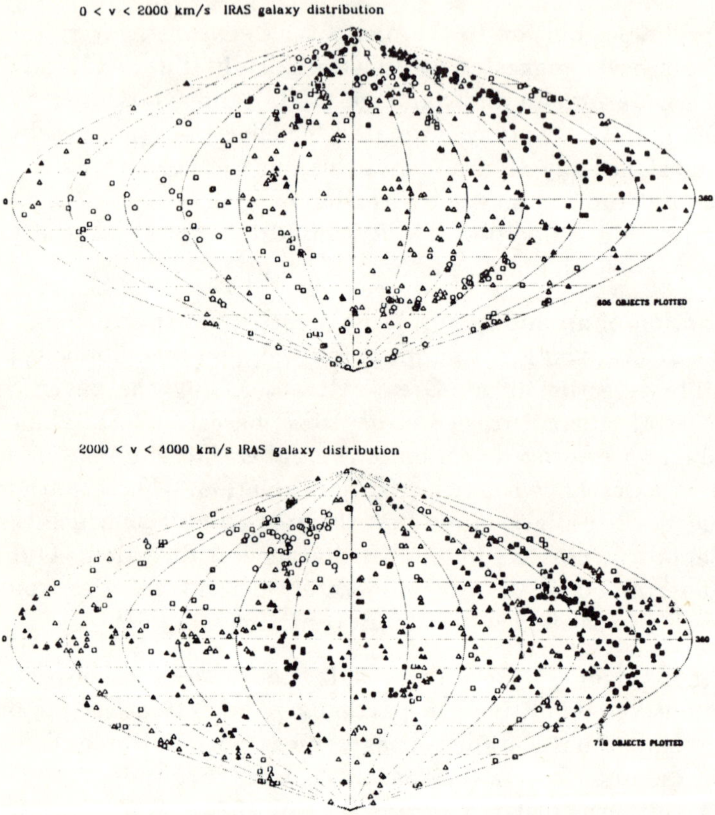

Figure 3. Equal area whole-sky plots in galactic coordinates of (a) top, the distribution of IRAS galaxies with velocities less than 2000 km s^{-1}, and (b) bottom, the same for galaxies with velocities between 2000 and 4000 km s^{-1}. In (a) the core of the LSC stands out near $l^{II} = 260°$ and $b^{II} = 75°$. In (b) Hydra-Centaurus and the region of the Great Attractor stand out near $l^{II} = 310°$ and $b^{II} = 15°$. (From Stauss and Davis 1988).

Finally, Davis, Yahil, Strauss, Tonry and Huchra have just completed a redshift survey of ~ 2500 IRAS galaxies for the express purpose of mapping mass fluctuations to predict the velocity field. The IRAS sample was chosen, despite the limitations mentioned above, because of its exceptional uniformity over the sky. Figure 3 shows the surface distributions of the IRAS galaxies in the 0-2000 km $s^{-1}$and the 2000-4000 km s^{-1}ranges. Both Yahil (1988) and Strauss and Davis (1988) have translated 3-D IRAS maps into a velocity field. Figure 4 shows contours of the distribution of galaxies (corrected for selection as a function of distance) in the supergalactic plane. Figure 5 shows the flow field predicted from the galaxy distribution.

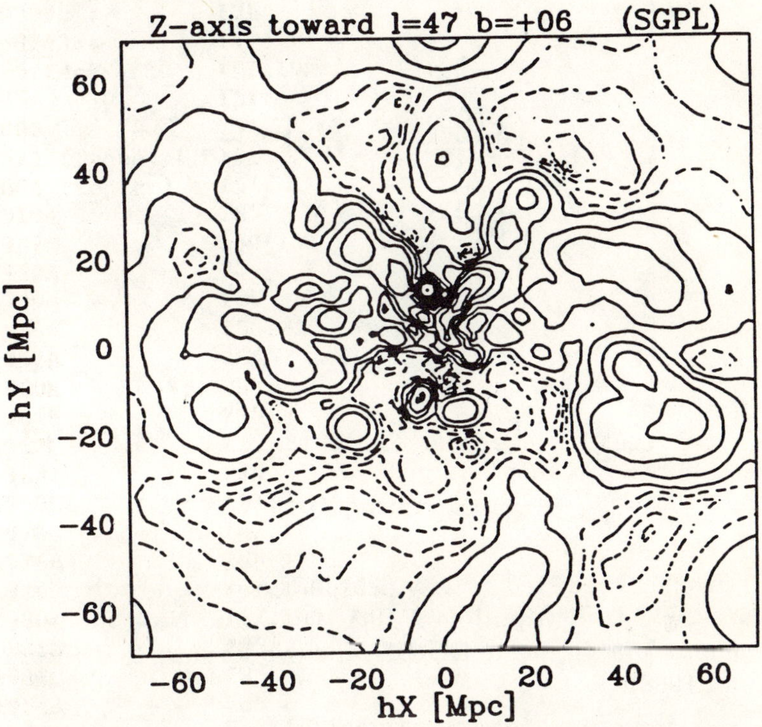

Figure 4. The density contours for the distribution of IRAS galaxies near the supergalactic plane. The LG is at the coordinate center (0,0), Virgo is the very dense region just above the LG (0,13), and the Hydra-Centaurus, Great Attractor complex is to the left (-40,20). The Perseus-Pisces complex is to the right (50,-20). (From Yahil 1988).

While there is qualitative agreement between the predicted velocity field and the Great Attractor model, they disagree in detail. In particular, the IRAS maps predict a much larger Virgo infall than is seen in the SS data. They also predict a significant motion towards the great supercluster in Perseus-Pisces, which has not been detected in the SS data. We're getting there, but we're not there yet!

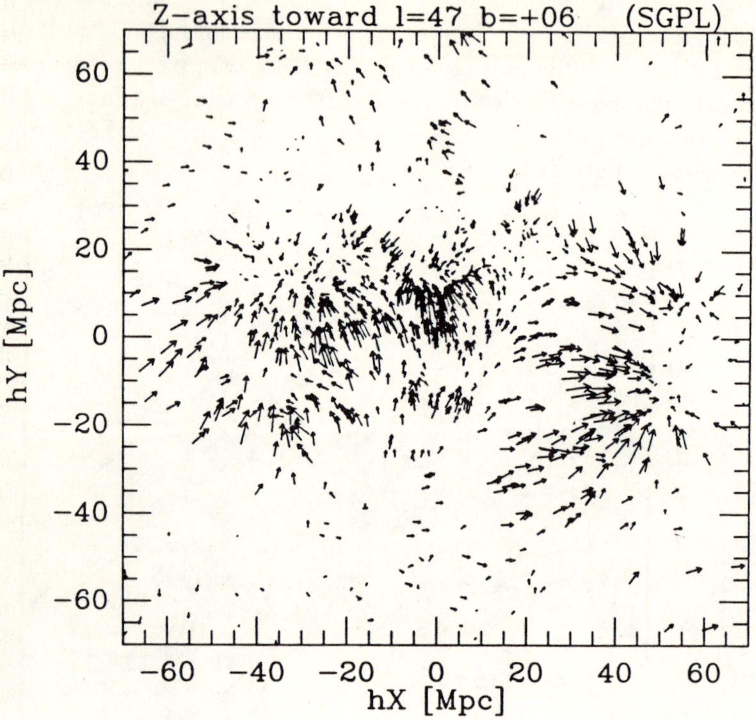

Figure 5. Predicted gravity field in the supergalactic plane from the 3-D distribution of IRAS galaxies. Note the large pull into *both* Virgo and HC-GA, and the flow into P-P. (From Yahil 1988).

III. THE LSC VELOCITY FIELD FROM IRTF

In determining the parameters for an interior flow model, we are fundamentally limited by the number of galaxies inside the LSC with accurately determined *relative* distances. We don't need to know the Hubble Constant, although knowing the flow is usually necessary to derive it later. To date, the best and most detailed determination of the infall pattern into Virgo is still that of AHMST (1982).

AHMST started with the (1) non-linear spherical infall model as developed by Schechter (1980), and (2) IRTF distances to 306 edge on spiral galaxies. The galaxies came from the semi-velocity limited catalog of Tully and Fisher, which has just recently been publised as the *Nearby Galaxies Catalog* (Tully 1988). The relative IRTF distances to individual edge on spirals are accurate to 20% to 25% (Aaronson and Mould 1983).

We then fit the model for the combination of pattern infall, rotation and the random motion of the LG. An example of a residual plot with residual velocity vectors superposed on the infall pattern is shown in Figure 6.

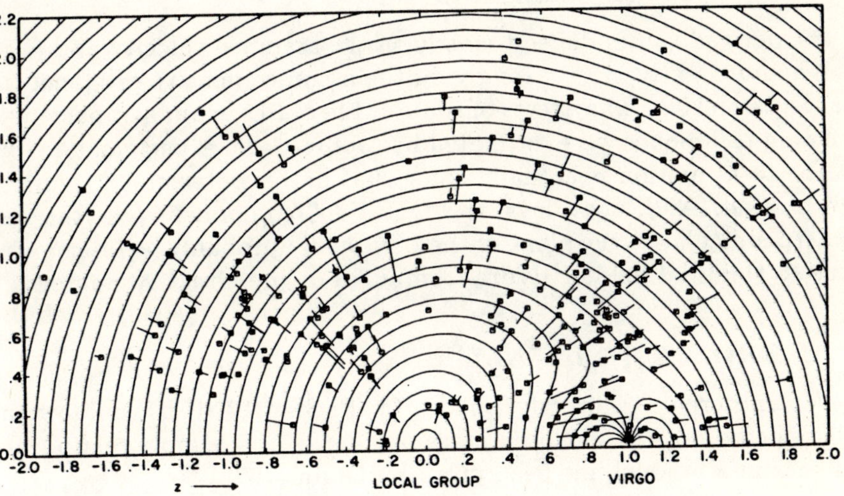

Figure 6. The velocity field of the LSC with an assumed infall velocity at the LG of $\sim$ 260 km s^{-1}. Superposed on the pattern are the velocity residuals from a simple model with that infall; positive residuals point away from the LG. (From AHMST).

Fits to the models were made both by predicting redshift from the observed distance and comparing with measured redshift, and by predicting line-width, given magnitude and redshift, and comparing with the observed line width. As pointed out by Schechter (1980), Tully (1988), and AHMST, the line-width residual scheme is somewhat less susceptible to problems with selection biases (eg. the Malmquist effect which plagues all magnitude limited samples) and its use is prefered. It is also important to stay away from the triple valued region in performing the fits.

The velocity components of interest were calculated for a range of assumptions about the overdensity inside the LG, about the density profile of the supercluster (falloff with radius), and about the actual heliocentric velocity of Virgo. Throughout we corrected velocities to the centroid of the local group via the standard IAU formula (300 sin(l) cos(b)). With reasonable estimates of the above parameters, adjusted to minimize the scatter about the simultaneously derived IRTF relation, the minimum χ^2 infall solution was:

$$\text{Pattern infall, } V_p = 250 \pm 64 \text{ km s}^{-1}$$
$$\text{Total motion to Virgo} = 331 \pm 41 \text{ km s}^{-1}$$
$$\text{Pattern rotation, } V_r = 180 \pm 58 \text{ km s}^{-1}$$

The derived values depend *weakly* on the assumed velocity and overdensity, and slightly more strongly on the assumed density profile of the Local Supercluster. The effects of some of the biases in our sample and in our fitting techniques were estimated by Monte-Carlo experiments. Note that the total motion is actually somewhat better determined than the pattern infall. And, nowadays, the "rotation" component plus the motion perpendicular to the plane of the LSC would probably be interpreted as due to the tidal field of Hydra-Centaurus or the Great Attractor.

IV. Ω FROM THE INFALL

We can derive Ω directly (Silk 1974; Peebles 1976) if we know the infall velocity and the enclosed mass overdensity. A simple relation for the linear infall model has been given by Gunn (1978):

$$\Omega^{0.66} \sim 3\left(\frac{V_p}{V_H}\right)\left(\frac{\delta\rho}{\rho}\right)^{-1}$$

where V_p = peculiar flow velocity with respect to Virgo, V_H is the the proper Hubble Velocity of Virgo = $V_p + V_{obs} + V_{rand}$, V_{obs} is V_h plus the correction to the local group, and $\delta\rho/\rho$ is the overdensity inside our radius. For non-linear models, simple corrections to this formula have been quoted by Villumsen and Davis (1986):

$$\Omega^{0.66} \sim 3\left(\frac{V_p}{V_H}\right)\left(\frac{\delta\rho}{\rho}\right)^{-1}\left(1+\frac{1}{3}\left(\frac{\delta\rho}{\rho}\right)\right)^{1/2}$$

and also by Yahil (1985):

$$\Omega^{0.66} \sim 3\left(\frac{V_p}{V_H}\right)\left(\frac{\delta\rho}{\rho}\right)^{-1}\left(1+\left(\frac{\delta\rho}{\rho}\right)\right)^{1/4}$$

These approximations to the exact solutions of the Friedmann equation are valid for small overdensities, or more precisely, far from the turnover radius. They can underestimate Ω when $\delta\rho/\rho$ is very large, but may overestimate Ω for flattened systems. The origin of the difference between these last two formulae have, for the moment, been lost in the depths of time, but Yahil's is correct for a supercluster with a density distribution similar to a de Vaucouleur's $R^{1/4}$ law (Regös and Geller 1988), and both equations give approximately the same results for the case examined here.

I want to stress again that the appropriate infall velocity, V_P, to use in the calculation of the supercluster mass is the *pattern* infall at the LG, **not** the total infall velocity, nor the total measured far-field motion. It also can be very dangerous to take the motion of an individual object rather than the full pattern, as has been pointed out by Bushouse *et al.* (1985), especially when the velocity field is very noisy (as in their models) and random motions are not corrected for. This is *not* a problem when the pattern infall is used.

The heliocentric velocity, V_h, of Virgo can be taken as either

$$1150 \pm 51\ km\ s^{-1} \quad \text{(JH)}\ *$$

$$1094 \pm 42\ km\ s^{-1} \quad \text{(JH - ST)}$$

$$1292 \pm 10\ km\ s^{-1} \quad \text{(M87)}$$

and corrected to Local Group Centroid by either

$$IAU_{corr} = -77\ km\ s^{-1}\ *$$

or the

$$STY_{corr} = -115\ km\ s^{-1}$$

The "STY" correction was adoped by Sandage and Tammann (1981) for the RSA and derived by Yahil, Tammann and Sandage (1977). The IAU correction is the compromise adopted by the IAU which includes both a galactic rotation component and the local random motion.

My preferred values (noted by *'s) give a Virgo velocity corrected to the local group of $V_{obs} = 1073 \pm 50$ km s^{-1}. We then can add in the peculiar and random component velocities from the IRTF solution, $V_{rand} = 81 \pm 30\ km\ s^{-1}$, and $V_P = 250 \pm 64\ km\ s^{-1}$, which gives $V_H = 1404\ km\ s^{-1}$. Finally, under the usual assumption that galaxies trace the mass, the enclosed overdensity, $\delta\rho/\rho$, has been estimated to lie between 2 and 3 (Davis *et al.* 1980; Yahil, Sandage and Tammann, 1980). The density falls off from Virgo with radius approximately as R^{-2} to R^{-3} (Jones 1976). Substituting into the second or third equations above we find

$$\Omega \sim 0.25\ to\ 0.35,$$

three or four times less than the closure density. This is larger than some of the earlier quoted values because we have approximately corrected to a non-linear model, and smaller than some earlier values because the pattern infall is only 250 km s^{-1}.

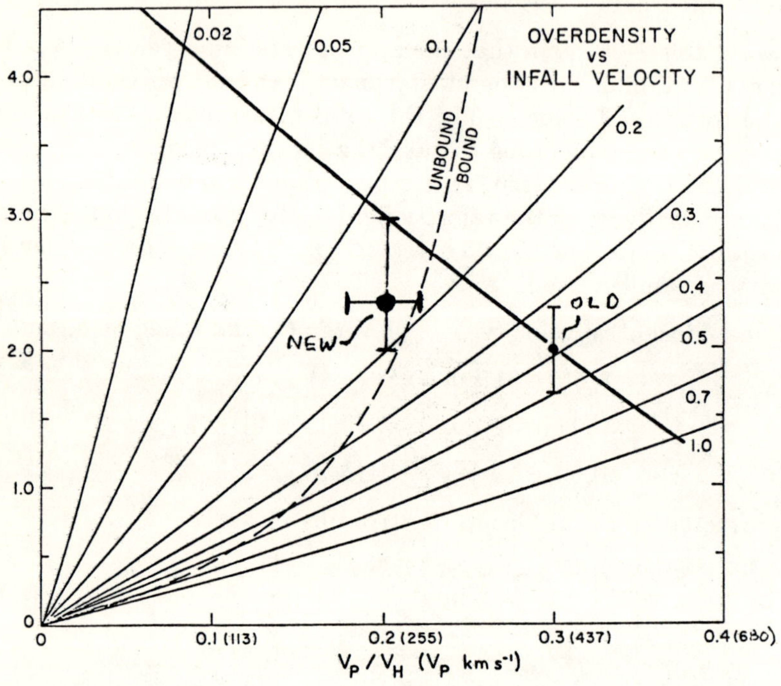

Figure 7. The over density-infall velocity plot of Silk for the full Friedmann infall solution. The dashed line separates the bound from the unbound solutions, the best current values are indicated. (Adapted from Davis *et al.* 1980).

Since the critcal mass-to-light ratio for the CfA survey is $\sim$ 1550 $M_\odot/L_\odot$ (Zwicky magnitudes, $H_0 = 100$), this implies that the $< M/L >$ of the material inside the LSC (inside the radius of the LG), is

$$< M/L >_{VF} \sim 400 \textit{ to } 500 \; M_\odot/L_\odot,$$

which is essentially the same $< M/L >$ found for galaxy clusters (Faber and Gallagher 1979; Geller 1988).

REMEMBER: This argument works backwards, too! That is to say, if the LSC is massive enough, $\Omega_{galaxies} \sim 0.3$, then V_P MUST be on the order of 250 km s^{-1}. This effect is demonstrated by Yahil's (1988) modeled pattern from the IRAS survey seen above. If Ω is 1, then the velocity field would have to be even more distorted unless the mass in the universe is very smoothly distributed on scales of 20 Mpc. This should provide an additional constraint on biased galaxy formation models.

One interesting sidelight to the study of Ω via the infall to Virgo was first pointed out by Silk (1974) and first experimentally fitted by Davis, *et al.* (1980). Solutions to the infall motion at the LG fall into two separate regions in the over-density/infall-velocity plane. Solutions in which the LG is not bound to the LSC lie to the left in figure 7 and bound solutions lie to the right. With the best values available in 1980, Davis *et al.* concluded that we were bound to the LSC and would fall into Virgo on a timescale of order 50 Billion years. With our lower solution for the infall velocity, the LG is comfortably unbound — you don't have to take out any cluster crash insurance just yet!

V. THE INFALL AND THE HUBBLE CONSTANT

Given a distance estimate for Virgo, D_{Virgo}, and the fully corrected Hubble, or cosmological velocity of Virgo, V_H, we can derive the Hubble Constant (neglecting, of course, other perverse flows).

Recent estimates of D_{Virgo}, for example, have been given by a variety of authors. Table 1 presents a representative summary.

Table 1 Virgo Distance Estimates

Name	(m-M)	D (Mpc)
de Vaucouleurs	30.70	13.8
Sandage and Tammann	31.70	21.9
Mould, Aaronson & Huchra *et al.* 1980	31.02	16.0
Pritchett & van den Bergh 1987	31.45	19.5
Pierce and Tully 1988	30.97	15.6
Dressler 1987	31.64	21.3
Harris 1988	31.7 (31.55)	21.9 (20.4)
Melnick 1988	30.9	15.1

Both de Vaucouleurs, and Sandage and Tammann have provided a large variety of estimates hovering around those required to give H_0 either 100 or 50 $\mathrm{km\,s^{-1}Mpc^{-1}}$. Mould *et al.* use the IRTF relation, Pierce and Tully use an I-band Tully-Fisher relation, Harris uses the turnover in the globular cluster luminosity function, Melnick uses a velocity dispersion/luminosity relation for extragalactic HII regions, and Dressler uses a $D_n - \sigma$ relation for the bulges of early type spirals to measure the distance to Virgo. The Harris values in parentheses are the distance modulus and distance that would result if the M31 globular clusters were used to calibrate the luminosity function turnover ($M_T \sim$ -6.85), rather than galactic globulars ($M_T \sim$ -7.0). The total range of currently quoted values runs from about 15 to 22 Mpc and is representative of the variance associated with the measuring techniques. This range is unlikely to converge until a few years after HST is launched.

The cosmological velocity of the cluster is the observed heliocentric velocity plus the correction to the center of the local group, plus the corrections for both pattern infall and random motion.

The lowest estimate of H_0 can be forced with the biggest distance and the lowest cosmological velocity:

$$\begin{aligned} D &\sim 23 \text{ Mpc}, \quad \text{and} \\ V_H &= 1052 - 115 + 290 = 1227 \ \text{ km s}^{-1} \\ \Rightarrow \text{H}_0 &= 53 \text{ km s}^{-1} \text{ Mpc}^{-1}. \end{aligned}$$

where I have added the quoted errors in each quantity in the appropriate direction (in this case, to lower the derived H_0). For this estimate, I have also used the STY correction to the local group, even though the AHMST flow solution was derived with the IAU correction.

The highest estimate can be forced by choosing the smallest distance and the largest velocity:

$$\begin{aligned} D &\sim 13.8 \text{ Mpc}, \quad \text{and} \\ V_H &= 1302 - 77 + 372 = 1597 \ \text{ km s}^{-1} \\ \Rightarrow \text{H}_0 &= 116 \text{ km s}^{-1} \text{ Mpc}^{-1}. \end{aligned}$$

The "best," consistent estimates I can come up with is based primarily on the IRTF distance plus zero points from infrared photometry of Cepheids in local group calibrators (eg. Freedman 1988), with the AHMST solution and (for consistency) the IAU correction for the LG centroid.

$$\begin{aligned} D &= 16.0 \text{ Mpc}, \quad \text{and} \\ V_H &= 1150 - 77 + 331 = 1404 \ \text{ km s}^{-1} \\ \Rightarrow \text{H}_0 &= 88 \text{ km s}^{-1} \text{ Mpc}^{-1}. \end{aligned}$$

Almost by definition, error bars on H_0 are meaningless, given the uncertainties in the distances of calibrators, biases in both data sets and observers, and the presence of large scale flows. H_0 might fall anywhere in the range above, or even outside it. Life would be easier for most theoretical cosmologists if the real value was near the low end.

VI. THE VIRGO CORE

A few years ago, as part of the extension of the CfA survey, I completed a redshift survey of the galaxies in the 6° radius circle centered on the Virgo Cluster. This work has been described in more detail elsewhere (Huchra 1985), but it is nonetheless important enough to update

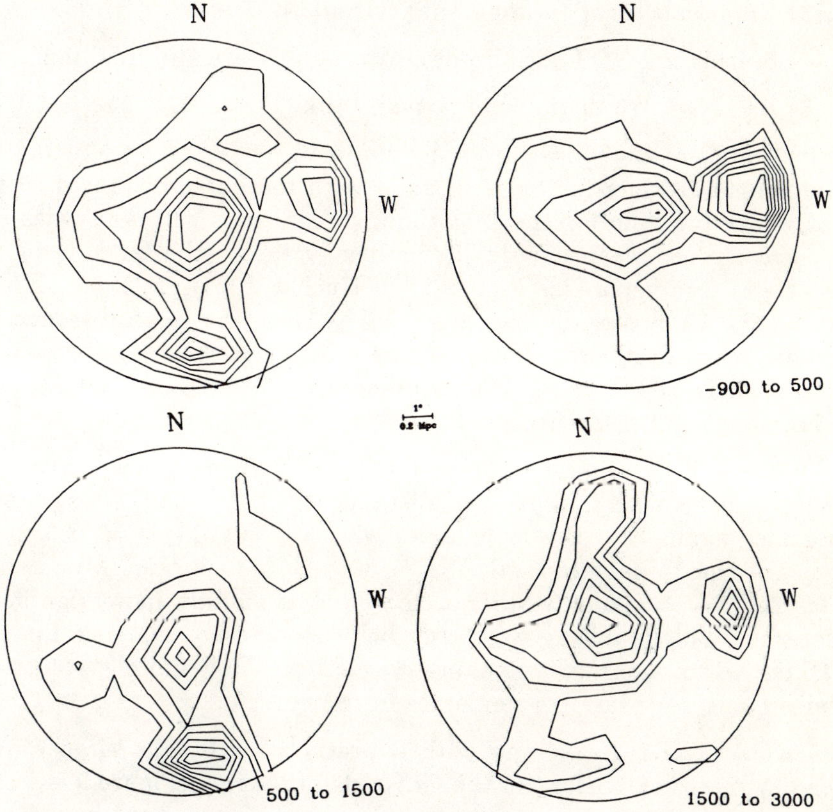

Figure 8. Surface distribution of (a) all galaxies with velocities $\leq$ 3000 km s^{-1}in the Virgo core, (b) the same for those galaxies with velocities less than 500 km s^{-1}, (c) galaxies between 500 and 1500 km s^{-1}, and (d) galaxies with velocities between 1500 and 3000 km s^{-1}. The clumpy nature of the cluster core is easily seen. (From Huchra 1985).

here.There are now 688 galaxies in the 6° circle with known redshifts, and 428 of these galaxies have redshifts $\leq$ 3000 km s^{-1}(and thus are possible cluster members). Figure 8 shows the isopleths for galaxies in different velocity ranges in the Virgo core.

We can redo the earlier analyses and we find that the important results have not changed:

(1) Virgo is clumpy, the 6° core is not dynamically relaxed.

(2) Virgo W at 2100 km s^{-1} is almost certainly background.

(3) Ellipticals form an apparently virialized core.

(4) Spirals are NOT virialized, and probably are still infalling.

(5) M/L $\sim$ 375, if the core is *bound* (like the M/L of the LSC!).

The infall of the spirals into the core can be tested by plotting velocity versus distance for those galaxies with measured distances. If the galaxies are infalling for the first time, the galaxies with velocities less than the cluster mean should preferentially lie on the far side of the cluster and vice versa (see Figure 11 of Huchra 1985). In 1985, I found that for the 12 cluster galaxies with $V_h \leq 1100$, the mean distance determined from IRTF was D = 17.7 $\pm$ 1.3 Mpc. For the 6 galaxies with $V_h \geq 1300$, D = 14.1 $\pm$ 1.0 Mpc, a difference, ΔD of 3.6 $\pm$ 1.6 Mpc, or a whopping big 2.25 σ effect.

Pierce and Tully (1988), have improved and increased the sample of Virgo spirals with independent distance estimates, so the test can be performed again. For the 14 galaxies with $V_h \leq 1100$, D = 18.3 $\pm$ 1.1 Mpc. For the 14 galaxies with $V_h \geq 1300$, D = 15.3 $\pm$ 0.7 Mpc, which yields ΔD = 3.0 $\pm$ 1.3 Mpc, a smaller, but slightly more significant difference. This is a long way from being conclusive evidence for core infall of the spirals, but it remains suggestive. We definitely need more good distances for late type galaxies in the core.

I would like to leave you with a pretty view of the Virgo cluster from our recent extensions of the CfA redshift survey. Figure 9 is a cone diagram of a 13° wedge centered on Virgo. The cut runs between 6° and 19° and from 9^h to 16^h, so it is almost perpendicular to the plane of the supercluster. The velocity limit is 4000 km s^{-1}. You can see that the Virgo cluster is relatively isolated in velocity space; the front of the Virgo-Coma Void can be seen just beyond the "finger of God" that represents the virial+infall motions in the cluster core. The Leo group lies to the right (at $\sim$ 1000 km $s^{-1}$and $11.^h5$), and the Virgo W complex at 2200 km $s^{-1}$and $12.^h3$ is seen as probably background to Virgo itself. Note that galaxies with negative velocities have not been plotted.

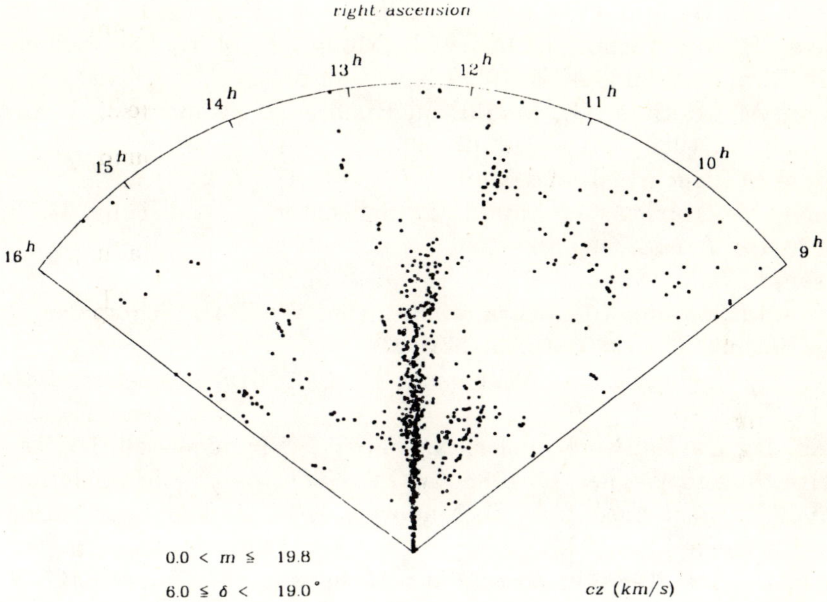

Figure 9. Cone diagram of the Virgo core region from the CfA Redshift Survey extensions. The center part of the strip, $8.°5 \leq \delta \leq 14.°5$, has nearly complete redshift information, the declination zones outside of that region are only about 40% complete.

ACKNOWLEDGEMENTS

I would like to thank my collaborators and friends, Marc Aaronson, Greg Bothun, Mark Cornell, Marc Davis, Jeremy Mould, Paul Schechter, Bob Schommer, Mike Strauss, John Tonry, Brent Tully, Amos Yahil, Dave Latham and Woody Sullivan, without whose efforts our recent studies of the Virgo cluster infall would never have been done. Especially important have been Marc Aaronson and Jeremy Mould. Until his untimely death last year, Marc Aaronson was the leader of our efforts to unlock the mysteries of the Hubble Flow. I would also like to thank Margaret Geller for useful discussions, Sidney van den Bergh for forcing me to write this up, and Barbara Elwell for help in assembling the manuscript. Typing services provided by Fang, Inc. This work has been supported by the Smithsonian Institution and by NASA grant NAGW-201.

REFERENCES

Aaronson, M., Bothun, G., Cornell, M., Dawe, J., Dickens, R., Hall, P., Sheng, H. M., Huchra, J., Lucey, J., Mould, J., Murray, J., Schommer, R. and Wright, A. E. 1988, *Ap. J.* in press.

Aaronson, M., Bothun, G., Mould, J., Huchra, J., Schommer, R. and Cornell, M. 1986, it Ap. J. **302**, 536.

Aaronson, M., Huchra, J. and Mould, J. 1979, *Ap. J.* **229**, 1.

Aaronson, M., Huchra, J., Mould, J., Schechter, P. and Tully, R. B. 1982, *Ap. J.* **258**, 64.

Aaronson, M. and Mould, J. 1983, *Ap. J.* **265**, 1.

Aaronson, M., Mould, J., Huchra, J., Sullivan, W. T. III, Schommer, R. and Bothun, G., 1980, *Ap. J.* **239**, 12.

Boughn, S., Cheng, E. and Wilkinson, D. 1981, *Ap. J. Letters* **243**, L113.

Burstein, D., Davies, R., Dressler, A., Faber, S., Lynden-Bell, D., Terlevich, R. and Wegner, G. 1986, in *Galaxy Distances and Deviations from Universal Expansion*, B. Madore and R. B. Tully, eds. (Dordrecht: Reidel), p. 123.

Burstein, D. *et al.* 1988, in *Large Scale Motions in the Universe*, G. V. Coyne and V. Rubin, eds. (Rome: Vatican Press).

Bushouse, H., Melott, A., Centrella, J. and Gallagher, J. 1985, *M. N. R. A. S.*, **217**, 1P.

Cheng, E., Saulson, P., Wilkinson, D. and Corey, B. 1979, *Ap. J. Letters* **232**, L139.

Davis, M. and Huchra, J. 1982, *Ap. J.* **254**, 437.

Davis, M. and Peebles, P. J. E. 1983, *Ann. Rev. Astron. and Ap.*, **21**, 109.

Davis, M., Tonry, J., Huchra, J. and Latham, D. 1980, *Ap. J. Letters* **238**, L113.

de Vaucouleurs, G. 1956, in *Vistas in Astronomy*, **Vol. 2**, A. Beer, ed. (New York: Pergammon), p. 1584.

de Vaucouleurs, G. 1958, *A. J.* **63**, 253.

de Vaucouleurs, G. 1972, in *I.A.U. Symposium No. 44, External Galaxies and Quasistellar Objects*, D. S. Evans, ed. (Dordrecht: Reidel), p. 353.

de Vaucouleurs, G. 1976, *Ap. J.* **205**, 13.

de Vaucouleurs, G. and de Vaucouleurs, A. 1964, *Reference Catalogue of Bright Galaxies*, (Austin: Univerity of Texas).

de Vaucouleurs, G. and de Vaucouleurs, A. 1973, *Astron. and Ap.* **28**, 109.

de Vaucouleurs, G. and Peters, W. 1968, *Nature* **220**, 868.

de Vaucouleurs, G. and Peters, W. 1981, *Ap. J.* **248**, 395.

de Vaucouleurs, G., Peters, W. Bottinelli, L., Gouguenheim, L. and

Paturel, G. 1981, *Ap. J.* **248**, 408.

Dressler, A. 1987, *Ap. J.* **317**, 1.

Dressler, A. Faber, S., Burstein, D., Davies, R., Lynden-Bell, D., Terlevich, R. and Wegner, G. 1987, *Ap. J. Letters* **313**, L37.

Fabbri, R., Guidi, I., Melchiorri, F., and Natale, V. 1980, *Phys. Rev. Letters* **44**, 1563.

Faber, S. and Gallagher, J. 1979, *Ann. Rev. Astron. and Ap.* **17**, 135.

Fixsen, D., Cheng, E. and Wilkinson D. 1983, *Phys. Rev. Letters* **50**, 670.

Freedman, W. L., 1988, this volume.

Geller, M. 1988, in *Large Scale Structures in the Universe*, 17th Annual Saas-Fee Course, L. Martinet and M Mayor, eds. (Geneva: Geneva Observatory), p. 69.

Gorenstein, M. and Smoot, G. 1981, *Ap. J.* **244**, 361.

Gunn, J. E. 1978, in *Observational Cosmology*, 8th Saas-Fee Course, Maeder, A., Martinet, L. and Tammann, G. eds. (Geneva: Geneva Observatory).

Han Ming Sheng and Mould, J. R. 1988, private communication.

Harris, W. 1988, this volume.

Hoffman, G., Olson, D. and Salpeter, E. 1980, *Ap. J.* **242**, 861.

Hoffman, Y. 1986, *Ap. J.* **308**, 493.

Huchra, J. 1985, in *The Virgo Cluster*, O. Richter and B. Binggeli, eds. (Garching: ESO), p181.

Kaiser, N. 1987, *M. N. R. A. S.* **227**, 1.

Jones, B. J. T. 1976, *M. N. R. A. S.* **174**, 429.

Lahav, O., Rowan-Robinson, M. and Lynden-Bell, D. 1988, preprint.

Lilje, P., Yahil, A., and Jones, B. 1986, *Ap. J.* **307**, 91.

Lubin, P., Epstein, G., and Smoot, G. 1983, *Phys. Rev. Letters* **50**, 616.

Lubin, P., Villela, T., Epstein, G., and Smoot, G. 1985, *Ap. J. Letters* **298**, L1.

Madore, B. and Tully, R. B. eds, 1986 *Galaxy Distances and Deviations from Universal Expansion*, (Dordrecht: Reidel).

Meiksin, A. and Davis, M. 1986, *A. J.* **91**, 191.

Melnick, J. 1988, this volume.

Mould, J. Aaronson, M. and Huchra, J. 1980, *Ap. J.* **238**, 458.

Ogorodnikov, K. F. 1952, *Vop. Kosmogony*, **1**, 150.

Ostriker, J., Peebles, P. J. E. and Yahil, A. 1974, *Ap. J. Letters* **193**, L1.

Peebles, P. J. E. 1976, *Ap. J.* **205**, 318.

Pierce, M. and Tully, R. B. 1988, *Ap. J.* **330**, 579.

Pritchet, CV. and Van den Bergh, S. 1987, *Ap. J.* **318**, 507.

Regös, E. and Geller, M. 1988, private communication.

Rubin, V. 1951, *A. J.* **56**, 47.

Rubin, V., Thonnard, N., Ford, W. K. and Roberts, M. 1976, *A. J.* **81**, 719.

Sandage, A. and Hardy, E. 1973, *Ap. J.* **183**, 743.

Sandage, A. and Tammann, G. 1976a, *Ap. J. Letters* **207**, L1.

Sandage, A. and Tammann, G. 1976b, *Ap. J.* **210**, 7.

Sandage, A. and Tammann, G. 1981, *Revised Shapley-Ames Catalog of Bright Galaxies*, (Washington: Carnegie Institution).

Sandage, A. and Tammann, G. 1988, *Ap. J.* **328**, 1.

Schechter, 1980, *A. J.* **85**, 801.

Sandage, A., Tammann, G. and Hardy, E. 1972, *Ap. J.* **172**, 253..

Shapley, H. and Ames, A. 1932, *Annals H. C. O.* **88**, No. 2.

Shapley, H. 1934, *M. N. R. A. S.* **94**, 791.

Shaya, E. 1984, *Ap. J.* **280**, 470.

Shaya, E. 1986, in *Galaxy Distances and Deviations from Universal Expansion*, B. Madore and R. B. Tully, eds. (Dordrecht: Reidel), p. 229.

Silk, J. 1974, *Ap. J.* **193**, 525.

Silk, J. 1977, *Astron. and Ap.* **59**, 53.

Smoot, G. and Lubin, P. 1979, *Ap. J. Letters* **234**, L83.

Stewart, J. and Sciama, D. 1967, *Nature* **216**, 748.

Strauss, M. and Davis, M. 1988, in *Large Scale Motions in the Universe*, G. V. Coyne and V. Rubin, eds. (Rome: Vatican Press).

Tammann, G. 1972, *Astron. and Ap.* **21**, 355.

Tammann, G. and Sandage, A. 1985, *Ap. J.* **294**, 81.

Tonry, J. and Davis, M. 1981, *Ap. J.* **246**, 680.

Tully, R. B. 1988, *Nearby Galaxies Catalog*, (Cambridge: Cambridge).

Tully, R. B. 1988, this volume.

Tully, R. B. and Fisher, J. R. 1977, *Astron. and Ap.* **54**, 661.

Tully, R. B. and Shaya, E. 1984, *Ap. J.* **281**, 31.

Villumsen, J. and Davis, M. 1986, *Ap. J.* **308**, 499.

Weiss, R. 1980, *Ann. Rev. Astron. and Ap.* **18**, 489.

Wilkinson, D. T. 1984, in *Large Scale Structure of the Universe, Cosmology and Fundamental Physics*, G. Setti and L. Van Horn, eds. (Geneva: CERN), p. 153.

Yahil, A. 1985, in *The Virgo Cluster*, O. Richter and B. Binggeli, eds. (Garching: ESO), p359.

Yahil, A. 1988, in *Large Scale Motions in the Universe*, G. V. Coyne and V. Rubin, eds. (Rome: Vatican Press).

Yahil, A., Tammann, G. and Sandage, A. 1977, *Ap. J.* **217**, 903.

Yahil, A., Sandage, A. and Tammann, G. 1980, *Ap. J.* **242**, 448.

Yahil, A., Walker, D. and Rowan-Robinson, M. 1986, *Ap. J. Letters* **301**, L1.

DISCUSSION

S. VAN DEN BERGH: Could you say something about the reason why you find a much higher value of the infall velocity than do Sandage and Tammann?

J. HUCHRA: Actually, the two values are almost within their quoted errors: 220 ± 50 for Sandage and Tammann's average and 250 ± 64 for AHMST. That's pretty good if you also include the fact that Sandage and Tammann's average includes four values from exactly the same data set.

B. TULLY: You mentioned that AHMST identified something modelled as rotation of the Local Supercluster and, indeed, de Vaucouleurs suggested such motion much earlier. The current interpretation would be that this is motion caused by the 'Great Attractor' and, perhaps, associated with the 'local velocity anomaly'.

J. HUCHRA: Yes, that is correct. Originally we called it rotation but a variety of authors, Lilje, Yahil and Jones and you and Shaya, for example, identify this component as probably due to the tidal torque of Hydra-Centaurus.

C. MANNING: Was your derivation of the over-density within the Virgo Cluster based on the assumption that light traces mass?

J. HUCHRA: Yes, by definition. Marc Davis has found it necessary to assume light does not trace matter in his recent simulations. This has an effect on the derived Ω.

THE DISTANCE OF THE VIRGO CLUSTER - A REVIEW

G. A. Tammann
Astronomisches Institut der
Universität Basel
European Southern Observatory, Garching

ABSTRACT. Recent distance determinations of the Virgo cluster yield converging results: from globular clusters (Harris, 1988) (m-M)=31.74±0.17, from novae (Pritchet and van den Bergh, 1987) 31.57±0.43, from supernovae (Tammann, 1987) 31.86±0.20, and from the D_n/σ relation (Dressler, 1987) 31.85±0.19. Previous discrepancies of the blue and/or infrared Tully-Fisher relation are due to selection effects; a nearly complete sample of 73 inclined spirals (Sb-Sm) is free of such selection ambiguities and requires 31.60±0.15 (Kraan-Korteweg et al., 1988). The formal mean of 31.73±0.09 (r=22 Mpc) is quite insensitive to the particular choice of the local calibrators. The modulus difference between subclusters A and B is found to be 0.00±0.08. - Relative distance indicators beyond Virgo have consistently led adherents of the long and short distance scales - independent of the true cluster recession velocity and our Virgocentric infall velocity - to the conclusion that H_o(global)xr(Virgo)≈1240+90 km s^{-1} and hence H_o(global)≈56 from all available evidence.

1. INTRODUCTION

At this time it is a pleasure to review the Virgo cluster distance because for the first time there are five independent distance determinations available, that agree so well, that a realistic value seems at hand. The history of the Virgo distance has been outlined by Richter (1985) and de Vaucouleurs (1986). Beginning with Hubble (1929), who put the cluster at 2 Mpc, the distance increased with time to reach an asymptotic value of roughly 22 Mpc during the last decade. Mainly due to the advent of the Tully-Fisher method in 1976 the data became, however, quite noisy, with distances ranging from ~13-24 Mpc. Thanks to the enormous progress over the last few years it is now possible to reduce this noise quite drastically.

In Section 2-6 the five independent routes to the cluster distance are discussed. The conclusions are given in Section 7.

2. GLOBULAR CLUSTERS

Distances of galaxies can be derived on the assumption that the luminosity function of their globular clusters peak at constant luminosity. For the Virgo cluster a reliable exploit of this method has long been unfeasable, because the turn-over point of the luminosity function occurs for the corresponding distance at an apparent magnitude as faint as ~25 mag. An additional problem is the successful treatment of the important fore- and background contamination. It marks the impressive progress of astronomical techniques that the photometry of Virgo globulars could now be pushed beyond the peak of the luminosity function (van den Bergh et al., 1985; Cohen, 1988a). The ultimate state of the art has been reached by Harris (1988). His thorough discussion covers also the crucial question whether the peak luminosity of the (outer) cluster population is universal, and he dispels much of the doubt about this matter. For four Virgo ellipticals he finds the peak of the luminosity function at $B=24.74\pm0.08$; with a Galactic peak luminosity of $M_B=-7.00\pm 0.15$ [assuming M_V(RR Lyr)=0.6] he obtains then a Virgo cluster modulus of 31.74 ± 0.17.

One of Harris' galaxies (M87) lies in the Virgo subcluster A and two (N4472, N4365) lie in subcluster B (cf. Binggeli et al., 1987). For the former he finds the peak luminosity at $B=24.99\pm0.12$, for the two latter at a mean of $B=24.78\pm0.09$. If anything this puts B in front of A (cf. Section 6).

3. NOVAE

The discovery of novae in the Virgo cluster by Pritchet and van den Bergh (1986) is an observational achievement that would have been thought to be unfeasable still a few years ago. They discovered nine novae, for six of which they could determine post-maximum decline rates d (mag.per day). The latter six novae lie in NGC 4472.

In M31 the relation between m_B(max) and log d is reasonably well determined from 57 novae that were observed by Arp (1956) and Rosino (1964, 1973). A comparison of the Virgo novae with the m_B(max)-log d diagram of M31 shows that the former are 6.8 ± 0.4 mag. fainter than their M31 counterparts. This magnitude difference carries a selection bias correction of -0.2 ± 0.1 as determined by the original authors from Monte Carlo simulations. The true distance modulus of M31 of $(m-M)=24.26\pm0.08$ from the infrared photometry of Cepheids (Welch et al., 1986; cf. also Pritchet and van den Bergh, 1988) is hardly at stake; also at this Conference the quoted moduli bracket this value to within less than ±0.1. However, what is needed here is not the true modulus but the apparent blue modulus $(m-M)_{AB}$. While Pritchet and van den Bergh (1987) implied $A_B=0.39$ (if the present true modulus is adopted), Sandage and Tammann (1988) have argued for a somewhat higher compromise value of $A_B=0.51\pm0.15$. Using the latter value one obtains $(m-M)_{AB}=24.77\pm0.17$ for M31. Stepping up this value by 6.8 ± 0.4 yields a Virgo modulus of $(m-M)=31.57\pm0.43$. Strictly speaking this value refers to subcluster B, because NGC 4472 is the first-ranked galaxy of this structure.

The prospects to tie the Virgo novae directly to the Galactic m_B(max)-log d relation and thus to obtain a truly independent distance determination are discussed by Cohen (1988b). When a Galactic calibration will become competitive, an additional bias problem may turn up because different selection mechanisms may work on the novae in the Galaxy and on novae seen in external galaxies. This makes the present comparison between two external galaxies (M31 and NGC 4472) particularly attractive. The only requirement is that the two novae samples are equally biased, whatever the true parent population of the m_B(max)-log d diagram may be. This parent distribution is still quite uncertain, and one may therefore wonder if the Monte Carlo bias correction of Pritchet and van den Bergh (1987) is indeed justified. Only additional Virgo novae with well observed decline rates will solve this ambiguity and will - more importantly - reduce the still rather large fitting error of 0.4 mag. This calls for further extended observing runs with large telescopes.

4. SUPERNOVAE

Supernovae of type Ia are good standard candles. The three best observed SNeIa follow a mean light curve to within <0.1 mag. in six different wavebands for at least 100 days past maximum (Leibundgut, 1988). The distinction between "fast" and "slow" SNeIa can be shown in most cases to be the result of poor photometry (Cadonau et al., 1985). Their spectra are of telling similarity but for a few exceptions. Differences of their expansion velocities at maximum are small enough to be difficult to establish beyond doubt (cf. Branch, 1988). The eleven SNeIa in the Virgo cluster with reasonably well defined B(max) have an observed scatter of

$\sigma(M)$=0.48 mag. (Table 1), most of which must be due to poor photometry - that in some cases is quite casual, - to the color-sensitive corrections for internal absorption, and to a lesser extent to front-to-back effects in the cluster. This lends support to the assumption that SNeIa are very useful standard candles indeed.

Table 1. SNeIa in the Virgo Cluster with B(max).

SN	Galaxy	Type	Subcluster	B°(max)
1919a	N4486	E	A	11.89
1939a	N4636	E/S0	B	12.69
1939b	N4621	E/S0	A	12.29
1957b	N4374	E	A	12.12
1960r	N4382	S0 pec	A	11.61*
1961h	N4564	E	A	11.75
1965i	N4753	S0 pec	A	12.53*
1976b	N4402	Sb	A	13.16*
1981b	N4536	Sc	B	11.96*
1983g	N4753	S0 pec	A	12.56*
1984a	N4419	Sab	A	11.96*
				12.23±0.14

*corrected for internal absorption

The mean value of B(max) of the five SNeIa in Table 1, that occurred in (nearly dust-free) E/S0 galaxies, is ⟨B(max)⟩=12.15±0.16. The six SNeIa in S0pec and spiral galaxies, which were corrected for their internal absorption on the assumption that $(B-V)_{max}$=-0.27 (Cadonau, 1986) and $A_B=E_{B-V}$ (Tammann, 1987), give ⟨B(max)⟩=12.30± 0.23. The satisfactory agreement of these two means justifies the adoption of an overall mean of ⟨B(max)⟩= 12.23±0.14. This should provide a good Virgo distance if in addition the absolute magnitude of SNeIa is known.

There are presently five routes to estimate M_B(max) of SNeIa, ranging from two SNe that occurred in galaxies whose distances are known from brightest stars to model predictions of deflagrating C/O white dwarfs. The different methods and their references are discussed elsewhere (Tammann, 1987) and are not repeated here. Each method carries a considerable uncertainty of typically ±0.6 mag., but the combined evidence gives a useful approximation of M_B(max)= -19.70±0.25. This yields with ⟨B(max)⟩=12.23±0.14 a Virgo cluster modulus of (m-M)=31.93±0.29.

Nine of the SNeIa in Table 1 have occurred in subcluster A, only two in subcluster B. Taken at face value they suggest a modulus difference $(m-M)_{A-B}$=-0.15±0.34.

In addition there are several direct distance determinations for the type II SN 1979c in the Virgo cluster spiral M100 (subcluster B). A consistent expansion parallax of $(m-M)=31.80\pm0.30$ was found by Panagia et al. (1980), Branch et al. (1981), and Kirshner (1985). VLBI measurements of the radiosphere of the same SN have led Bartel (1988) to a distance modulus of 31.71 ± 0.70.

The weighted Virgo modulus from SNeIa and from the one SN II becomes then 31.86 ± 0.20. It is clear that any new, well observed SN in the Virgo cluster, which should produce about one SN per year, will further improve the result. Eventually the route via SNe will become a highly competitive method for the distance of the Virgo cluster.

5. THE D_n- σ RELATION

Dressler et al. (1987) have shown that there is a tight correlation between the diameter D_n and the aperture-dependent velocity dispersion σ for elliptical galaxies. Here D_n is the diameter of a circle that encompasses a specified mean surface brightness. This photometrically defined diameter depends on the adopted absorption and the luminosity profile of the E galaxy. The diameter correction given by Lynden-Bell et al. (1988) is $\Delta\log D_n=0.32\Delta A_B$, which translates into a distance effect of $\Delta(m-M)=1.6\Delta A_B$. The absorption has paradoxially thus a stronger effect on this diameter distance than on a luminosity distance. With this caveat it is possible to derive exceptionally good <u>relative</u> distances from the D_n- σ relation, because its scatter amounts for Virgo ellipticals to only $\sigma(\log D_n)=0.059$, corresponding to $\sigma(M_B)=0.30$ mag. However, lacking a suitable local calibrator the relation cannot be used to derive absolute distances.

It is therefore most remarkable that Dressler (1987) could extend the D_n-σ relation to bulges of S0 and early-type spirals. To avoid serious problems with disc light and internal absorption he redefined D_n to a brighter mean surface brightness, viz. 19.75 B mag.arcsec.$^{-2}$. From a sample of 26 Virgo galaxies, and omitting two of them, he established the following relation (cf. Fig.1)

$$\log D_n=1.333 \log \sigma-(1.572\pm0.014) \qquad (1)$$

Some of the scatter about this relation is due to the (considerable) errors of the velocity dispersion σ and to front-to-back effects. If it is read entirely in $\log D_n$, it still amounts to only $\sigma(\log D_n)=0.068$, corresponding to $\sigma(m-M)=0.34$ mag. This suggests that the internal dust problems have been dealt with successfully. A more serious problem are the two omitted galaxies, NGC 4382 and NGC 4417. Are they just the tip of an iceberg? Do the remaining 56 Virgo members of type S0-Sb [Binggeli et al. (1985)], which are fainter on average by 1.3 mag., have indeed smaller D_n and σ values, or do they contaminate the mean relation of Fig.1? The danger is not so severe because the three calibrator galaxies used below match roughly the luminosity of the galaxies of the Dressler sample. Moreover the brighter half of that sample yields the same Virgo modulus as the fainter one to within 0.07±0.14 mag. This shows that the relation of eq.(1) is fortunately quite insensitive to absolute magnitude.

A Virgo modulus can be derived from eq.(1) if galaxies with known D_n and σ and known distance are available. This is the case for three spirals, viz. the Galaxy, M31, and M81; their relevant data, which are set out in Table 2 (cf. also Fig.1), need some comments. The data for the Galactic bulge are taken from Terndrup (1988), those of M31 and M81 from Dressler (1987). The Galaxy may be of somewhat later type than the S0-Sb sample, but Dressler (1987) has shown that also the Sbc galaxy NGC 4501 fits well the mean relation. In principle the Virgo cluster distance has to be known to match the velocity dispersion σ of the calibrators to the aperture size (5"x5") that was applied for Virgo. However, the remaining mismatch should introduce only negligible systematic errors (cf. Dressler, 1987). The distances of the Galactic bulge, of M31 and M81 are from Tern-

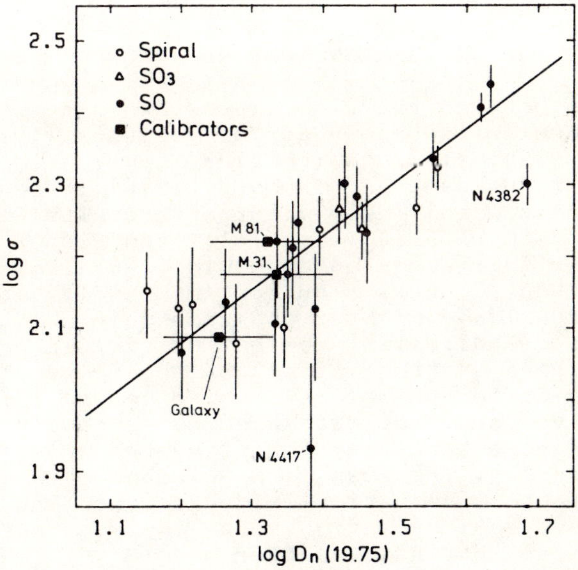

Fig.1. Dressler's (1987) D_n-σ relation for the bulges of Virgo galaxies. The three calibrating galaxies are shown for a best fitting Virgo modulus of 31.87.

drup (1987), Welch et al. (1986), and Freedman (1988) as well as Jacoby (1988), respectively. The diameter D_n of the M31 bulge was reduced to a compromise front absorption of A_B=0.44±0.10 (McClure and Racine, 1969).

The three calibrating bulges fit eq.(1) best with a distance modulus of the Virgo cluster of 31.85±0.18 (cf.Table 2). Taking into account the zero point error of eq.(1) one obtains (m-M)=31.85±0.19.

Table 2. The Calibration of the D_n-σ Relation

Calibrat.	$(m-M)^\circ$	log D_n	σ	log D_n	(m-M)	$(m-M)_{Virgo}$
Gal.Bulge	14.46±.20	4.741±.040	124±9	3.522±.059	17.61±.30	32.07±.36
M31	24.26±.15	2.862±.035[1]	150±5	1.533±.040	7.67±.20	31.93±.25
M81	27.65±.30	2.148±.027[2]	166±8	0.762±.039	3.81±.19	31.46±.36
						31.85±.18

[1] A_B=0.44±0.10
[2] A_B=0.15±0.08 (Burstein and Heiles 1978)

Twenty-one galaxies of the Dressler sample lie in subcluster A, only three in B. The formal distance difference between the two uneven sets is $\Delta(m-M)_{A-B}$=0.03±0.13.

6. THE 21cm LINE WIDTH-MAGNITUDE RELATION

The relation between the 21cm line width and luminosity of disc galaxies is also known as Tully-Fisher (1977a) relation or TF relation. Since its first application to the Virgo cluster (Sandage and Tammann, 1976) it has been used by many authors with widely divergent results. While some authors found (m-M)$\lesssim$31.0 (e.g. Tully and Fisher, 1977b; Visvanathan, 1981; Aaronson an Mould, 1983; Bottinelli et al., 1984, 1988), others required (m-M)$\gtrsim$31.5 (Sandage and Tammann, 1976, 1984; Richter and Huchtmeier, 1984). Different calibrators, different corrections for internal absorption (which have a strong effect on the slope of the TF relation), and different colors (e.g. blue and infrared) were used to derive the Virgo distance, but an analysis of these differences shows that they cannot be the sole reason for the discrepancies. The suspicion that the main reason for the unsatisfactory situation is due to sampling effects was elaborated by Teerikorpi (1987); he implied that, if the TF relation has an intrinsic scatter of $\sigma(M)\gtrsim$0.5 mag., one has to consider complete samples very deep into the cluster luminosity function.

It has been proposed (Schechter, 1980; Kraan-Korteweg et al., 1986) to use the inverse TF relation (logΔv_{21}

versus M) in order to avoid the selection bias (Malmquist effect) of magnitude-limited samples. This route has recently led Tully (1988a, b) to suggest a low Virgo distance. However, the elimination of the Malmquist bias has its price: the slope <u>and</u> the intercept of the regression line of the inverse TF relation are inherently different from one sample to another. It is therefore

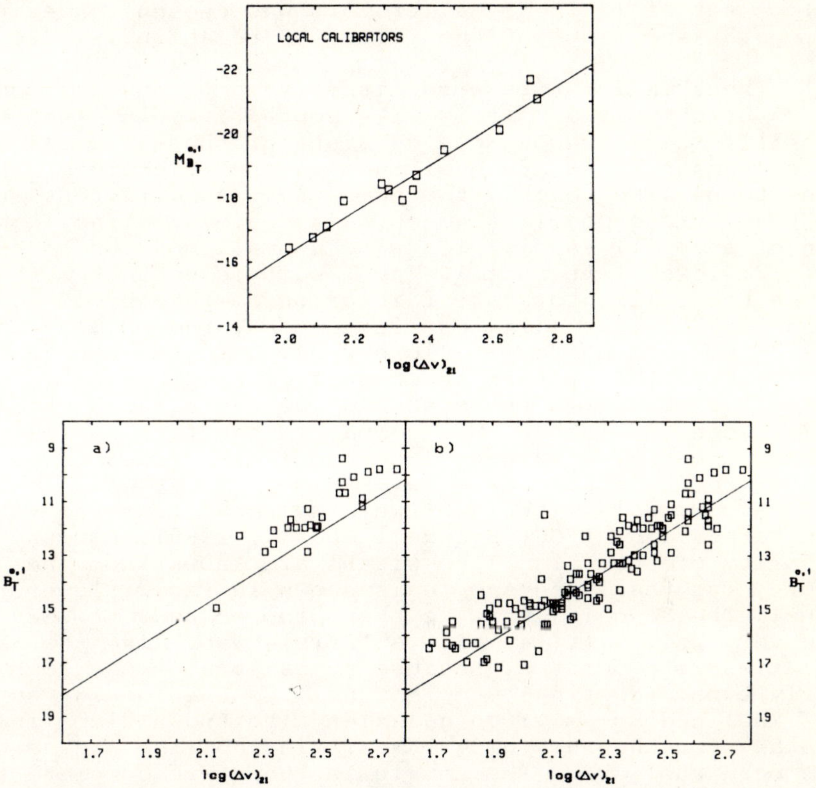

Fig.2. Upper panel: the blue Tully-Fisher relation for the calibrating galaxies. Lower left panel: the Tully-Fisher relation of 25 bright Virgo galaxies. (The sample happens to agree with the set of Virgo members for which H-magnitudes are available). This sample or similar ones were frequently directly compared to the local calibrators. A naive comparison yields a highly biased modulus of ~30.9. The full drawn line is taken from the lower right panel: here a (nearly) <u>complete</u> sample of 87 Virgo Sa-Sm spirals with inclinations >45° is shown. The full drawn line is the best fit to the subset of 73 Sb-Sm galaxies and requires - if compared to the local calibrators - a modulus of 31.60.

impossible to combine in any meaningful way the inverse regression of a (magnitude-limited) sample of distant galaxies with a (distance-limited) sample of nearby calibrators.

The effect of selection bias on the TF distance of the Virgo cluster has recently been studied in detail by Kraan-Korteweg et al. (1988). They proceeded essentially in three steps:

(1) A set of 13 local calibrators was chosen to define the zeropoint and slope of the (blue) TF relation (Fig.2, upper panel).

(2) These calibrators were directly compared to a sample of 25 bright Virgo spirals with appropriate inclinations ($i>45°$). This Virgo sample consists of those spirals for which H-magnitudes are published in the literature; it happens to be also roughly the same that led different authors to small Virgo moduli. Indeed, also here, a direct comparison of step (1) and (2) leads to a small modulus of ~30.9 (Fig.2, lower left panel). The result is untrustworthy because the calibrators are fainter on average by $\Delta M=1.7$ mag. than the 25 Virgo galaxies (if a mean Virgo modulus of ~31.7 is assumed from Sections 2-5). Moreover the former are systematically fainter in surface brightness by $\Delta m'_{25}=0.25$ mag. (in the system of de Vaucouleurs et al., 1976) than the latter. One compares therefore apples with pears. Clearly, a larger Virgo sample is needed to ban the dangers of selection effects.

(3) All spirals with $i>45°$ were selected from the Virgo Cluster Catalog (VCC; Binggeli et al., 1985). For the vast majority of them 21cm line widths are known from the literature. The resulting sample is shown in Fig.2, lower right panel. The Sa galaxies were then dropped because they follow very poorly, if at all, a TF relation. Also the Sab galaxies were omitted, because no calibrator has such an early type. The final sample consisted then of 73 (out of 89) inclined Sb-Sm Virgo galaxies with known 21cm line widths. For the remaining 16 galaxies no line widths have yet been published. The sample is therefore complete to 82%, and it is considered to be sufficiently representative and unbiased that it is called here "complete". A comparison of this complete sample with the calibrators in step (1) yields a Virgo modulus of 31.60±0.15.

The very large difference between the distances resulting from step (2) and (3) may be surprising. As a comparison of the two TF relations in the lower panel of Fig.2 shows, the reason for the difference is the strong bias of the 25 galaxies in step (2). This sample contains at any given line width particularly bright galaxies, which results in an artificially small scatter $\sigma(M)$ and in a regression line which is too bright by ~0.7 mag. The complete sample must define a much better regression line and it reveals also the true scatter of $\sigma(M)=0.7$ mag. of the (blue) TF relation. This scatter is not the result of a front-to-back effect. Such an effect would be expected to be particularly strong between subclusters A and B. A separation of the complete sample according to these subclusters indicates a distance difference of $\Delta(m-M)_{A-B}=-0.31\pm0.21$, but the scatter $\sigma(M)$ is not significantly reduced for the two subsamples. That the observed scatter is real is also supported by field galaxies for which Roberts (1987) and Sandage (1988) have determined $\sigma(M)$ to be $\sim$1.0 mag.

Kraan-Korteweg et al. (1988) have found a pronounced dependence of the TF relation on the galaxy type. Its actual size depends on the adopted correction for internal absorption, but it constitutes a problem which has to be dealt with. One could introduce the type as an additional free parameter into the TF relation. However, it was preferred instead to match the observable parameters of the calibrators and the complete sample as closely as possible. The match in mean (number-coded) type T and line width is shown in Table 3. The two sets of galaxy agree also well in absolute magnitude and surface brightness. It may seem surprising that the local calibrators and the complete sample, which leave little freedom of manipulation, yield such a good match in parameter space, but actually it is exactly what one expects from fair samples of galaxies.

Table 3. The Match of the Local Calibrators and the 82% Complete Virgo Sb-Sm Sample

	Calibrators	Virgo Sample
$\langle T \rangle$	6.2	6.0
$\langle \log\Delta v_{21} \rangle$	2.37	2.37

The quoted TF distance of the Virgo cluster rests on blue magnitudes, because infrared H-magnitudes are available only for a small subset of the complete sample. Yet by means of the tight relation between line width and (B-H)-color, Kraan-Korteweg et al. (1988) could show that also the infrared data are entirely consistent with a Virgo modulus of 31.60, - in fact any other value would question the universality of the TF relation. The infrared TF relation may have a slightly smaller dispersion $\sigma(M)$ than the blue relation (cf. also Sandage, 1988), but this possible advantage is offset by its steeper slope. - For further details on the TF distance of the Virgo cluster the reader is referred to Kraan-Korteweg et al. (1988).

The remaining incompleteness of the "complete sample" has probably little influence on the result. The missing 16 galaxies would decrease the mean magnitude of the total sample by 0.12 mag. The Virgo modulus would increase by the same amount _if_ the missing galaxies had on average the same line width. Since they have rather smaller line widths the effect on distance is likely to cancel in first order. A more severe problem is posed by the possibility that cluster spirals follow a systematically different TF relation as field galaxies (i.e. local calibrators). Van den Bergh (1984) and Gratton (1988) have argued that the gas and dust deficiency of the Virgo spirals may lead to an underestimate of the distance of 0.2-0.3 mag. If this is taken at face value the mean Virgo distance adopted in Section 7 has to be increased by 0.08 mag.

7. CONCLUSIONS

The five distance determinations of the Virgo cluster of Sections 2-6 are compiled in Table 4. The agreement of the individual values is very good. They give a weighted mean of $(m-M)=31.73\pm0.09$, equivalent to 22.2 ± 0.9 Mpc.

Table 4. The Virgo Modulus from 5 Methods

	$(m-M)^{\circ}$	Galaxy Type	Calibrators
Globular Clusters	31.74±0.17	E	RR Lyr
Novae	31.57±0.43	E	M31
Supernovae	31.86±0.20	E,S0,S	diff.methods
$D_n-\sigma$ relation	31.85±0.19	S0,S	Galaxy,M31,M81
21cm line widths	31.60±0.15	S	13 nearby gal.
	31.73±0.09		

The weight of this result is enhanced by the fact that the different avenues employ different types of galaxies, from E's to spirals. Moreover they rest on a variety of calibrators (cf. Table 4). While the globular clusters rely only on population II, the novae depend but on M31 whose distance is particularly well known. The $D_n-\sigma$ relation uses in addition the Galactic bulge and M81, which, however, enter with lower weight. The supernova distance comprises a number of calibrations including purely physical methods. Merely the 21cm line width distance depends on a larger number of local calibrator galaxies. Of the 13 such galaxies eight were discussed during this Conference by different authors. Their mean distances are 0.12 mag. smaller on average than the ones adopted for the calibration of the 21cm line width-magnitude relation. Such a small shift will change the overall mean of the Virgo cluster distance by only 2%. It is quite surprising that the cluster distance depends now so little on the local calibrators.

Possible sources of systematic errors were mentioned in Sections 2-6. The close agreement of the five distance determinations in Table 4 leaves, however, little room for systematic errors because the assumption seems too contrived that the independent methods should all suffer systematic errors of the same size (above $\sim$0.3 mag.) and sign. The external error of the adopted Virgo modulus is therefore likely to be smaller than three times the standard deviation, i.e. $3\sigma < 0.30$ mag. This estimate allows also for a possible foreground absorption of <0.10 mag; throughout this paper a value of $A_B=0.00$ was adopted as best approximation (cf. Sandage and Visvanathan, 1978).

Table 5. The Relative Distance of Subcluster A and B.

Method	$(m-M)_{A-B}$
Globular Clusters	0.21±0.15
Supernovae	-0.15±0.34
$D_n-\sigma$ relation	-0.03±0.13
21cm line widths	-0.31±0.21
first-ranked gal.	0.30±0.35
	+0.00±0.08

A longstanding question is the relative distance of subclusters A and B. Four determinations were given in the previous Sections. They are shown in Table 5 together with the relative distance that is indicated by the two respective first-ranked galaxies, M87 and NGC 4472 (Sandage and Tammann, 1987). The weighted mean difference is $\Delta(m-M)_{AB}=+0.00\pm0.08$. This shows that the two subclusters are at the same distance to within ±4%. In that case they should also have the same mean velocities. Indeed the velocity difference of 98+116 km s^{-1} (Binggeli et al., 1985) is only insignificantly different from zero.

With a mean cluster velocity of v=1196±67 km s^{-1} (for A and B, corrected for a Virgocentric infall velocity of v_{VC}=220 km s^{-1}; Binggeli et al., 1985) and r(Virgo)=22.2±2.7(3σ)Mpc the Hubble constant becomes H_o=53.9±7.2 km s^{-1}Mpc^{-1} at the distance of the cluster. The decisive question remains how close this solution is to the global value of H_o. Many authors have carried the distance scale beyond the Virgo cluster using a variety of (relative) distance indicators like first-ranked cluster galaxies, SNeIa, and the Tully-Fisher relation. A selection of relevant solutions is given in Table 6. The list is not exhaustive, but gives a representative overview. The most surprising result is that, although there are wide variations in r(Virgo) and H_o(global) [at roughly v_o 5000 km s^{-1}] between different authors, the product of r(Virgo)x H_o(global) is the same to within ±15%; this holds even for Hubble's (1936) early estimate. This shows that the Virgo cluster is tied so intimately into the far-out distance scale, irrespective of its observed velocity and the Virgocentric infall velocity, that its distance is truly fundamental for the determination of H_o(global). Thus with r(Virgo)x H_o(global)=1240±90(3σ) km s^{-1} (from Table 6) and r(Virgo)=22.2±3.3(3σ) Mpc everybody's H_o(global) becomes 55.9±9.2(3σ) km $s^{-1}Mpc^{-1}$. A weighted mean - admittably based on personal judgement - favors a product of 1150±50 km s^{-1} and hence H_o(global)=51.8±5.1 km $s^{-1}Mpc^{-1}$.

Table 6. The Virgo Distance and the Global Value of H_o from Different Authors.

Author(s)	r(Virgo)	H_o(global)	rH_o
Hubble (1936)	2.2	530	1170
Sandage and Tammann (1974)	19.5	57	1110
Visvanathan (1983)	18.8	74	1390
Tammann and Sandage (1985)	21.6	50	1080
de Vaucouleurs (1986) (mean for E+S Cloud)	13	100	1300
Aaronson et al.(1986)	14.6	90	1310
Dressler (1987)	20.0	67	1340
Tammann (1987)	20.9	55	1150
Bottinelli et al.(1988)	16.6	72.5	1200
Pierce and Tully (1988) Tully (1988)	15.6	90	1400
Rowan-Robinson (1988)	19	66	1250
Kraan-Korteweg et al.(1988)	21.6	56	1210
Melnik (1988)	15.1	86	1300
			1240+30

Acknowledgement

Sincere thanks are due to Dr.A.Sandage who has accompanied me through the genesis of this paper. The support of the Swiss National Science Foundation is gratefully acknowledged. I thank Mrs.M.Saladin and D.Cerrito for the editorial work on this manuscript.

References

Aaronson, M., and Mould, J. 1983, *Ap.J.***265**, 1.
Aaronson, M., Bothun, G., Mould, J., Huchra, J., Schommer, R.A., and Cornell, M.E. 1986, *Ap.J.***302**, 536.
Arp, H.C. 1956, *A.J.***61**, 15.
Bartel, N. 1988, in: *Supernova Shells and Their Birth Events*, ed.W.Kundt, Berlin: Springer, (in press).
Binggeli, B., Sandage, A., and Tammann, G.A. 1985, *A.J.***90**, 1681.
Binggeli, B., Tammann, G.A., and Sandage, A. 1987, *A.J.***94**, 251.
Bottinelli, L., Gouguenheim, L., Paturel, G., and de Vaucouleurs, G. 1984, *Astron.Astrophys.Suppl.***56**, 381.
Bottinelli, L., Gouguenheim, L., and Teerikorpi, P. 1988, *Astron.Astrophys.***196**, 17.
Branch, D. 1988, this Conference.
Branch, D., Falk, S.W., McCall, M.L., Rybski, P., Uomoto, A.K., and Wills, B.J. 1981, *Ap.J.***244**, 780.
Burstein, D., and Heiles, C. 1978, *Ap.J.***225**, 40.
Cadonau, R. 1986, Ph.D.Thesis, Basel Univ.
Cadonau, R., Sandage, A., and Tammann, G.A. 1985, in: *Supernovae as Distance Indicators*, ed.N.Bartel, Berlin: Springer, p.151.
Cohen, J.G. 1988a, *A.J.***95**, 682.
Cohen, J.G. 1988b, this Conference.
de Vaucouleurs, G., de Vaucouleurs, A., and Corwin, H.G. 1976, *Second Reference Catalogue of Bright Galaxies*, Austin: University of Texas Press.
de Vaucouleurs, G. 1986, in: *Cosmologia di Gamow, Enrico Fermi School, Course 86*, ed.F.Melchiorri and R.Ruffini, Amsterdam: North-Holland, p.1.
Dressler, A. 1987, *Ap.J.***317**, 1.
Dressler, A., Lynden-Dell, D., Burstein, D., Davies, R.L., Faber, S.M., Terlevich, R., and Wegner, G. 1987, *Ap.J.* **313**, 42.
Freedman, W.L. 1988, this Conference.
Gratton, L. 1988, preprint.
Harris, W.E. 1988, this Conference.
Hubble, E. 1929, *Proc.Nat.Acad.Sci.***15**, 168.
Hubble, E. 1936, *The Realm of the Nebulae*, Oxford University Press.
Jacoby, G.H. 1988, this Conference.
Kirshner, R.P. 1985, in: *Supernovae as Distance Indicators*, ed.N.Bartel, Berlin: Springer, p.171.
Kraan-Korteweg, R.C., Cameron, L.M., and Tammann, G.A. 1986, in: *Galaxy Distances and Deviations from Universal Expansion*, ed.B.F.Madore and R.B.Tully, Dordrecht: Reidel, p.65.

Kraan-Korteweg, R.C., Cameron, L.M., and Tammann, G.A. 1988, *Ap.J.* (in press).
Leibundgut, B. 1988, to be published.
Lynden-Bell, D., Faber, S.M., Burstein, D., Davies, R.L., Dressler, A., Terlevich, R.J., and Wegner, G. 1988, *Ap.J.*326, 19.
McClure, R.D., and Racine, R. 1969, *A.J.*74, 1000.
Melnik, J. 1988, this Conference.
Panagia, N. et al. 1980, *Mon.Not.R.astr.Soc.*192, 861.
Pierce, M.J., and Tully, R.B. 1988, *Ap.J.*330, 579.
Pritchet, C.J., and van den Bergh, S. 1987, *Ap.J.*318, 507.
Pritchet, C.J., and van den Bergh, S. 1988, *Ap.J.*331, 135.
Richter, O.-G. 1985, *ESO Workshop on the Virgo Cluster*, ed.O.-G.Richter and B.Binggeli, p.409.
Richter, O.-G., and Huchtmeier, W.K. 1984, *Astron.Astrophys.*132, 253.
Rosino, L. 1964, *Ann.d'Ap.*27, 498.
Rosino, L. 1973, *Astron.Astrophys.Suppl.*9, 347.
Roberts, M.S. 1987, *Observatory* **107**, 103.
Rowan-Robinson, M. 1988, *Space Science Rev.* (in press).
Sandage, A. 1988, *Ap.J.* (in press).
Sandage, A., and Tammann, G.A. 1974, *Ap.J.*194, 559.
Sandage, A., and Tammann, G.A. 1976, *Ap.J.*210, 7.
Sandage, A., and Tammann, G.A. 1984, *Nature* **307**, 326.
Sandage, A., and Tammann, G.A. 1987, *A Revised Shapley-Ames Catalog of Bright Galaxies*, 2nd ed., Washington: Carnegie Institution of Washington.
Sandage, A., and Tammann, G.A. 1988, *Ap.J.*328, 1.
Sandage, A., and Visvanathan, N. 1978, *Ap.J.*223, 707.
Schechter, P.L. 1980, *A.J.*85, 801.
Tammann, G.A. 1987, *Observational Cosmology*, eds.A.Hewitt and G.R.Burbidge, I.A.U.Symp.No.124, p.151.
Tammann, G.A., and Sandage, A. 1985, *Ap.J.*294, 81.
Teerikorpi, P. 1987, *Astron.Astrophys.*173, 39.
Terndrup, D.M. 1988, poster presented at this Conference.
Tully, R.B. 1988a, *Nature* **334**, 209.
Tully, R.B. 1988b, this Conference.
Tully, R.B., and Fisher, J.R. 1977a, *Astron.Astrophys.*54, 661.
Tully, R.B., and Fisher, J.R. 1977b, *I.A.U.Coll.*37, 95.
van den Bergh, S. 1984, *A.J.*89, 608.
van den Bergh, S., Pritchet, C., and Grillmair, C. 1985, *A.J.*90, 595.
Visvanathan, N. 1981, *Astron.Astrophys.*2, 67.
Visvanathan, N. 1983, *Ap.J.*275, 430.
Welch, D.L., McAlary, C.W., McLaren, R.A., and Madore, B.F. 1986, *Ap.J.*305, 583.

DISCUSSION

M.J.Pierce. Could you comment on the possibility that both the large dispersion and systematic deviation of the "extended" sample are due to errors in photographic photometry, as the CCD photometry B.Tully and I have obtained indicates a much smaller dispersion?

If you attribute the large dispersion in the "extended" Virgo sample to the lack of Malmquist bias, how do you explain the fact that the local calibrators (which are presumably unbiased) show such small dispersion?

G.A.Tammann. You know that for all distance determinations the paramount problem is the objective definition of samples. The next step is then to collect the best possible observations for such samples. The present observational data for the complete sample may not yet be fully satisfactory, but you see the large scatter and zero point shift even for galaxies brighter than 13 mag. (Fig.2); this cannot be due to errors. Just the fact that the complete sample has large scatter and is systematically shifted points to a real effect. But I agree that better data are extremely welcome to test for second-order effects. - The apparently small dispersion of the local calibrators has ~20% chance to stem from a parent population with $\sigma(M)=0.7$ mag.

R.B.Tully. Following up on Pierce's question, he and I became aware that galaxies at the distance of the Virgo cluster with Zwicky magnitudes of 14.5 mag. have CCD-measured magnitudes of 13.3. The large scatter and systematic shift of the luminosity-line width relation that you find compared with other data sets could be caused by such a systematic effect on magnitudes. Our Ursa Major cluster sample is essentially complete to 13.3 mag., well into the domain where you are finding the large scatter, and our dispersion is 0.3 mag.

G.A.Tammann. So far small scatter has been shown only for "convenient" galaxy samples. Compared to objective analyses (e.g. Roberts, 1987; Sandage, 1988) our scatter of $\sigma(M)=0.7$ mag. is still small. But I think, we all agree that we must begin to work on complete samples and just obtain additional observations if we feel that they are needed. The Virgo Cluster Catalog (complete to 18 mag.) was so far unique to define a complete sample of spirals. Now the Fornax cluster is also accessible (Ferguson and Sandage, 1988, preprint). I am afraid that our knowledge of the UMa cluster population is still fairly dim. Even so the bright UMa galaxies, whose observational parameters you do not doubt, are fully consistent with a large distance of $(m-M)=31.8$ (Kraan-Korteweg et al., 1988).

The Distances to the Virgo and Ursa Major Clusters and a Determination of H_o

Michael J. Pierce
Dominion Astrophysical Observatory
5071 W. Saanich Rd.
Victoria, BC V8X 4M6 CANADA

This report describes the result of an investigation of the luminosity–line-width relations (Tully and Fisher 1977, A. A. **54**, 661; hereafter TF relations) as a distance indicator. Particular emphasis has been paid to questions as to the suitability of applying the TF relations over a wide range of environments (i.e. from the relatively high densities found in clusters to the lower densities characteristic of the local calibrators in the field). This report is a synopsis of that presented by Pierce and Tully (1988, Ap. J. **330**, 579).

The data consists of B, R, and I CCD images of complete samples within the Virgo ($B_T < 12.0$) and Ursa Major ($B_T < 13.3$) Clusters. In addition, 3 nearby calibrating galaxies were also observed: M31, M33, and NGC 2403. The CCD data was suplemented with infrared magnitudes from Aaronson *et al.* (1982, Ap. J. Suppl. **50**, 241). Details of the reduction proceedure can be found in Pierce and Tully (1988). The HI line-widths were obtained from a

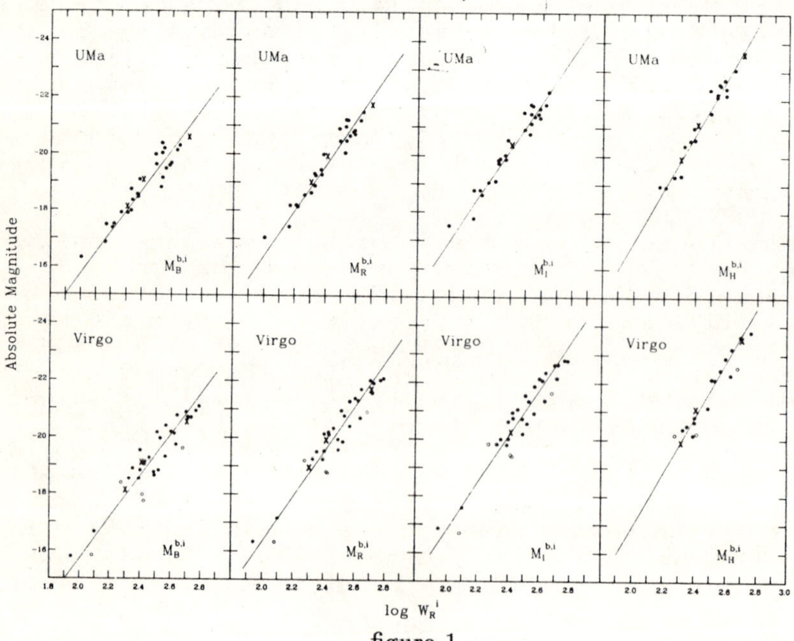

figure 1

variety of sources and were corrected for inclination along with the magnitudes (Tully and Fouque 1985, Ap. J. Suppl., **58**, 67).

In figure 1 the TF relations are presented for the two clusters on an absolute scale. The heavy crosses are the nearby calibrators [M33 (24.2), NGC 2403 (27.5), and M31 (24.4) in order of increasing luminosity along with adopted distances. See Pierce and Tully (1988) for sources]. The solid line is the regression in line-width which avoids any concerns of Malmquist or incompletion biases (see Tully, this volume). There is evidence for a slight morphological type dependence of the TF relations but only at the $1-2\sigma$ level. There is no difference in slope for the two samples despite the differing densities and crossing times for the two clusters. Thus there is good evidence that the TF relations for cluster and field environments are similar. The R, I, and H dispersions are only 0.30 mag for Ursa Major. The intrinsic dispersion could be as low as 0.25 mag. as the depth within Ursa Major is significant. The dispersion in Virgo is higher (0.40 mag.) and is presumed to be due to projection contamination as the most deviant galaxies have the distance-velocity signature characteristic of infall (see figure 2). The derived distances for the two clusters are:

$$(m-M)_{UMa} = 30.95 \pm 0.16(15.5 \pm 1.2Mpc)$$
$$(m-M)_{Vir} = 30.96 \pm 0.20(15.6 \pm 1.5Mpc)$$

with the small number of local calibrators dominating the error. The relative geometry of the two clusters is shown in figure 3.

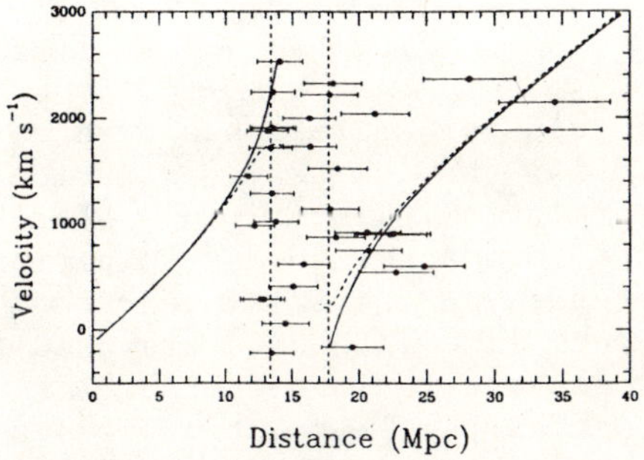

figure 2

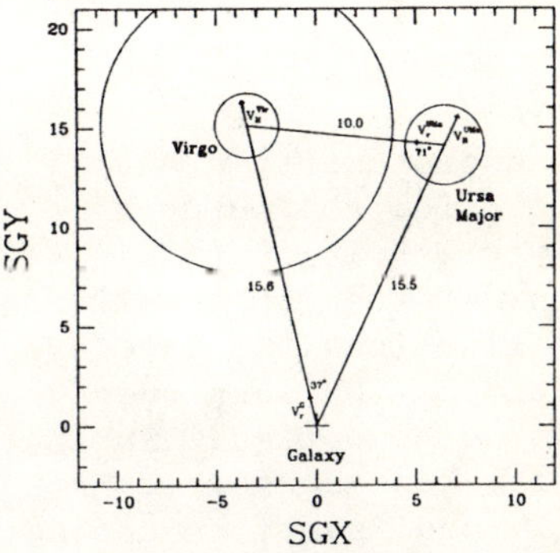

figure 3

Adopting a Virgocentric retardation of 300 km s^{-1} at the Local Group (Aaronson *et al.* 1982, Ap. J., **258**, 64) and the model of Tully and Shaya (1984, Ap. J., **281**, 31) allows estimates of H$_o$. For Virgo:

$$H_o = (1316 \pm 120)/(15.6 \pm 1.5) = 84 \pm 11 \text{ km s}^{-1} \text{ Mpc}^{-1}$$

For Ursa Major:

$$H_o = (1324 \pm 115)/(15.5 \pm 1.2) = 85 \pm 10 \text{ km s}^{-1} \text{ Mpc}^{-1}$$

The dominent source of error in the velocities is the uncertainty in the Virgocentric retardation at the Local Group. We are currently attempting to improve the situation with a CCD survey of several hundred nearby field galaxies.

CALIBRATION OF THE CCD TULLY-FISHER RELATION

Jeffrey A. Willick
Space Sciences Laboratory and Dept. of Physics
University of California, Berkeley
Berkeley, California 94720 USA

ABSTRACT CCD images of galaxies permit one to carry out accurate galaxy photometry using well-defined apertures. They provide, in addition, surface photometric data which yield accurate inclination estimates for spirals based on isophote ellipticity. Such images have been used to develop a more accurate version of the luminosity-21 cm linewidth correlation for spirals. When line of sight depth within galaxy clusters is accounted for by noting redshift-distance correlations among galaxies belonging to a given cluster, the scatter in the luminosity-21 cm linewidth relation appears to be as small as 0.23 mag.

CCD red photometry of nearly 150 cluster spirals has been carried out for the purpose of studying the luminosity-rotation velocity, or Tully-Fisher (TF), relation. The imaging information afforded by the CCD data allows accurate determination of inclinations, surface brightnesses, and total integrated magnitudes. The magnitude and 21 cm linewidth data are used to determine an optimal expression for the luminosity-rotation velocity dependence. It is found that the TF relation is distinctly nonlinear, in the sense that the magnitude-log linewidth relation requires a quadratic term to accurately describe the data. The slope of the relation is therefore not constant; however, for average spiral galaxy rotation velocities (log linewidth roughly 2.50), the effective slope is very nearly 7.5, corresponding to luminosity varying as the third power of the rotation velocity.

The problem of bias associated with the sample's magnitude limit (the "cluster Malmquist effect") has been dealt with by a novel technique that uses expressions originally derived by Teerikorpi. In this scheme the data are fit to a preliminary model, the parameters of the relation estimated, and initial estimates of magnitude bias derived; the data are then corrected for bias, the model fit to the data once again, and so on until convergence of the parameter estimates is achieved. Numerical simulations have been carried out and have verified that the method does indeed yield unbiased estimates of the parameters. Comparison of the real data with simulations shows the level of bias present to be small in any case, and unlikely to significantly influence the distance estimates.

The rms error in predicted absolute magnitude which characterizes the "CCD Tully-Fisher relation" may be as low as 0.20 mag, although more data will be required to confirm this low value. Such a small scatter makes it possible to resolve line-of-sight dispersion within individual clusters. For eight of the clusters studied, it has been possible to show that the distances within clusters are positively correlated with radial velocity, demonstrating that the spiral galaxies in these clusters are not members of virialized cluster cores. A phenomenological model is developed to account for the dependence of distance on velocity within these clusters; the final calibration of the TF parameters utilizes this model and achieves a significant reduction in scatter. The Virgo cluster does not at all conform to the model; rather than a positive distance-radial velocity correlation being observed, the Virgo galaxies show a general anticorrelation between distance and radial velocity, with some exceptions. This complexity of structure is hinted at within some other clusters as well; all this suggests that galaxy clusters should be used carefully as calibrators of the Tully-Fisher relation.

The model also yields distances to, and velocities of, cluster "dynamical cores"; the ratio of these provides estimates of the Hubble constant for each cluster. The absolute calibration for these distances is provided by the Ursa Major cluster, for which Pierce and Tully (this conference) found D=15.5 Mpc using a similar observational technique. The six cluster in the model with redshifts in excess of 4500 km/sec yield very similar results for the Hubble constant, with a mean value of 91 km/sec/Mpc. This value is fully corrected for bias, should be little influenced by peculiar velocities since it is based on distant clusters, and carries a small formal error owing to the very small scatter of the distance indicator. It is likely to be significantly revised only if the primary calibration, the assumed distance to the Ursa Major cluster, is changed.

GLOBULAR CLUSTERS IN M104 (NGC 4594) AS DISTANCE INDICATORS

Terry J. Bridges and David A. Hanes
Physics Department, Queen's University
Kingston, Ontario, Canada

We have recently (April 1988) acquired deep CCD images in several fields of the spheroid of M104 (NGC 4594, the Sombrero galaxy) using a double-density 1024x640 RCA CCD at the prime focus of the Canada-France- Hawaii Telescope (CFHT). This Sa galaxy has a prominent bulge and a demonstrably large cluster population (more than a thousand in total, according to Harris, Harris and Harris 1984 A.J. 89, 216). It lies in the southern extension of the Virgo cluster but may be a foreground object.

Our deep photometry, only just begun at this writing, will reach to beyond B = 25 mag in frames which total 7200 seconds exposure in 0.8-0.9 arcsec seeing. (We are using the powerful DAOPHOT data reduction system.) For likely values of the M104 distance modulus, our photometry will almost certainly reach beyond the turnover of the globular cluster luminosity function. Our objectives are:

(i) to carry out deep photometry of the globular cluster population and to derive the luminosity function to the turnover or beyond;

(ii) to determine the specific frequency (number of clusters per unit galaxian magnitude);

(iii) to study the spatial distribution of globular clusters, especially as that compares to the underlying spheroid luminosity profile;

(iv) to establish the existence (or not) of a central deficiency of globular clusters;

(v) to assess the universality of the globular cluster luminosity function;

and (vi) to derive a distance for M104.

THE USE OF GIANT EXTRAGALACTIC HII REGIONS AS DISTANCE INDICATOR: NEW RESULTS FROM A CATALOG OF HII REGIONS IN THE GALAXY NGC 4321

R. ARSENAULT[1], J. BOULESTEIX[2], Y. GEORGELIN[2], J.-R. ROY[3]

[1] European Southern Observatory, Karl-Schwarzschild-Str.2, D-8046 Garching b. München, FRG

[2] Observatoire de Marseille, 2 Place Le Verrier, 13248 Marseille Cedex 4, France

[3] Département de Physique, Université Laval, Québec, Canada

ABSTRACT A comprehensive set of observations has been obtained on the spiral galaxy NGC 4321 (M100) at the CFHT using CIGALE, a Fabry-Pérot interferometer coupled with a photon counting camera. Those observations have been used to get Hα fluxes, Hα diameters and velocity width for 108 HII regions in M100. Those parameters show no correlation between diameter and velocity width nor between flux and velocity width.

The CIGALE data has been used to catalog the HII regions in NGC 4321. The original aspect of this catalog is that the Hα line profile of each HII region has been obtained by integrating the HII region flux in each wavelength map. The catalog provides HII regions diameters (D), fluxes (F) and velocity widths (W). This data is used to check for correlations between diameter, luminosity and velocity width among the HII regions of NGC 4321, correlations which have been observed in Giant Extragalactic HII Regions (GEHR) [1],[2],[3]. These relations can, in principle, be used as distance indicator.

The plots of log F versus log W and log D versus log W do not show any trend between those 2 variables. The correlation coefficients are 0 in both cases. Selecting the best points (highest signal to noise ratio line profiles) one can distinguish boundaries in the D-W plane and similarly in the F-W. There is an upper limit to the velocity width (94 % of the HII regions have a velocity width smaller than 35 km/s). An upper envelope can also be seen in both plots having a slope of 1.65 in the D-W plane and 3.3 in the F-W plane. There is clearly not a cause to effect relation between D and W nor between F and W in the HII regions of NGC 4321. We do not believe that GEHRs are another class of HII regions, since their parameters (diameter, flux, velocity width) increase smoothly from the smallest value to the largest; there is no discontinuity in the parameters distributions. There are more indications in [4] that GEHRs are not

fundamentally different from smaller HII regions.
It is not too surprising to find no correlations between diameter (or flux) and velocity width for the HII regions of NGC 4321. In normal HII regions one expects many physical mechanisms which acting simultaneously would probably destroy any scaling of those parameters. Such physical mechanisms have been presented in previous papers and are: stellar winds, supernovae explosions, champain flows, Kolmogoroff turbulence, virial equilibrium.
We believe that the relations D-W and F-W (or L-W) in GEHRs have a statistical basis. The criterion: "largest HII regions" allows one to pick up the objects located near the upper physical limit, i.e. having the largest diameter and velocity width. In luminous galaxies one will likely find objects near this physical limit. However, less luminous galaxies contain less likely the largest (or brightest) HII regions located at the physical limit. The smaller (or less luminous) although largest HII regions in less luminous galaxies will span a given range of velocity widths for their diameter (or luminosity). Therefore, the probability to pick up a "largest" HII regions with velocity width smaller than 35 km/s is higher in less luminous galaxies. This can produce the relations observed in GEHRs but those will obviously be scattered and their slope and zero-point will depend on the sample choice and size. The same kind of reasoning also explains the relation between the galaxy luminosity and the mean velocity width of its 3 largest HII regions [5].
Therefore we conclude that there is evidence that the relations between diameter, and velocity width and between luminosity and velocity width in GEHRs have a statistical basis. The distance indicator potential of GEHRs is thus questionable. These results are very interesting from the point of view of the physics of HII regions.

REFERENCES

1. Melnick, J., Terlevich, R., Moles, M., Garcia-Pelayo, J.-M., 1987, M.N.R.A.S. 226, 849.
2. Arsenault, R., and Roy, J.-R., 1987, Astron. Astrophys., in press.
3. Hippelein, H., 1986, Astron. Astrophys. 160, 374.
4. Kennicutt, R.C, 1984, Astrophys. J. 287, 116.
5. Roy, J.-R., and Arsenault, R., 1986, Astrophys. J. 302, 579.

VI

DEVIATIONS FROM THE HUBBLE FLOW

ON BEST DISTANCE ESTIMATORS AND GALAXY STREAMING

D. LYNDEN-BELL, DAVID BURSTEIN, ROGER L. DAVIES, ALAN DRESSLER, S.M. FABER, ROBERTO TERLEVICH & GARY WEGNER
Institute of Astronomy, Madingley Road, Cambridge CB3 0HA

ABSTRACT We determine the best distance indicator that can be formed by linearly combining n logarithmic observables measured for each member of a cluster. The σ, D_n relationship can hardly be improved by using all 8 of our observables.

The evidence for galaxy streaming from both the elliptical galaxy survey and the best AHM data is portrayed in new diagrams. These show the motions that led us first to bulk streaming and then to the Great Attractor model. It may be significant that the best fitting great attractor model places it at the edge of the good data. The redshift surveys indicate a large concentration of galaxies there but both IRAS and optical dipoles indicate that at least half of the Local Group's motion is more probably generated by the gravity of closer galaxies. However, those methods ignore matter masked by the galactic plane so the result might change when hidden parts of the great attractor are included.

1.HOW TO FIND THE BEST DISTANCE INDICATOR

1.1 Outline.

In the elliptical galaxy survey (Faber *et al.* 1989) we measured 8 observables for each galaxy: Central velocity dispersion σ, Magnesium line strength index Mg, three differently defined angular diameters D_n, A_e and D_{FW}, the total magnitude B_T, the B-V colour and the axial ratio a/b. Of these D_n, A_e, D_{FW} and B_T depend on distance in the sense that they would change if the same galaxy were observed from further away. However, one may always use log ratios such as $\log(D_{FW}/D_n)$, $\log(A_e/D_n)$ and $B_T + 5\ell n D_n$ which are distance independent and thus leave only one distance-dependent variable, $\log D_n$. To simplify notation we call it y and all the remaining logarithmic variables log σ, $\log(D_{FW}/D_n)$ etc. we denote collectively by a vector array $\mathbf{x}$. Our first problem is to find that linear combination of the $\mathbf{x}$ that is best correlated with y for the members of some well-observed cluster such as Coma or Virgo. We shall then show that the accuracy of the resultant distances is independent of our initial choice of D_n as the distance-independent variable.

To determine distance as accurately as possible, we ask what linear combination of the $\mathbf{x}$ leads to the correlation with least dispersion in the distance-dependent variable y. To do this for N objects in m clusters (currently 1) and n components of $\mathbf{x}$, we minimise

$$S = \frac{1}{N-n-m} \sum_{cluster\ members} (y - \mathbf{a}\cdot\mathbf{x} - c)^2 \tag{1}$$

over all choices of the linear coefficients $\mathbf{a}$ and the constant c. Actually we eliminate c by using the deviations of y and $\mathbf{x}$ from their means over the cluster members $\overline{y}$ and $\overline{\mathbf{x}}$. Thus we minimise

$$S(\mathbf{a}) = \frac{1}{N_e} \sum (y - \overline{y} - \mathbf{a}\cdot(\mathbf{x} - \overline{\mathbf{x}}))^2 \tag{2}$$

where

$$N_e = N - n - m \tag{3}$$

Notice that we may now extend the sum over any set of well observed clusters rather than concentrating on only one since the distance dependence of y is eliminated in $y - \overline{y}$.

Minimising over all choices of $\mathbf{a}$ we obtain

$$\partial S/\partial\mathbf{a} = 0 = \mathbf{Y} - \mathbf{a}\cdot\mathbf{M} \tag{4}$$

where

$$\mathbf{Y} = \frac{1}{N_e}\sum (y - \overline{y})(\mathbf{x} - \overline{\mathbf{x}}) \tag{5}$$

and $\mathbf{M}$ is the matrix

$$\mathbf{M} = \frac{1}{N_e}\sum (\mathbf{x} - \overline{\mathbf{x}})(\mathbf{x} - \overline{\mathbf{x}}) \tag{6}$$

Thus $\mathbf{a}$ is given by

$$\mathbf{M}^{-1}\cdot\mathbf{Y} \tag{7}$$

The accuracy with which these $\mathbf{a}$ are determined from the cluster data is given by the inverse of $\mathbf{M} = \partial^2 S/\partial\mathbf{a}\partial\mathbf{a}$. The principal values of $\mathbf{M}^{-1}S$ give the mean squares of the errors in the principal axis directions.

Using the minimising $\mathbf{a}$, our best distance indicator is given by

$$log\ r = \mathbf{a}\cdot\mathbf{x} - y + const \tag{8}$$

When distance ratios are needed, the value of the constant is not needed. For absolute distances the constant must be determined by applying formula (8) to a collection of calibrating objects whose distances are known.

Although our prescription (2) gives the distance indicator with least dispersion for a thoroughly observed cluster, it is often true that clusters are not observed so as to be complete in $\mathbf{x}$ no matter what y may be, but are observed in a sample that is complete to a given value of y. When this is so, it is best to use the other regression line which gives the means of $\mathbf{a}\cdot\mathbf{x}$ at given values of y. While this distance indicator will be less accurate it will be less subject to Malmquist bias and is therefore preferred.

1.2 Distance Error Independent of Initial Choice.

We could have chosen a different distance-dependent variable as our y by taking $-1/5B_T$ or $\log D_e$ in place of $\log D_n$. The most general y that depends on distance like -log r is of the form $y^* = y - \mathbf{b} \cdot \mathbf{x}$.

Both $-1/5B_T$ and $\log D_e$ are of that form of course. We need to show that we get the same distance indicator whether we begin with y or with y^*. Our method minimises S^* over all $\mathbf{a}^*$ and

$$S^*(\mathbf{a}^*) = \frac{1}{N_e} \sum [y^* - \mathbf{a}^* \cdot \mathbf{x} - c]^2 = \frac{1}{N_e} \sum [y - \mathbf{a}' \cdot \mathbf{x} - c]^2 = S(\mathbf{a}')$$

where $\mathbf{a}' = \mathbf{a}^* + \mathbf{b}$.

It is now clear that minimising S^* over all $\mathbf{a}^*$ will lead to the same minimum as minimising $S(\mathbf{a}')$ over all $\mathbf{a}'$. Furthermore the final distance variable determined will be the same $\mathbf{a} \cdot \mathbf{x} - y$ since the minimising $\mathbf{a}'$ is just our old minimising $\mathbf{a}$.

The fact that "any y will do" leads to an amusing conundrum. Suppose we chose for our y, a distance-dependent quantity that had been so badly measured that its values were totally useless and suppose that another choice of y, y^* would give us a well measured and almost perfect distance indicator. How can it be that plotting a useless quantity against the distance independent $\mathbf{a} \cdot \mathbf{x}$ can give as accurate distances as plotting the good y^*. This conundrum is easy to unravel. The large scatter in the original y is mirrored by the large scatter in $y - y^*$ which is 'one' of the $\mathbf{x}$ of the first formulation. Our choice of the best $\mathbf{a}$ ensures that all this scatter is along the line of correlation so our poor choice of y is not reflected in the breadth of the correlation, in fact the *net* coefficient of the bad measureable in the correlation (8) will be zero.

The above example suggests that the most sensible y to choose is that distance dependent primary measurable that has the largest net coefficient in the final relationship y - $\mathbf{a} \cdot \mathbf{x}$ = const. Furthermore, it suggests that it may be desirable to leave out variables with low coefficients altogether if their inclusion only serves to increase n and thereby increase S. In the application that follows, variables with ill-determined small coefficients have been progressively omitted.

1.3 Application to our Virgo and Coma data.

Table 1 gives the 'best' coefficients $\mathbf{a}$ and their errors $\Delta\mathbf{a}$ and the resultant error in log r when all 8 variables are progressively eliminated. In Table 1 $\mu_e = B_T - 5logA_e$ and $\mu_n = B_T - 5logD_n$.

TABLE 1 Best distance coefficients for Coma + Virgo

log σ a_1 $\pm\Delta a_1$	$\mu_{e/5}$ a_2 $\pm$	log(D_{FW}/D_n) a_3 $\pm$	Mg_2 a_4 $\pm$	B-V a_5 $\pm$	$\mu_{n/5}$ a_6 $\pm$	log a/b a_7 $\pm$	error in log r
1.15	.169	-.405	-.532	.426	-.147	-.087	.0644
±.090	.070	.188	.730	.474	.329	.144	
1.16	.250	-.381	-.567	.370			.0625
±.089	.076	.174	.705	.425			
1.18	.225	-.371	-.374				.0620
±.089	.075	.173	.621				
1.13	**.228**	**-.398**					**.0613**
±.088	**.075**	**.172**					
1.16	.115						.0677
±.101	.084						
1.18							.0672
±.093							

The best indicator found is

$$log\ r = -\ logD_n + (1.13 \pm .088) log\ \sigma + (.228 \pm .075)\mu_e/5 + \\ + (-.398 \pm .172) log(D_{FW}/D_n)$$

but the log distance error of .0613 is not much better than the simple expression

$$log\ r = -logD_n + (1.18 \pm .093) log\ \sigma$$

whose corresponding error is .0672 so we have used the latter for simplicity (Dressler *et al.* 1987, Lynden-Bell 1986, Dressler 1988). The above Table clearly indicated that no improvement can be made by using the axial ratios of elliptical galaxies nor by involving B-V or Mg_2 and log σ (see Burstein *et al.* 1987).

2. GALAXY STREAMING AND THE GREAT ATTRACTOR

Our studies of the motions of elliptical galaxies (Lynden-Bell *et al.* 1988, Lynden-Bell 1987) led us to deduce that they were streaming at some 521 km/s towards $\ell = 307$, b = 9. A somewhat better fit to the data was found by putting a great attractor centred in that direction at a distance of 4350 ± 350 km/s. When we allow the direction to the great attractor to vary a marginally better fit still is found at 4200 ± 300 km/s toward $\ell = 309 \pm 10$, b = 18 ± 10 (Faber & Burstein 1988). We believe it to be no coincidence that on looking near there we found the largest concentration in the sky of galaxies of diameter greater than one arc minute (centred on $\ell = 312$, b = 25, see Figure 1). Could we only see through

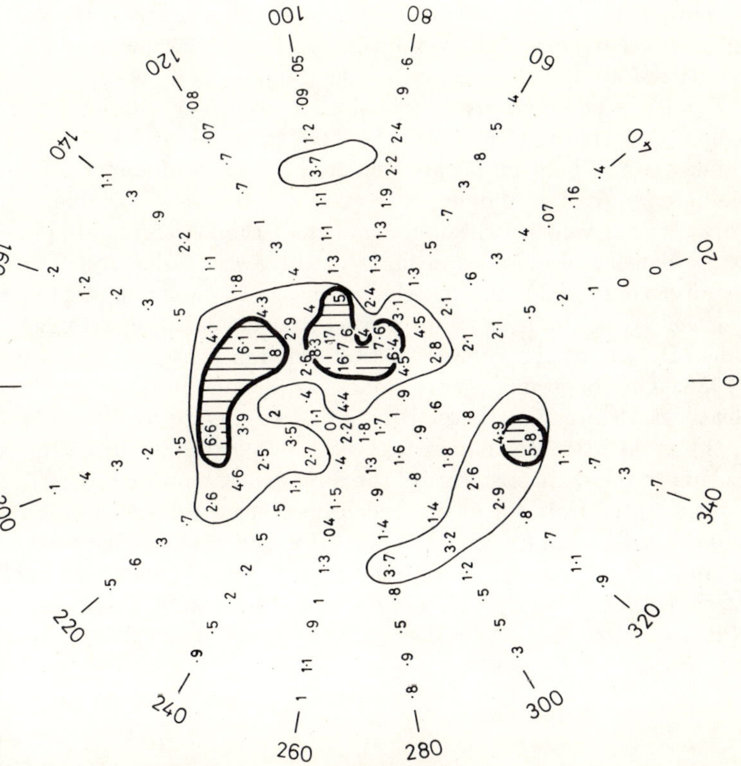

Figure 2

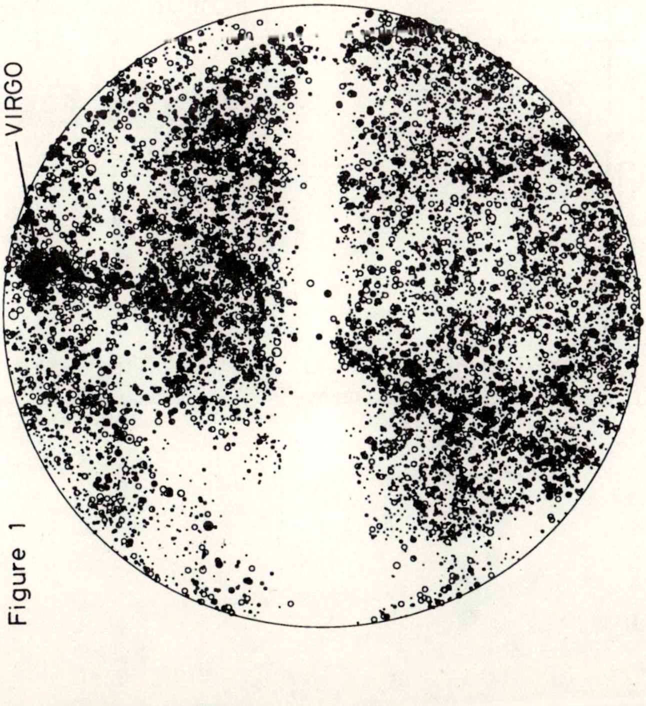

Figure 1

the galactic murk its true centre might well appear at lower galactic latitude. There is now strong evidence that the structure extends across the galactic plane joining the concentration Telescopium Pavo. Figure 1 shows the hemisphere of the sky centred on the intersection of the galactic and supergalactic planes (ℓ = 327, b = 0) in an equal area projection. Galaxies of diameter > 1.1′ on the ESO scale are plotted together with the UGC galaxies, and MCG galaxies fill in the missing strip. The symbols are weighted by major diameter. We have used such diameter limited catalogues to correct for the very variable completeness in Huchra's radial velocity catalogue and map the supergalactic plane. Figure 2 plots the density of galaxies in units of the average density. The density is averaged over cones of 30° total angle centred on the main supergalactic band of galaxies. Density contours are interpolated from cones 20° apart in supergalactic longitude. Details of the completeness correction are given in Lynden-Bell & Lahav 1988. The cones average over much larger volumes at greater distances so large densities at larger distances correspond to much greater masses. In spite of this, the great attractor region SGL = 160-200 stands out. Virgo may be a dense outlier to the main structure of the Supergalaxy in Centaurus-Telescopium. Figure 3 is adapted from the galaxy redshift surveys of Dressler (1988), da Costa *et al.* (1986) and Fairall (1988) which are all in this region. The vertical lines are at 5000 km/s. All three demonstrate the great excess of galaxies in the 2500-6000 km/s interval. Fairall's survey is all South of the Galactic Plane and da Costa's is all North of it, so this establishes the continuity in redshift of the structures seen in projection in Figure 1.

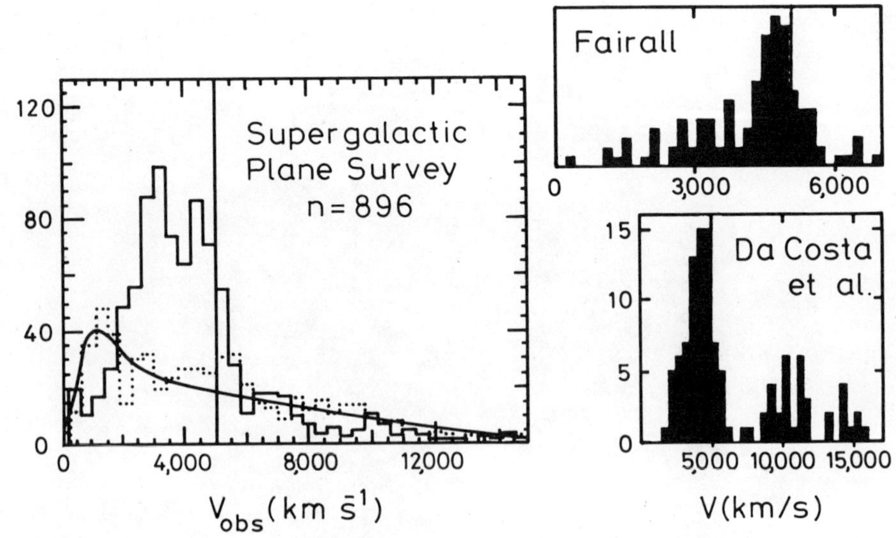

Figure 3

Figures 4, 5, 6 plot the elliptical galaxy survey with the best AHM data in double ended cones of 30° semi-angle about three primary directions: in the supergalactic plane towards the great attractor $\ell = 315$, b = 18, SGL = 162, SGB = 0, Figure 4: Perpendicular to that but in the supergalactic plane, Figure 5, $\ell = 156$, b = 71, SGL = 72, SGB = 0 (this includes the Virgo Cluster whose members we have plotted individually but at the mean distance): Towards the supergalactic pole $\ell = 47$, b = 6 (Figure 6), but here the paucity of galaxies forced us to open the cone to a semi-angle of 45°. In each diagram the excess radial velocity over and above pure Hubble flow in the cosmic-microwave-background frame of rest is plotted vertically and the coordinate along the cone's axis is plotted horizontally. Excess velocities in the backward pointing ends of the double cones have had their signs reversed. Thus the plots represent excess velocity in the direction of the primary cone as a function of coordinate in that direction. A bulk streaming motion in the direction of the cone would be represented by all points being systematically higher than zero. A lower-than-Hubble expansion rate would give a negative slope while a super-Hubble expansion rate would give a positive slope. It is clear that the points in Figure 4 are systematically high and that there is a large dispersion of peculiar velocity in the general region of the great attractor. However, there is a paucity of points beyond it which according to the model should show negative peculiar velocities. Evidently the program has fitted the model in just such a way that the lack of strong negative velocities is accounted for by a lack of data. The finding of negative velocity residuals beyond the main outward flow in this direction remains a major challenge to observers trying to tie down the cause of the observed galaxy streaming seen in both the elliptical and AHM data. There is some evidence of a lower than Hubble expansion in Figure 5 over and above the 'up-down-up' signature that can be seen in Virgo itself around R = 1300 km/s. Such a slower than Hubble expansion could be caused by the tidal field of the great attractor but it is hard to separate the effect from Virgo infall with precision. Little can be seen from the wider cone towards the supergalactic poles, although it is strange to see all three AHM cluster points lying high.

3. OPTICAL AND IRAS DIPOLES

Both the gravitational field and the light flux from a galaxy fall as r^{-2}. If doubling the mass of a galaxy also doubles its light then its gravity field at a point will still be proportional to its light flux there. Adding many such contrubutions vectorially we find that the net extragalactic light flux past the Local Group should be proportional to the gravity field on it. That in turn should generate the motion of the Local Group relative to the CMB. Early tests of this by Huchra using the Shapley Ames catalogue and by those analysing IRAS sources have found rough agreement in direction but only at the 25-30° level which is rather too loose to convince doubters of the direct connection claimed (Yahil *et al.* 1986, Meiksin & Davis 1986, Lahav 1987).

The new colour selected IRAS galaxy samples of Meurs and Harmon (1988) was constructed to see objects closer to the galactic plane. However, when we

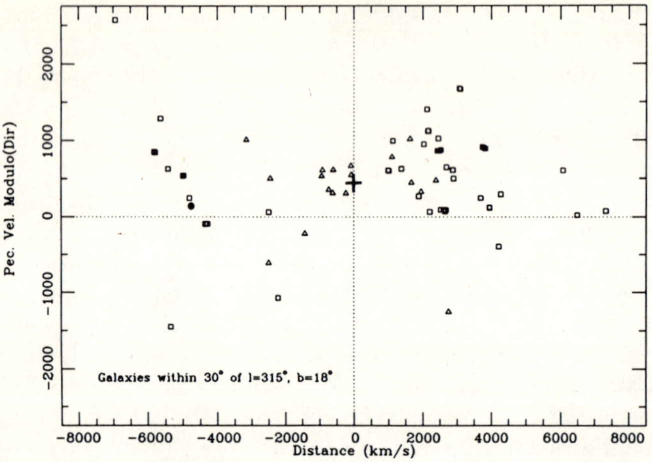

Figure 4

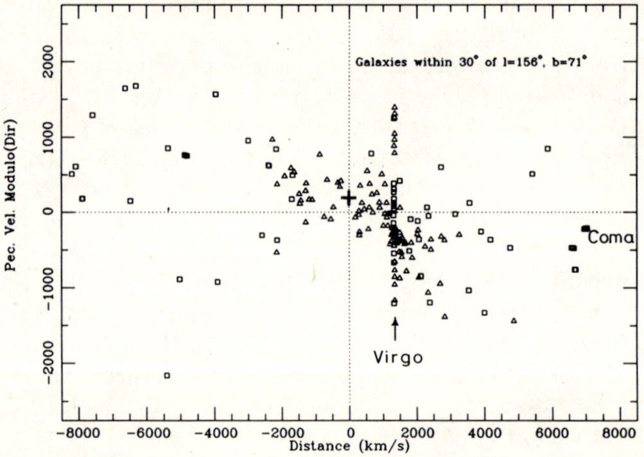

Figure 5

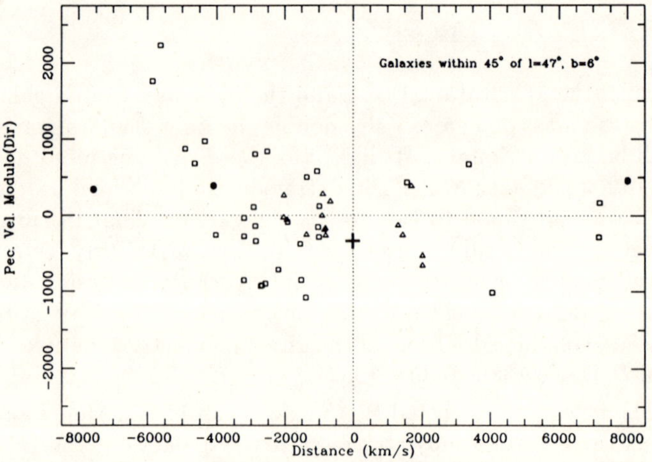

Figure 6

persuaded Harmon to measure the dipole from it (Harmon *et al.* 1987) both number and flux weighted dipole directions were significantly nearer (8° and 6°) the CMB dipole relative to the Local Group than were the previously measured IRAS dipoles. Lahav's calibration of the diameter functions of both the ESO and UGC galaxy catalogues from redshift surveys shows that ESO measurements of diameter are 1.13 ± 0.05 times larger than UGC diameter measurements (Lahav *et al.* 1988). This new calibration has given an optical dipole pointing to ℓ = 261, b = 29 remarkably close to the ℓ = 269, b = 28 motion of the Local Group relative to the CMB. His calibration is confirmed by the ratio of the small diameter number counts in the ESO and UGC polar caps each of which tend towards a uniform distribution. The two densities would be brought into coincidence if ESO diameter measurements were on average 1.10 ± 0.02 times larger than UGC measurements.

The good agreement of the dipole directions suggest that the motion of the Local Group is indeed gravitationally induced by material intimately connected with the observed galaxies. This is evidence that light does trace mass but detailed point by point comparisons (Yahil 1988, Davis & Strauss 1988) are needed. However, analysing the origin of the observed optical dipole we find half of it comes from galaxies with diameters greater than 4′ (Lynden-Bell & Lahav 1988). This suggests that the gravity field on the Local Group comes from only half the distance of the Great Attractor. In assessing this it is important to realise that the gravity field on the Local Group includes that which causes the Virgo infall and the 'push' from the Local Void which we find gives rise to the component of the Local Group's motion across the Supergalactic plane. Furthermore, the light dipole includes no contribution from the part of the great attractor hidden behind the galactic plane. Indeed, neither IRAS nor optical dipoles have taken account of such a contribution. Such problems are now under active scrutiny but leaving them aside, it looks as though much of the gravity that causes the motion of the Locap Group is more local. Whereas that tentative result will no doubt be welcomed by cold dark matter enthusiasts, it is only fair to point out that the data of Figure 5 show a surprisingly large scale uniformity with no clear infall from the back of anything nearer than 4000 km/s, at which point the motion still looks positive!

REFERENCES

Burstein, D. Davies, R.L., Dressler, A., Faber, S.M., Lynden-Bell, Terlevich, R.J., and Wegner, G. 1988, *Towards Understanding Galaxies at High Redshift*, eds. R.G. Kron & A. Renzini (Kluwer:Boston), p.17.

Da Costa, L.N., Nunes, M.A., Pellegrini, P.S., Willmer, C. 1986, A.J. **91**, 6.

Davis, M., and Strauss, M. 1988, Vatican Symposium 1987, *Large Scale Motion in the Universe*, eds. G. Coyne and V.C. Rubin.

Dressler, A. 1987, Ap. J., **317**, 1.

Dressler, A. 1988, Ap. J., **329**, 519.

Dressler, A., Lynden-Bell, D., Burstein, D., Davies, R.L., Faber, S.M., Terlevich, R.J., and Wegner, G. 1987, Ap. J., **313**, 42.

Faber, S.M., Burstein, D., Davies, R.L., Dressler, A., Lynden-Bell, D., Terlevich, R.J., and Wegner, G. 1989, Ap. J. Suppl., in press.

Faber, S.M., and Burstein, D. 1988, Vatican Symposium 1987 *Large Scale Motion in the Universe*, eds. G. Coyne and V.C. Rubin.

Fairall, A.P. 1988, M.N.R.A.S. **230**, 69.

Harmon, R.T., Lahav, O., and Meurs, E.J.A. 1987, M.N.R.A.S., **228**, 5P.

Lahav, O. 1987, M.N.R.A.S. **225**, 213.

Lahav, O., Rowan-Robinson, M., and Lynden-Bell, D. 1988, M.N.R.A.S., submitted.

Lynden-Bell, D. 1986, Q.J.R.A.S., **27**, 319.

Lynden-Bell, D. 1987, Q.J.R.A.S., **28**, 186.

Lynden-Bell, D., and Lahav, O. 1988, Vatican Symposium 1987 *Large Scale Motions in the Universe*, eds. G. Coyne and V.C. Rubin.

Lynden-Bell, D., Faber, S.M., Burstein, D., Davies, R.L., Dressler, A., Terlevich, R.J., and Wegner, G. 1988, Ap. J. **326**, 19.

Meiksin, A., and Davis, M. 1986, A.J. **91**, 191.

Meurs, E.J.A., and Harmon, R.T. 1988, A.& A., submitted.

Yahil, A. 1988, Vatican Symposium 1987, *Large Scales Motions in the Universe*, eds. G. Coyne, and V.C. Rubin.

Yahil, A., Walker, D., and Rowan-Robinson, M. 1986, Ap. J.**301**, L1.

DISCUSSION

B. TULLY: In response to Judy Cohen's question, I have argued that the plane of the Local Supercluster can be followed to 15,000-20,000 km/s with its center being located in the Pisces-Cetus region.

D. LYNDEN-BELL: Yes, that is even further than the most distant members of our survey. We think there is evidence that the centre of the supergalactic plane distribution is in Centaurus, with Virgo off on one side.

J. COHEN: Where is the end of the supercluster? Why does the "great attractor" appear to be in the supergalactic plane?

D. LYNDEN-BELL: I think the "Great Attractor" is probably the centre of the structure that makes the supergalactic plane and that we lie near a prominent outlier called Virgo. Why galaxies are planarly distributed near her is a most intriguing question. I doubt if the supergalaxy extends beyond 6,000 km/s but that may be merely where the good catalogues to run out. I have no answer as to why Perseus and Coma lie in the supergalactic plane too!

T. SHANKS: How do you reconcile your evidence for structure on scales of 40-50h^{-1} Mpc with the galaxy correlation function results in 17^m redshift surveys which show very little evidence for galaxy clustering on any scale above 10h^{-1} Mpc?

D. LYNDEN-BELL: All we would claim is that the large scale motions exist. The only evidence on density that we have unearthed are those represented by the galaxy surveys and those found by the redshift surveys.

T. SHANKS: The caveat about the correlation function result is that if large-scale structure existed on scales >150h^{-1} Mpc we may have normalised it away. If this is the case it could even be easier to find consistency with your original bulk motion solution.

EFFECT OF THE LOCAL VELOCITY ANOMALY ON THE HUBBLE CONSTANT

R. BRENT TULLY
Institute for Astronomy, University of Hawaii, Honolulu, Hawaii, USA 96822

ABSTRACT It is argued that the luminosity - line width distance estimator can be used to determine the distances to individual field galaxies in a way that avoids Malmquist bias. A new calibration of the luminosity - line width relation based on CCD photometric observations of almost complete cluster samples is presented. The scatter at near-infrared wavelengths appears to be only ~ 0.25 mag. It has been suggested that the scatter in the luminosity - line width relation is large, the apparent increase in the Hubble ratio with distance is an artifact of Malmquist bias, and H_0 has a value in the vicinity of 50 km/s/Mpc. It is concluded here that the scatter in this distance estimator is small, Malmquist bias can be avoided, the apparent increase in the Hubble ratio with distance we still see is due to a local velocity anomaly, and $H_0 \sim 90$ km/s/Mpc.

INTRODUCTION

Three issues will be raised: (a) there will be a discussion of Malmquist bias and how to avoid it, (b) calibrations of B, R, and I band versions of the luminosity - line width relation for the Virgo and Ursa Major clusters will be presented, and (c) it will be demonstrated that galaxies in the cloud that we live in and the nearest adjacent cloud have systematically low velocities for their distances compared with galaxies in all the other nearby clouds. If these results are valid then it follows that the global value of the Hubble Constant is high. This article provides a summary of more extensive discussions of the same issues by Pierce and Tully (1988) and Tully (1988a) and an earlier review by Tully (1988b).

MALMQUIST BIAS

There has been a lot of discussion of Malmquist bias lately (Teerikorpi 1984; Giraud 1985, 1987; Bottinelli et al. 1986, 1987, Feast 1987, Tammann 1987, Sandage 1988, Lynden-Bell et al. 1988). Reasonable conclusions from what has been said are (a) one must be wary of systematic underestimation of galaxy distances, with the systematic effects themselves functions of distance, and (b) to statistically correct for biases one needs a lot of information regarding the properties of the distance estimator relations and the clumpiness in the distribution of galaxies, perhaps more information than is available. It is argued here that the problem can be formulated in a simpler fashion, such that distances can be estimated to individual galaxies free of bias over all ranges of distance.

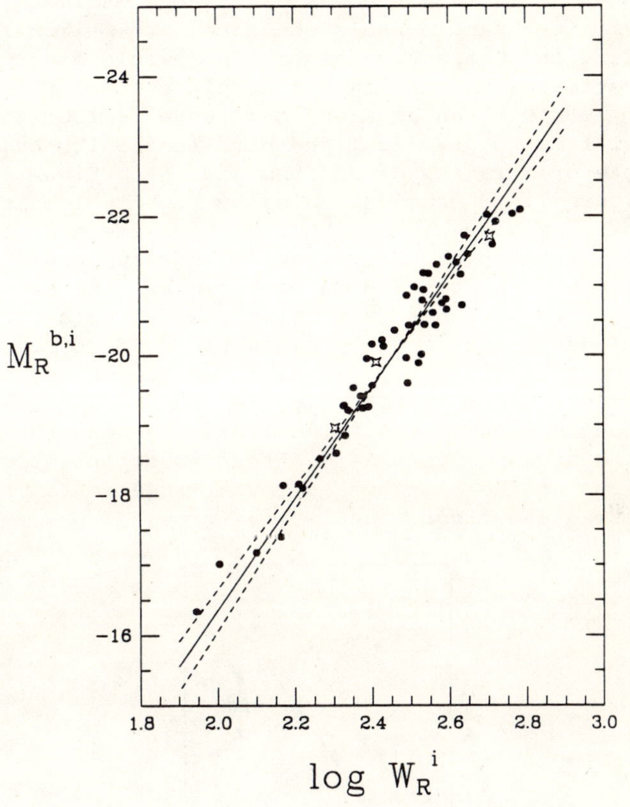

Fig 1. Luminosity-line width relation for the combined Virgo and Ursa Major clusters. The shallow dashed line is derived from the regression with uncertainties in luminosity. The steep dashed line represents an arbitrary relation that would generate a negative Malmquist bias. The solid line is derived from the regression with uncertainties in line width and provides the bias-free estimator. The absolute magnitude of the relationship is established by fitting to the three local calibrator systems: M31, M33 and NGC 2403 (open stars).

A key point to appreciate is that it is possible to formulate distance-estimator relations which generate *anti-Malmquist bias;* ie, which provide systematic *over*estimates of galaxy distances. This point can be understood with recourse to Figure 1, which is an R band luminosity - HI line profile width plot for galaxies drawn from the Virgo and Ursa Major clusters. The flatter dashed line is defined by the regression with uncertainties in luminosity and if that relation is used as a distance estimator on a magnitude-limited sample, it generates distances that suffer from Malmquist bias. By contrast, the steeper dashed line which has no physical significance, if it were used as a distance estimator on a magnitude-limited sample would generate anti-Malmquist bias. Both the flat and the steep dashed lines would give unbiased distance estimates in the mean if complete samples were available. But with an apparent magnitude limited sample, the tendency will be to draw from the high-luminosity end of a field sample of galaxies drawn from a distribution like the one seen in Figure 1. These intrinsically high luminosity galaxies will tend to lie *above* the flatter dashed line and hence be mistakenly taken for fainter galaxies and have their distances underestimated -the standard Malmquist bias. However, by construction, the steeper dashed line was drawn so steep that at the luminous galaxy end systems would tend to lie *below* this line. Using this relation, galaxies usually would be presumed to be brighter than they actually are and distances would generally be underestimated. The lesson is that the amplitude and even the sign of systematic errors is dependent on the assumed slope of the distance estimator relation. There must be an intermediate slope that is bias free.

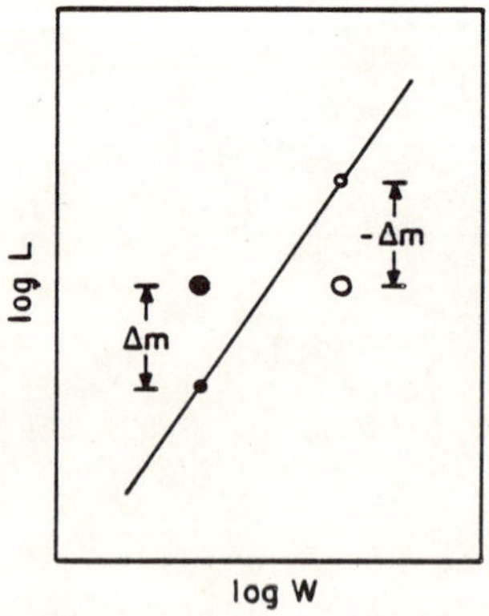

Fig 2. Symbolic luminosity-line width relation. If the straight line is derived from the regression with uncertainties in line width, then there is equal probability of drawing either the large closed circle or the large open circle from a magnitude-limited sample. The errors associated with drawing either of these objects are equal and opposite in sign.

It is the regression with errors in the distance-independent line width parameter that provides the bias-free slope. The reason should be clear from the cartoon Figure 2. The straight line is supposed to be the regression with errors in line width. If the system represented by the large filled circle is selected from a magnitude-limited sample then it will erroneously be assumed to have the intrinsic luminosity of the small filled circle on the mean relationship and an error of Δm will be made in the derived distance modulus. However, by construction of the regression, statistically the sample would contain the system represented by the open circle which would have the same apparent magnitude at the same distance and hence be as likely to be observed. In this case, the assumed luminosity is indicated by the small open circle and the error in distance modulus would be $-\Delta m$. Distance estimates would be uncertain by an amount characterized by the dispersion of the relationship, but there would be no significant systematic error.

This argument has required four assumptions that have not been made explicit. Two that seem reasonable as approximations given the state of our knowledge are that the luminosity-line width relation is linear in a log-log plot and that there is a normal distribution in log line width. A third assumption is that profile line widths are measured in an unbiased fashion. This condition was not necessarily true in the past and even in the present, but it is now possible to amass samples of spirals free of line width bias within, say, a distance corresponding to 3000 km/s with present radio receivers. Finally, the fourth assumption that must be made is that the relationship is being calibrated using systems which are statistically the same as systems that are the subsequent object of study. In practise, we are asking that spiral galaxies in the Virgo and Ursa Major clusters be representative of spirals in typically less crowded regions. Our best argument in support of this proposition comes from the Ursa Major Cluster sample. The density of galaxies in this cluster is modest, relative velocities are small such that crossing times are half the age of the universe, and galaxy types are overwhelmingly late and gas-rich. It is in this cluster that we measure the lowest dispersion to the luminosity - line width relation. Except for a slightly larger dispersion, there is no apparent difference between the Virgo Cluster relation and this one in Ursa Major.

The key to the avoidance of Malmquist bias, then, is the determination of the regression on line width in a representative sample that is complete over the intrinsic magnitude interval of the sample that will subsequently be studied. At present, we have a sample drawn from the Ursa Major Cluster that is almost complete to $B_T = 13.3$ ($M_B = -17.7$) combined with a sample drawn from the Virgo Cluster that is almost complete to $B_T = 12.0$ ($M_B = -19.0$). These samples are being substantially enlarged by observations of fainter systems so the present discussion is preliminary. However, the Ursa Major sample already goes faint enough to be applicable to our field sample.

It has been appreciated that the amplitude of Malmquist bias depends on a particular distance estimator's dispersion, which may be disputed, and it also depends in an ambiguous way on the clumpiness of the galaxy distribution. The claim being made here is that unbiased distance estimates can be made to individual galaxies in the field *whatever the dispersion* and *independent of structure* in the distribution of galaxies.

[At the conference, Lynden-Bell pointed out that even if distances are unbiased, because there are more galaxies truly at large distances with error thrown inward than galaxies truly at small distances with errors thrown outward, the observed median will be lower than the true median in a way dependent on the distribution of galaxies, but simple simulations suggest the effect will be insignificant.]

NEW CALIBRATIONS OF LUMINOSITY - LINE WIDTH RELATIONS

The work described here is amplified upon by Pierce (1988a) in this same volume. There have been a number of concerns raised regarding the utility of luminosity-line width relations as distance estimators. We now have B, R, and I band CCD observations of roughly 30 useful galaxies in each of the Virgo and Ursa Major clusters (the R band data for the two clusters combined is illustrated in Fig. 1). Most indications at present support the proposition that the intrinsic dispersion of luminosity-line width relations is small and that the relations provide good distance estimates. What follows are point-by-point evaluations of some of the issues that have become controversial, beginning with some of the most contentious.

(a) There have been discussions concerning the true intrinsic dispersion of the luminosity-line width relations. Values in the range 0.7 - 1.0 mag have been suggested by Richter and Huchtmeier (1984) and Sandage (1988). Yet analyses based on more homogeneous data have repeatedly resulted in values in the range 0.4 - 0.5 mag (Tully and Fisher 1977; Bottinelli et al. 1980; Aaronson and Mould 1983; Aaronson et al 1986). Our newest results (Pierce and Tully 1988) suggest a dispersion of only 0.25 mag at near-infrared passbands, after accounting for probable cluster depth effects. These results are in line with other results based on CCD photometry (Bothun and Mould 1987; Willick 1988). By contrast, Kraan-Korteweg, Cameron, and Tammann (1988) and, at this conference, Tammann (1988) contend that the Virgo Cluster scatter really is more like 0.7 mag at B band if the sample is complete to rather faint levels and, moreover, that there is a systematic offset from the luminosity-line width relation defined by bright galaxies. The reconciliation of these contrasting claims is probably the most pressing issue in this business. It is plausible that the non-photometric magnitudes that Kraan-Korteweg et al. were forced to use have both substantial scatter and systematic errors. Unfortunately, we do not yet have enough reduced CCD photometry of the fainter galaxies in the Kraan-Korteweg et al. sample to test this hypothesis. But the critical observations have probably already been made. Pierce and I now have CCD measurements of 60 Virgo spirals with HI information, double the number used in this discussion. The new material substantially overlaps the Kraan-Korteweg sample and should provide a clarification of the situation.

(b) It was suggested by Roberts (1978) and Rubin et al. (1985) that there are substantial type dependencies in luminosity-line width relations. The effect has always been much more marginal in more homogeneous or more closely volume-limited data sets (de Vaucouleurs et al. 1982; Aaronson and Mould 1983). With our almost-complete CCD samples we see a 2-sigma effect at B band and 1-sigma effects in the near-infrared bands. It should be appreciated that the Rubin et al. sample was specifically chosen to span a large luminosity range within each morphological type and, to that end, the sample was very successful. However, it is consequently an entirely *inappropriate* sample to use to look for the distance-dependent relationships that would characterize complete, volume-limited samples.

(c) Aaronson and Mould (1983) reported the detection of slight curvature in the H band luminosity-line width relationship and that eventuality would not come as a surprise. However, Bottinelli and Gouguenheim have contended in private that if one uses a measure of the maximum in the rotation curve as the line width parameter rather than the width at the 20 percentile intensity then there is no curvature and our most recent work confirms that result for practical purposes. Pierce (1988b) is arguing there may be a slight break from a linear relationship corresponding to a more pronounced break in a plot of luminosity vs. effective radius.

(d) There has been controversy over the slope of the luminosity-line width relations, which has particular relevance to an understanding of the physical basis of the correlation. As Bottinelli et al. (1980) and Tully, Mould, and Aaronson (1982) have pointed out, there is a significant dependency of the slope on wavelength. It has been appreciated that the details depend on the precise definition of parameters and we favor a definition of line width that characterizes the maximum rotation velocity, V_{max} (Tully and Fouqué 1985). A result of our CCD work is that the slope depends sensitively on the photometric aperture that is chosen and, with total magnitudes, the relation reaches an asymptotic value in the infrared of

$L \propto V_{max}^{3.2}$. The slope of the relationship is smaller than the usual value of 4 that theories try to explain.

(e) We found that the directly measured scatter in the near-infrared in our Virgo sample at ~ 0.4 mag is greater than the ~ 0.3 mag in Ursa Major. It is possible that the greater scatter in Virgo is due to environmental effects but it is at least as plausible that the increased scatter is due to a depth effect, with galaxies somewhat to the foreground and background of Virgo caught in the act of falling into the cluster. The line-of-sight to Ursa Major seems to be fortuitously clean.

(f) Early on, it was feared that absorption corrections to extremely edge-on systems would be a major source of uncertainty (Sandage and Tammann 1976) but it has been evident for almost as long that scatter in deprojected luminosity-line width relations are small for edge-on systems and the more important deprojection problem is the correction to line widths toward face-on orientations (Fisher and Tully 1977). A big improvement has come with the CCD photometry because inclinations can now be measured with an estimated accuracy of $\pm$ 3° (rather than $\pm$ 7° with previous data). With this improvement, we are willing to consider galaxies as face-on as 30°.

H_0 INCREASING WITH DISTANCE: MALMQUIST BIAS OR LOCAL VELOCITY ANOMALY?

The phenomenon of H_0 apparently increasing with distance over the range zero to 2000 km/s has been demonstrated by Giraud (1985), Bottinelli et al. (1986), Tammann (1987), and Sandage (1988). The general concensus has been that this phenomenon was an artifact of Malmquist bias, though Giraud (1987) felt that this is only part of the story. Certainly, the techniques discussed by the above authors were susceptible to Malmquist bias. The phenomenon has the signature of Malmquist bias, may be partially due to the bias, but is it all bias?

Figure 3 suggests that it is not. To construct this figure, distances were determined for galaxies in the Aaronson et al. (1982) sample using the bias-free luminosity-line width distance estimator in the H band discussed in a previous section. The Hubble ratio, velocity/distance, was calculated for the individual systems (if a galaxy is in a group, the group barycentric velocity is used). Then the galaxies were binned according to membership in structures I call clouds (Tully 1987, 1988c). The mean value of the Hubble ratio for all members of a common cloud is determined and if at least five estimates contribute within a single cloud then the mean value is plotted in Figure 3. Panel (a) in this figure is based on strictly observed velocities and panel (b) is based on velocities modified by the specific spherical Virgocentric retardation model discussed by Tully and Shaya (1984).

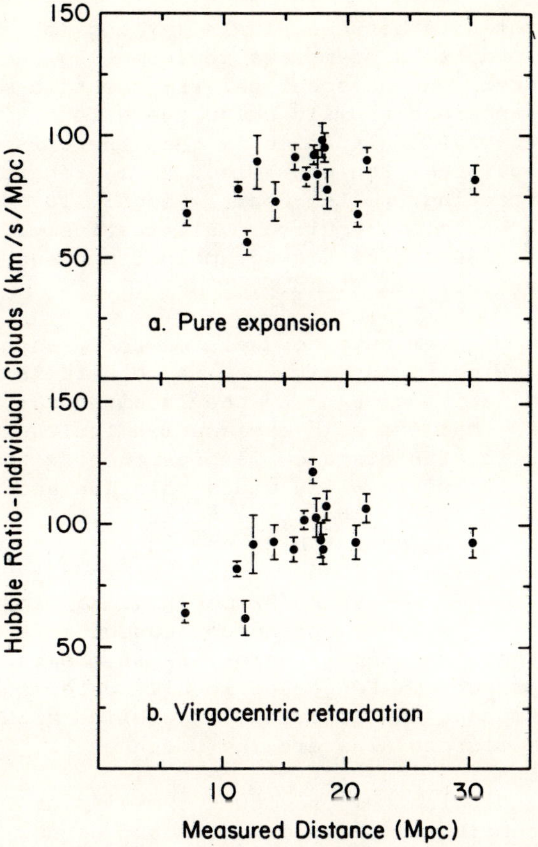

Fig. 3. Hubble ratio for individual clouds of galaxies, plotted as a function of the distances of the clouds. (a) Velocities as observed. (b) Velocities modified in accordance with a specific model of the mass distribution.

In each of the cases illustrated in Figure 3, two conspicuously low points are associated with the Coma-Sculptor Cloud that we live in and its companion Leo Spur. It is to be appreciated that most galaxies within 800 km/s (excluding those in the Virgo Cluster) lie within these two features, which is to say there is a major contribution from the members of these clouds to samples dominated by bright galaxies. In our presentation, the contributions from all those nearby galaxies is reduced to only two points and the anomalous nature of the Hubble ratio determined from these regions becomes reasonably clear. In fact, de Vaucouleurs (1958; more recently, de Vaucouleurs and Peters 1968, 1986: compare Fig. 2 in latter reference with Fig. 3 here) has been aware of this local anomaly and originally tried to explain it in terms of rotation of the Local Supercluster. It would appear that the first hint of the phenomenon was given by Rubin (1951).

The differences between panels (a) and (b) demonstrate a sensitivity to details of models representing the distribution of nearby matter, but no spherical Virgocentric mass model can diminish the apparent anomaly below the effect seen in panel (a). Our tentative explanation is that there is enough mass within our local Coma-Sculptor Cloud to be retarding expansion and to be attracting the Leo Spur. Roughly 10^{14} $M_{\odot}$ and M/L ~ 200-300 $M_{\odot}/L_{\odot}$ would be required. Alternatively, some of the effect might arise out of a quadrupole component of the Great Attractor gravitational field.

It would be premature to dwell on the possible explanation of the local velocity anomaly. At issue here is the influence on the determination of H_0. If the local anomaly is denied and it is accepted that the sole explanation for the increase of the Hubble ratio with distance is bias then, with the zero-point for the luminosity-line width relation accepted here, one would conclude H_0 is in the low sixties. If it is accepted that we can avoid Malmquist bias and the results of Figure 3 are taken at face value then H_0 must be greater than the free expansion value of H_0 ~ 80 km/s/Mpc (Fig. 3a) and, depending on the actual mass distribution and consequent bulk motions in the vicinity of the Local Supercluster, H_0 in the range 85 - 95 km/s/Mpc is probable. Hence the locally determined value of the Hubble Constant can be nicely reconciled with the value obtained from distant clusters (Aaronson et al. 1986).

There is growing evidence that there is at least a tenth the closure density in clumped matter and we now know that galaxies live in filamentary structures surrounded by voids, so the kinds of motions being suggested could be anticipated. Presumably most cosmic observers live in overdense regions, so from a cosmic perspective a typical situation will be a measured increase of the Hubble ratio with distance from the local point of reference. It would be an error to quickly dismiss such an apparent effect as solely an artifact of bias.

Acknowledgements

My collaborator on the CCD photometry program is Mike Pierce. Ed Shaya is participating in an effort to develop a more refined model of the local velocity anomaly. This research is supported by a grant from the US National Science Foundation.

REFERENCES

Aaronson, M., Bothun, G., Mould, J. R., Huchra, J. P., Schommer, R. A., and Cornell, M. E. 1986, Ap. J., **302**, 536.

Aaronson, M., Huchra, J. P., Mould, J. R., Tully, R. B., Fisher, J. R., van Woerden, H., Goss, W. M., Chamaraux, P., Mebold, U., Siegman, B., Berriman, G., and Persson, S. E. 1982, Ap. J. Suppl., **50**, 241.

Aaronson, M., and Mould, J. R. 1983, Ap. J., **265**, 1.

Bothun, G., and Mould, J. R. 1987, Ap. J., **313**, 629.

Bottinelli, L., Fouqué, P., Gouguenheim, L., Paturel, G., and Teerikorpi, P. 1987, Astron. Astrophys., **181**, 1.

Bottinelli, L., Gouguenheim, L., Paturel, G., and de Vaucouleurs, G. 1980, Ap. J., **242**, L153.

Bottinelli, L., Gouguenheim, L., Paturel, G., and Teerikorpi, P. 1986, Astron. Astrophys., **156**, 157.

de Vaucouleurs, G. 1958, A. J., **63**, 253.

de Vaucouleurs, G., Buta, R., Bottinelli, L., Gouguenheim, L., and Paturel, G. 1982, Ap. J., **254**, 8.

de Vaucouleurs, G., and Peters, W. L. 1968, Nature, **220**, 868.

de Vaucouleurs, G., and Peters, W. L. 1986, Ap. J., **303**, 19.

Feast, M. W. 1987, Observatory, **107**, 185.

Fisher, J. R., and Tully, R. B. 1977, Comments on Astrophys., **7**, 85.

Giraud, E. 1985, Astron. Astrophys., **153**, 125.

Giraud, E. 1987, Astron. Astrophys., **174**, 23.

Kraan-Korteweg, R. C., Cameron, L. M., and Tammann, G. A. 1988, Ap. J., **331**, (Aug. 15).

Lynden-Bell, D., Faber, S. M., Burstein, D., Davies, R. L., Dressler, A., Terlevich, R. J., and Wegner, G. 1988, Ap. J., **326**, 19.

Pierce, M. J. 1988a, Pub. A.S.P., **100**, (this issue).

Pierce, M. J. 1988b, Ph. D. Thesis, University of Hawaii.

Pierce, M. J., and Tully, R. B. 1988, Ap. J., **330**, 579.

Richter, O. G., and Huchtmeier, W. K. 1984, Astron. Astrophys., **132**, 253.

Roberts, M. S. 1978, A. J., **83**, 1026.

Rubin, V. C. 1951, A. J., **56**, 47.

Rubin, V. C., Burstein, D., Ford, W. K. Jr., and Thonnard, N. 1985, Ap. J., **289**, 81.

Sandage, A. 1988, Ap. J., **331**, (Aug 15).

Sandage, A., and Tammann, G. A. 1976, Ap. J., **210**, 7.

Tammann, G. A. 1987, in "Observational Cosmology", IAU Symp. 124, eds. A. Hewitt and G. R. Burbidge, p. 151.

Tammann, G. A. 1988. Pub. A. S. P., **100**, (this issue).

Teerikorpi, P. 1984, Astron. Astrophys., **141**, 407.

Tully, R. B. 1987, Ap. J., **321**, 280.

Tully, R. B. 1988a, Nature, **334**, 209.

Tully, R. B. 1988b, in "Vatican Study Week: Large Scale Streaming Motions in the Universe", eds. V. C. Rubin and G. V. Coyne (Princeton: Princeton University Press).

Tully, R. B. 1988c, "Nearby Galaxies Catalog", (Cambridge: Cambridge University Press).

Tully, R. B., and Fisher, J. R. 1977, Astron. Astrophys., **54**, 661.

Tully, R. B., and Fouqué, P. 1985, Ap. J. Suppl., **58**, 67.

Tully, R. B., Mould, J. R., and Aaronson, M. 1982, Ap. J., **257**, 527.

Tully, R. B., and Shaya, E. J. 1984, Ap. J., **281**, 31.

Willick, J. 1988, Pub. A. S. P., **100**, (this issue).

DISCUSSION

T. SHANKS: How do you reconcile the small scatter seen in your Ursa Major and Virgo Tully-Fisher relations with the much larger scatter seen in the Aaronson et al. IR Local Sueprcluster sample?

B. TULLY: Burstein separated the IR sample into "good" and "bad" categories based on whether or not the H band aperture correction was based on reliable B band photometric information. If he uses the "good" subsample with a kinematic model that accounts for the Great Attractor, and Virgo Cluter, and a local anomaly then he finds scatter comparable with what I am now reporting.

S. DJORGOVSKI: If you evaluate the Tully-Fisher relation independently in the two clusters, what difference do you get for the slopes?

B. TULLY: No significant difference.

J. MOULD: There are dynamical models for the Local Supercluster available now based on the distribution of IRAS galaxies. Do these offer any support for your local velocity anomaly?

B. TULLY: I have not looked into that matter in detail. I think the IRAS sample is too coarse on such a local scale. With Ed Shaya, we are trying to do a similar analysis but focused on the local region and using what we know about the distrubiton of optically-identified galaxies.

THE UNIVERSALITY(?) OF DISTANCE-INDICATOR RELATIONS

S. DJORGOVSKI, R. DE CARVALHO, and M.-S. HAN
California Institute of Technology
Pasadena, CA 91125, USA.

ABSTRACT. We examine the origin of distance-indicator relations for galaxies, and the possibility that they may vary with the large-scale environment. The relations reflect formative and evolutionary processes of galaxies, and contain some information about them. They are expected to depend on the environment in a complex manner. There are hints that both the Tully-Fisher relation for the spirals, and the $D_n - \sigma$ relation for the ellipticals may depend on the parent cluster properties such as richness, velocity dispersion, or galaxy number density, but these tentative dependences are hard to separate from the selection effects. If the distance-indicators vary from one cluster to another, some of the large peculiar motions claimed in the literature are partly spurious. Until the environmental effects on distance-indicator relations are better understood, their use for the mapping of large-scale velocity fields and measurements of the far-field Hubble constant may be premature.

The distance-indicator relations, e.g, Tully-Fisher (TF), Faber-Jackson (FJ), the "fundamental plane" of elliptical galaxies and bulges (FP), etc., play a crucial role in detremining the H_0, and are essential for the measurements of large-scale velocity fields. In order for the relations to be usable for these purposes, they have to be universal, that is, same in all large-scale structure (LSS) environments, at a high level of repeatibility. (Recall that at the redshift of 5000 km s^{-1}, a 10% distance error translates into a false peculiar velocity of 500 km s^{-1}.) The distance-indicator relations are tools in exploring the LSS and the associated Hubble flow and deviations from it, and like all tools, they must have their limits of accuracy. Our purpose here is to explore such limits, by looking at the origin of distance-indicator relations, their sensitivity to the formative and evolutionary processes for galaxies in different environments, and to compare the relations derived for individual clusters using the available large and homogeneous data bases. Whether the distance-indicator relations are universal or not, they contain valuable information about the galaxy formation.

The physics underlying the distance-indicator relations is actually very simple, the same for all of them, and clearly universal. The astrophysics implied by their existence is complex, and highly nontrivial; it contains practically all the puzzles of galaxy formation and evolution.

For any galaxy bound by the newtonian gravity the following energy equation must hold:

$$\frac{GM}{\langle R \rangle} = k_E \frac{\langle V^2 \rangle}{2} \tag{1}$$

where G is the gravitational constant, M is the galaxy mass, $\langle R \rangle$ a suitable mean radius defined so that the left side of the Eq.(1) is the potential energy, $\langle V^2 \rangle/2$ a mean kinetic energy per unit mass, and k_E the virialization constant: $k_E > 1$ for a bound system, and $k_E = 2$ for a virialized one. Clearly, the k_E is determined by the overal energy dissipation during the formation of a galaxy, or at least its visible components.

Let an operationally defined, observed radius of a galaxy derived in a non-isophotal way (e.g., the core radius in a Hubble or King model fit, the de Vaucouleurs' r_e, or the e-folding radius of an exponential disk) be:

$$R = k_R \langle R \rangle \tag{2}$$

The parameter k_R thus reflects the density structure of a galaxy. Similarly, let an operationally defined, observed velocity scale of some sort (e.g., the maximum rotational velocity of a spiral, or the central velocity dispersion of an elliptical) be defined as:

$$V^2 = k_V \langle V^2 \rangle \tag{3}$$

The parameter k_V reflects the kinematical structure of a galaxy, e.g., the velocity dispersion tensor with all of its anisotropies in elliptical, the breakdown of the kinetic energy between the circular and random motions, etc. Finally, let the I be an operationally defined mean surface brightness of a galaxy, and the relation between the luminosity L, I, and R:

$$L = k_L I R^2 \tag{4}$$

The parameter k_L thus reflects the luminosity structure of a galaxy in the given bandpass, which in general is not the same as its density structure. Obviously, the parameters k_R, k_V, and k_L depend on the operational definitions of observables, whereas k_E does not, and it represents an intrinsic quality of a galaxy.

With these definitions, it is easy to derive from Eq.(1):

$$R = K_{SR} V^2 I^{-1} (M/L)^{-1} \tag{5}$$

and

$$L = K_{SL} V^4 I^{-1} (M/L)^{-2} \tag{6}$$

where (M/L) is the global mass-to-light ratio of a galaxy, and the combined structural parameters K_{Sx} are:

$$K_{SR} = \frac{k_E}{2G k_R k_L k_V} \tag{7}$$

and

$$K_{SL} = \frac{k_E^2}{4G^2 k_R^2 k_L k_V^2} \tag{8}$$

Let us call the Eqs.(5) and (6) *the generalized distance indicator relations*, combining the distance-dependent quantities R and L with the distance-independent expressions on the right-hand sides. In practice, the Eq.(5) comes in the form of the FP relation,

$$R \sim V^A I^B \tag{9}$$

where the observed values of coefficients A and B are about 1.4 and −0.9 (Djorgovski & Davis 1987). The FP relation also comes in the guise of the $D_n - \sigma$ relation proposed by Dressler *et al.* (1987). As shown by these authors, the two forms are equivalent, providing that all galaxies considered have the same integrated light profiles (which may, but do not have to correspond to the de Vaucouleurs' profile), or in our terminology, that k_R and k_L are the same for all galaxies (see also Phillipps 1988b). The Eq.(6) appears in the guises of TF and FJ relations,

$$L \sim V^C \tag{10}$$

where the observed values of coefficient C are generally close to 4. Some other distance-indicator relations proposed in the literature can be traced to these forms, and the underlying physics is the same. †

If the isophotally defined radii or magnitudes are used instead, that effectively ammounts to absorbing the surface brightness term into the R or L. In fact, that is commonly done for the TF relation, and also with the FP in its $D_n - \sigma$ incarnation.

In order for the distance-indicator relations of these forms to work, the products $K_{SR}(M/L)^{-1}$ and $K_{SL}(M/L)^{-2}$ have to be *power-law functions of the explicite variables R, L, I, or V alone.* The deviations of the observed coefficients A, B and C from their "canonical" values of 2, −1, and 4, define these functions. If these functions are not pure power-laws, the distance-indicator relations become nonlinear. The cosmic scatter and possible variations in the intercepts of the relations may reflect the dependence on other, unaccounted for variables, e.g., the properties of the LSS environment. This defines the limits of applicability of distance-indicator relations.

All of the constituent parameters of K_{SR}, K_{SL}, and (M/L) can depend on the formative and evolutionary histories of galaxies. Our present understanding of galaxy formation is that it generally consists of a series of dissipative merging and infall processes, most of which can be conditioned by the large-scale environment (e.g., Silk & Norman 1981; Silk 1987; etc.). Some of the relevant processes may include: mergers and violent relaxation; gradual infall and formation of disks; tidal shocks, tidal torquing, and the origin of angular momenta; dark halo stripping, and gas sweeping; biasing of the star formation by the neighboring protogalaxies via their radiation fields and galactic winds; metallicity enrichment and gas retention in protocluster cores, and its effects on the stellar IMF; cooling flows and abnormal IMF's; etc. The environment thus affects not only the supply of the galaxy-building material, but also the dynamical structure of the final products (for example, the velocity anisotropies

† The derivation of the $L \sim V^4$ relation from the Gott-Rees (1975) hierarchy by Faber (1982), and the constraint on the initial density perturbation spectrum index from there may be incorrect, as it ignores some of the terms in a more complete Eq.(6). Inclusion of the missing terms leads to a virial theorem tautology. Unlike the dissipationless, mostly unrelaxed LSS, galaxies do not retain any memory of the initial density perturbation spectrum: their properties are dertermined by the dynamical relaxation and merging processes, and by dissipation (cooling), which takes them out of the clustering hierarchy, and defines them as units.

of ellipticals are dynamical fossiles of the past mergers), the primordial IMF and thus the stellar (M/L), the ratio of the dark and visible matterial, etc. Most of these phenomena are generic, i.e., independent of the exact scheme of the LSS formation (top-down vs. bottom-up, CDM, or other scenarios).

Galaxian properties are thus expected to depend on the LSS environment, and this is indeed observed. The most prominent effect is the morphology-density relation (Dressler 1980; Postman & Geller 1984), which is closely related to the morphology-clustering relation (Davis & Geller 1976) and morphology-LSS topology relation (Giovanelli *et al.* 1986). Related to that is the surface brightness-clustering relation (Davis & Djorgovski 1985; Djorgovski & Davis 1986; but see also Bothun *et al.* 1986). There are indications that the galaxy luminosity function varies with the environment (Binggeli 1987; Giovanelli & Haynes 1988). Finally, there are controversial, but intriguing alignment effects (Djorgovski 1987a, and references therein). The crucial question is, are there large-scale dependences of galaxian properties relevant for the distance-indicator relations within a given Hubble type, or in the case of the TF relation, within the whole sequence of spirals?

The (M/L) is one of the principal parameters of interest here, and perhaps the easiest one to track down and measure. Large-scale variations in the (M/L) are naturally expected in many, and perhaps all scenarios of LSS and galaxy formation (Hoffman *et al.* 1982; Silk 1988). Such variations are indeed one of the cornerstones of the whole idea of biased galaxy formation (see, e.g., Dekel & Rees 1987 for a review). The data on clusters are hard to come by, but there are some indications that the (M/L) ratios of galaxy clusters vary by a factor of $3-10$, and perhaps depend on the cluster mass (N. Bahcall 1981; Schectman & Dressler 1988). If there are such variations in (M/L), the essential question is, are they coupled to the individual galaxies? The (M/L) appears to vary along the Hubble sequence (Tinsley 1981; Rubin *et al.* 1982). That can translate into a type-dependence of the TF relation, and from there, via morphology-density relation, into a dependence on the cluster properties. If one assumes a constancy of the K_{SR} parameter for ellipticals, then from the observed equation of the FP, a dependence of (M/L) on mass follows, with a considerable residual scatter (Faber *et al.* 1987; Djorgovski 1987b); it is not yet clear if there is a dependence on the environment, but such test is in principle possible. Variations of (M/L) by a factor of 2 even within a single Hubble type are easily allowed by the present data.

The mean surface brightness I reflects the star formation history of a galaxy, and thus merger-caused starbursts or gas sweeping. There is some disagreement as whether the effects of sweeping are evident in the properties of cluster spirals (Bothun *et al.* 1984, 1985b; Giovanelli *et al.* 1986), but that is clearly a possibility. Phillipps (1988a) finds a systematic difference in surface brightness between the H I deficient and normal spirals in the Virgo cluster.

The parameter k_E is directly related to the degree of energy dissipation (cooling) during the formation of a galaxy, through the inverse Compton mechanism in the initial collapse phase, the cooling of protostars through the Lyα, H_2, or hyperfine transition lines of first metals, or other reprocessed radiation, and through the redistribution of orbital energy of stars in merging proto-galactic fragments (violent relaxation). The later process need not be complete

$(1 < k_E \leq 2)$, but it may also be possible for some galactic cores to be supervirialized $(k_E > 2)$. In the well-known cooling diagram for protogalaxies (e.g., Rees & Ostriker 1977; Blumenthal *et al.* 1984; Silk 1985) the position of a galaxy is related to the degree of dissipation, which also determines its morphology. The Hubble sequence can be viewed as a dissipation sequence (Sandage 1986), and even within a given Hubble type, there may be a spread in k_E: the FP of ellipticals and bulges is almost parallel to the velocity dispersion – surface brightness plane, which is just a representation of the cooling diagram, viz., the temperature – density plane. The k_E may vary by tens of percent, even within a given Hubble type, but this variation would be very hard to detect directly, or separate from a variation in (M/L).

The parameters k_R and k_L depend on the luminosity and mass density structure of a galaxy. It is an observed fact that the surface brightness profiles of galaxies of all Hubble types, including the ellipticals, have different shapes (cf. Kent 1984, or Djorgovski 1985). The "mass types" of spirals do not correlate well with their Hubble types (Burstein 1982). The interplay between the (M/L) and velocity anisotropy makes it hard to constrain the mass distribution in ellipticals, even if the complete surface brightness and projected velocity dispersion information is available (Binney 1982). A wide range of models is possible for the disks as well, from "maximum disk", to "maximum halo" (see, e.g., Kent 1986). And even if one can start with a perfectly homologous family of galaxies, mergers and tidal encounters would modify the k_R and k_L. Some homogenization of the samples may be attempted by selecting galaxies by the shape of their surface brightness profiles: this was done with some success with the TF relation by Bothun & Mould (1987).

The parameter k_V is affected by the anisotropies of the velocity dispersion tensor, and the redistribution of kinetic energy between the ordered and chaotic motions in ellipticals, or by the shape of the rotation curve in spirals. Just like the density structures, the velocity structures of galaxies bear imprints of past mergers and tidal interactions. This is especially true for ellipticals, whose shape parameters do not correlate at all with the FP parameters, or mutually (Djorgovski & Davis 1987; Djorgovski 1987c). Two otherwise identical ellipticals with different amounts of radial anisotropy can have very different projected central velocity dispersion: Merritt (1988) shows that such effects can produce errors in (M/L) estimates (using the King-Minkowski method) of up to 150%.

Given all this, it is perhaps astonishing that there are working distance-indicator relations at all. Even if all of the observed cosmic scatter and peculiar velocities are attributed to the vartiations in K_{SR}, K_{SL}, and (M/L), the compound variations in these parameters cannot be much larger than 20%, say. Given that some fraction of the observed peculiar velocities must be gravitationally induced, the true "cosmic" variation must be even smaller. This reflects a remarkable and unexpected uniformity in the ways of manufacturing of galaxies. Possibly there are some fortuitous cancelations, but more likely, there are some as yet undiscovered regulative (feedback?) mechanisms operating in galaxy formation. Further studies of variations and scatter in distance-indicator relations may shed some light (if not mass) on this astrophysical mine.

Let us now examine some of the available data on distance indicator relations in galaxy clusters, in attempt to look for variations. This is a difficult problem, and it requires large and homogeneous sets of good-quality data. We use the set of Arecibo clusters from Bothun *et al.* (1985a) to examine the TF relation, and the set of clusters from Dressler *et al.* (1987) and Lynden-Bell *et al.* 1988, to examine the $D_n - \sigma$ relation. We can compare the slopes of relations between different clusters, but not the intercepts, which are distance-dependent. In order to check for any variation in intercepts, one needs an independent distance indicator which is at least as good, and there are none available at this time. We can, however, compare the relations derived for the cluster cores and envelopes, as the distances cancel exactly. Finally, we can test for the curvature in these relations, by plotting the slope against the redshift. Any one of these effects (variation of the slopes, intercepts, or the curvature) can generate false peculiar velocities, if an universal relation is assumed, and fitted to the data.

We use the results on 10 clusters from Lynden-Bell *et al.* (1988) in order to search for possible variations in the $D_n - \sigma$ relation. This study is mainly limited by the scarcity of clusters with more than 10 measured galaxies (only 6 clusters). The cluster properties which we use include the richness class, and the velocity dispersion, which is related to the depth of the gravitational potential of the cluster as a whole. We have used the "isophotal" diameters D_n, and the central velocity dispersions, σ, from Burstein *et al.* (1987) and Davies *et al.* (1987). The global solutions for the fundamental plane, viz., the slope and intercepts of the $D_n - \sigma$ relation, for the six more populous clusters, were taken from Lynden-Bell *et al.* (1988).

We find weak correlations of the $D_n - \sigma$ relation slope with the cluster velocity dispersion and richness class, as illustrated in Figure 1. We then divided the sample of galaxies in each cluster according to the projected radius from the cluster center, and evaluated the $D_n - \sigma$ relation for the inner and the outer

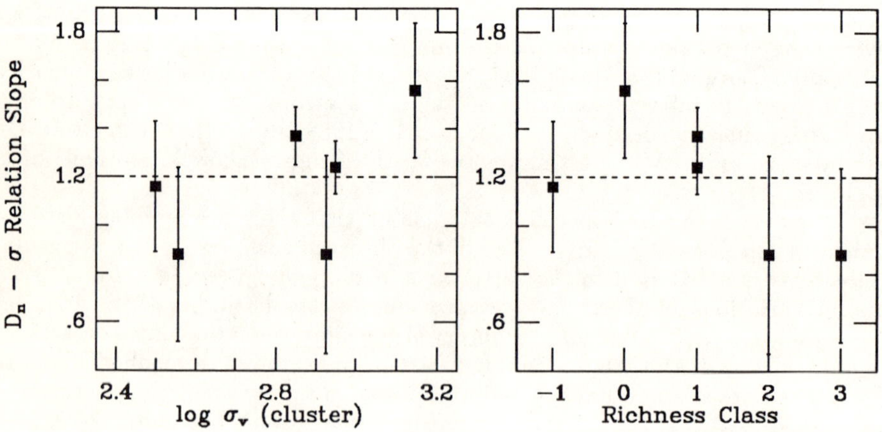

Figure 1. Dependence of the slope of $D_n - \sigma$ relation on the cluster properties, for the Lynden-Bell *et al.* (1988) clusters with more than 10 galaxies measured. The dashed lines indicate the "universal slope" proposed by those authors.

part of each cluster. We computed a minimum perpendicular distance least-squares fit, with all points weighted equally (unfortunately, there are no published error bars for D_n). Figure 2 shows the distribution of the $D_n - \sigma$ relation parameters for the inner and outer regions of the 6 clusters. The larger spread of the slopes for the inner sample is quite remarkable, and one would expect just the *opposite* effect if the outer sample was polluted by the non-cluster-members. However, the galaxy encounters should have more effect on galaxian properties in the inner regions. An alternative way to examine this effect is to correlate the *differences* in the slopes and intercepts (which are distance-independent)

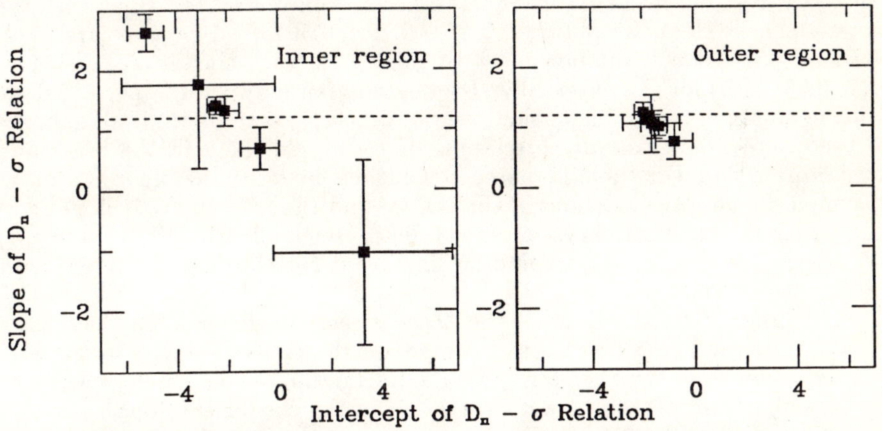

Figure 2. Parameters of the $D_n - \sigma$ relation solutions for the inner and the outer 50% of galaxies in the same 6 populous clusters as in Fig. 1.

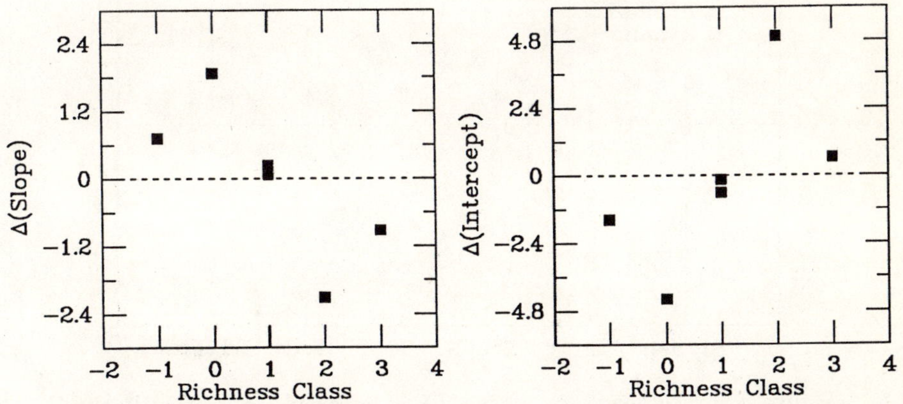

Figure 3. The differences (Inner − Outer) in $D_n - \sigma$ relation solutions for the the same 6 populous clusters as in Figs. 1 and 2, plotted against the cluster richness class.

for the inner and outer parts of each cluster with the cluster properties, as in Figure 1. The dependences of slope and intercept differences on the cluster richness class are shown in Figure 3; we find no perceptible dependence on the cluster velocity dispersion. The correlation between the intercept differences (Inner – Outer) and richness class is may be only a consequence of the slope difference variation. Any differences in intercepts, whether intrinsic or caused by the difference in slopes, would introduce a systematic error in relative distances, and thus a false component of the peculiar velocity.

To quantify these correlations, we have determined parametric and non-parametric correlation coefficients, listed below as: (Linear Regression; Spearman Rank; Kendall Rank). For the relation between the $D_n - \sigma$ slope with the cluster richness class, we obtain (–0.65; –0.79; –0.50) if all galaxies are used; (–0.66; –0.95; –0.65) if the inner 50% of galaxies in each cluster are used; and (0.17; 0.03; 0.10) for the outer 50%. For the relation between the $D_n - \sigma$ slope with the cluster velocity dispersion, we obtain (0.57; 0.48; 0.41) if all galaxies are used, and no significant correlations if we divide them in the inner and outer subsamples. For the differences in solutions between the inner and outer subsamples, computed as (Inner – Outer), we find the following: Δ(Slope) correlates with the cluster richness class as (–0.75; –0.96; –0.69), and Δ(Intercept) as (0.64; 0.81; 0.69). Again, we find no significant correlations with the cluster velocity dispersion.

To summarize, the slope of the $D_n - \sigma$ relation depends on the cluster richness class, and to a much lesser degree, on the cluster velocity dispersion. This dependence is stronger for the galaxies in the cluster cores, where the interactions dominate, and is weak or absent in the cluster envelopes.

We now consider the possible dependence of the TF relation on the morphological type. If the TF relation measures the properties of the disks only,

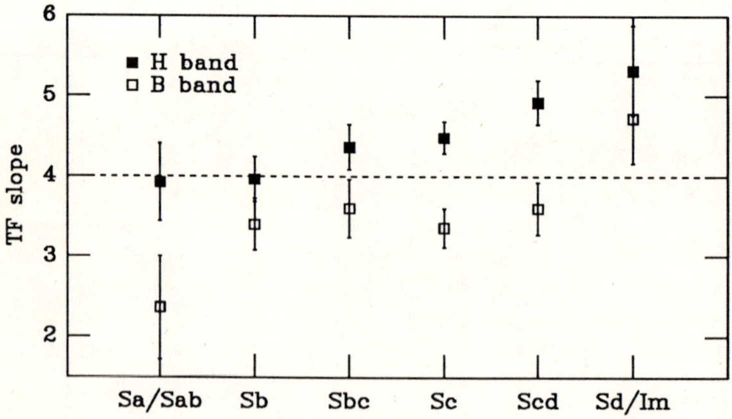

Figure 4. Dependence of the TF relation slope in the B and H bands on the Hubble type. The data were taken directly from Aaronson & Mould (1983). This dependence should translate into a dependence on environmental density, via the morphology–density relation.

a variation of the B/D ratio, the M/L, or the disk surface brightness along the Hubble sequence would induce a variation of the TF slope. In an early study, Roberts (1978) found a strong dependence of the TF relation on galaxy morphology. Then Rubin *et al.* (1982) detected a difference in TF slopes for the Sb and Sc galaxies. Aaronson and Mould (1983) claimed no significant dependence of the TF slope with the morphological type, but even their own data suggest that such dependence does exist (Figure 5). If the TF slope depends systematically on the galaxy morphology, it will then also depend on the cluster density, according to the morphology – environmental density correlation (Dressler 1980).

Kraan-Korteweg (1983) studied the *distance-independent* relation between the H I 21 cm line width and the IR surface brightness, and found a variation between different clusters. This was explored further by Bothun *et al.* (1985b), who also investigated the color – luminosity relation for galaxies in clusters, and found that it too may vary.

In order to examine the possible dependence of the TF relation on the cluster properties, we use the sample of 10 Arecibo clusters from Bothun *et al.* (1985a). A detailed description of the sample selection and the reliability of the observational parameters is found in the original paper. We performed a double linear regression TF relation fit to each of the ten clusters, by assigning a typical error of 0.04^m in H magnitude and of 20 km s^{-1} in line width as adopted from Table 3 of Bothun *et al.* (1985a). In order to guard against the pollution by non-cluster-members, we impose a relative velocity cutoff at 2.5 times the cluster velocity dispersion away from the mean cluster redshift.

We tested the dependence of the TF slope on three different environment indicators, namely, the Abell richness class, the velocity dispersion and the morphological type population. We also examine the correlation with the cluster redshift, which may be indicative of a curvature in the TF relation. The variations of the TF slope against these environment indicators are shown in Figure 4. There is a significant correlation of the TF slope with each of these variables. In our notation from above, we find the following correlation coefficients: for the dependence of the TF slope on the cluster richness class, (0.81; 0.49; 0.60) if all clusters are considered, and (0.83; 0.42; 0.62) if only the clusters with more than 10 galaxies measured are used. For the dependence on the cluster velocity dispersion, (0.69; 0.77; 0.69) for all clusters, and (0.82; 0.96; 0.90) for the clusters with more than 10 galaxies measured. We note that the correlations improve if the more reliable clusters (those with more than 10 galaxies in the sample) are used.

These correlations suggest that the TF relation has a shallower slope in rich or high-density clusters. The question now is whether this trend is intrinsic and due to a real and systematic variation of galaxian properties, or due to some bias and sample selection effects?

For the dependence of the TF slope on the cluster redshift, we obtain the following correlation coefficients: (0.71; 0.88; 0.73) for all clusters, and (0.93; 0.96; 0.90) for those with more than 10 galaxies in the sample. This effect is easily understood as due to the curvature in the TF relation, whereby one samples different portions of the luminosity function in clusters at different redshifts. This was already noticed by Aaronson *et al.* (1986), who tried a nonlinear fit to the data. The curvature of the TF relation can be produced by the non-circular motions in the spiral disks (Bottinelli *et al.* 1983), or could be partly contributed by the dependence of metallicity on disk luminosity (Bothun *et al.* 1984).

The environmental effects shown in Figure 5 might then be caused by this sample selection effect, *if* there is a good correlation between redshifts and the cluster parameters. We find a week correlation between redshift and the environmental parameters (one tends to pick richer and denser clusters at higher redshifts). However, the curvature effects cannot easily explain all of the environmental effects found above.

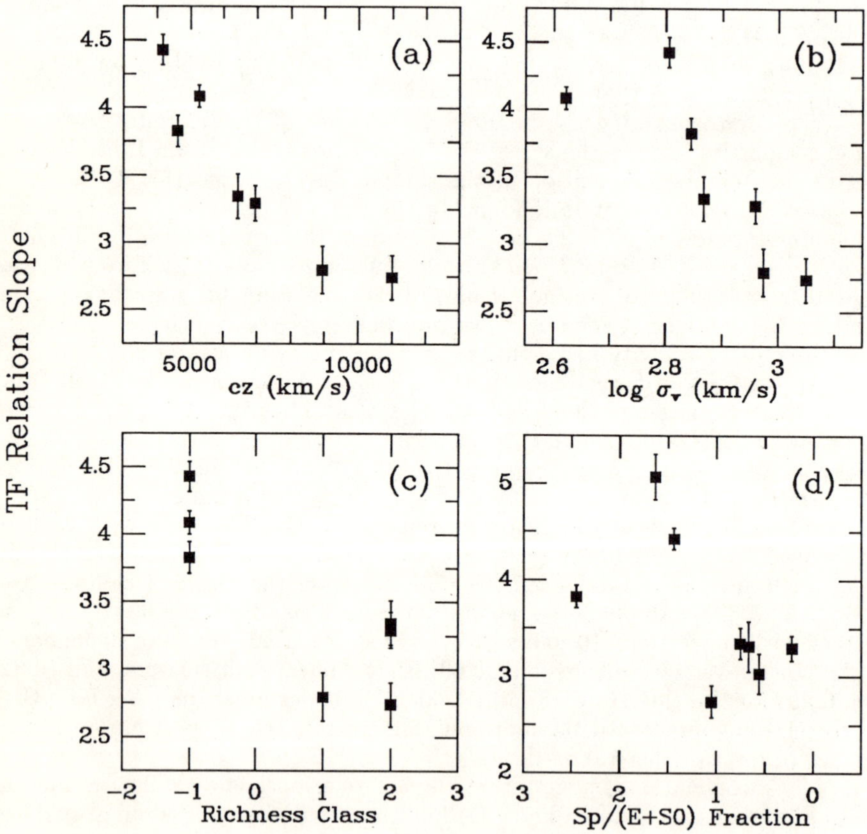

Figure 5. Dependence of the *H* band TF relation slope on: (a) cluster redshift, indicating a curvature in the relation; (b) cluster velocity dispersion; (c) cluster richness class; and (d) morphological composition, defined as a ratio of the fraction of spirals and the fraction of E and S0 galaxies. The data on Arecibo clusters are from Bothun *et al.* (1985a). Only the clusters with more than 10 galaxies were used in panels (a) – (c), and all clusters with available data in panel (d).

In a recent series of papers, Whitmore *et al.* (1988), Rubin *et al.* (1988), and Forbes & Whitmore (1988), explore the shapes of the rotation curves of spirals as a function of the environmental density (cluster *vs.* field, and as a function of position within a given cluster). They find a clear trend which can be most easily understood if the galaxies in denser environments have stripped or truncated halos; this trend is valid for spirals of all Hubble types and all Rubin-Burstein mass types. The implication is that the velocity widths for galaxies of identical luminosities may be systematically lower in denser environments, and thus the intercepts of the TF relation will differ as well.

The Malmquist bias effect in the TF relation has been extensively studied during the past few years (see Bottinelli *et al.* 1988 and references therein). The cluster sample we are dealing with is basically volume limited, the conventional Malmquist bias might not apply here. However as being pointed out by Teerikorpi (1986), cluster samples could suffer a similar biasing effect, due to a horizontal cut off the fainter tail of the TF relation. Such cut through TF relation with non-zero dispersion does result a slightly shallower slope, but the amount of this effect could not be too significant since the cluster sample is sufficiently deep in magnitude and the scatter in the TF relation is small (Aaronson *et al.* 1986). In fact, the sample we used here was actually selected according to the 21 cm line-width rather than the H magnitude (Bothun *et al.* 1985a), so that the incompleteness at the fainter part of the TF relation is more likely to be due to a vertical cut, and the relevant bias is to steepen the slope. Thus, this kind of bias is unlikely to produce the apparent environmental effect. Other possible sample bias such as HI signal to noise ratio and HI flux can probably be ruled out (Aaronson *et al.* 1986).

We therefore conclude tentatively that the correlations between the TF slope and the environmental parameters shown in Figure 5 are partly due to the curvature of the TF relation, and partly to either some as yet unknown sample biasing effects or a true environmental dependence. The later may well reflect the differences in formative and evolutionary processes for spiral galaxies in different environments.

In any case, we feel that further investigations of the universality of distance-indicator relations are both desirable and necessary. Their careless application can easily generate spurious effects and complicate further the measurements of the H_0. The burden of proof should be on those who claim that their empirical standard rulers are good to a couple of percent or less. On the positive side, the distance-indicator relations clearly contain valuable and perhaps unique information about the diversity of processes of galaxy formation and interaction of galaxies with their environment.

REFERENCES

Aaronson, M., and Mould, J. 1983, *Ap.J.* **265**, 1.

Aaronson, M., *et al.* 1986, *Ap.J.* **302**, 536.

Bahcall, N. 1981, *Ap.J.* **247**, 787.

Binggeli, B. 1987, in "Nearly Normal Galaxies", S. Faber (ed.), p. 195. New York: Springer.

Binney, J. 1982, in "Morphology and Dynamics of Galaxies", Saas-Fee course. Geneva: Observatoire de Geneve.

Blumenthal, G., *et al.* 1984, *Nature* **311**, 517.
Bothun, G., *et al.* 1984, *Ap.J.* **278**, 475.
Bothun, G., *et al.* 1985a, *Ap.J.Suppl.Ser.* **57**, 423.
Bothun, G., *et al.* 1985b, *Ap.J.* **291**, 586.
Bothun, G., *et al.* 1986, *Ap.J.* **308**, 510.
Bothun, G., and Mould, J. 1987, *Ap.J.* **313**, 629.
Bottinelli, L., *et al.* 1983, *Astr.Ap.* **118**, 4.
Bottinelli, L., *et al.* 1988, *Ap.J.* **328**, 4.
Burstein, D. 1982, *Ap.J.* **253**, 539.
Burstein, D., *et al.* 1987, *Ap.J.Suppl.Ser.* **64**, 601.
Davies, R., *et al.* 1987, *Ap.J.Suppl.Ser.* **64**, 581.
Davis, M., and Djorgovski, S. 1985, *Ap.J.* **299**, 15.
Davis, M., and Geller, M. 1976, *Ap.J.* **208**, 13.
Dekel, A., and Rees, M. 1987, *Nature* **326**, 455.
Djorgovski, S. 1985, Ph.D. thesis, University of California, Berkeley.
Djorgovski, S. 1987a, in "Nearly Normal Galaxies", S. Faber (ed.), p. 227. New York: Springer.
Djorgovski, S. 1987b, in "Starbursts and Galaxy Evolution", T.X. Thuan, T. Montmerle, and J. Tran Thanh Van (eds.), p. 549. Paris: Editions Frontières.
Djorgovski, S. 1987c, in "Structure and Dynamics of Elliptical Galaxies", T. de Zeeuw (ed.), p. 79. Dordrecht: Reidel.
Djorgovski, S., and Davis, M. 1986, in "Galaxy Distances and Deviations from Universal Expansion", B. Madore and R.B. Tully (eds.), p. 291. Dordrecht: Reidel.
Djorgovski, S., and Davis, M. 1987, *Ap.J.* **313**, 59.
Dressler, A. 1980, *Ap.J.* **236**, 351.
Dressler, A., *et al.* 1987, *Ap.J.* **313**, 42.
Faber, S. 1982, in "Astrophysical Cosmology", H. Bruck, G. Coyne, and M. Longair (eds.), p. 191. Vatican: Specola Vaticana.
Faber, S. 1987, in "Nearly Normal Galaxies", S. Faber (ed.), p. 175. New York: Springer.
Forbes, D., and Whitmore, B. 1988, preprint.
Giovanelli, R., Haynes, M., and Chincarini, G. 1986, *Ap.J.* **300**, 77.
Giovanelli, R., and Haynes, M. 1988, in "Towards Understanding Galaxies at Large Redshifts", R. Kron and A. Renzini (eds.), p. 133. Dordrecht: Reidel.
Gott, J.R., and Rees, M. 1975, *Astr.Ap.* **45**, 365.
Hoffman, Y., Shaham, J., and Shaviv, G. 1982, *Ap.J.* **262**, 413.
Kent, S. 1984, *Ap.J.Suppl.Ser.* **56**, 105.
Kent, S. 1986, *A.J.* **91**, 1301.
Kraan-Korteweg, R.C., 1983. *Astr.Ap.*, **125**, 109.
Lynden-Bell, D., *et al.* 1988, *Ap.J.* **326**, 16.
Merrit, D. 1988, *A.J.* **95**, 496.
Phillipps, S. 1988a, *Astr.Ap.* **194**, 77.
Phillipps, S. 1988b, *M.N.R.A.S.* **233**, 561.
Postman, M., and Geller, M. 1984, *Ap.J.* **281**, 95.
Rees, M. and Ostriker, J. 1977, *M.N.R.A.S.* **179**, 541.
Roberts, M. 1978, *A.J.* **83**, 1026.

Rubin, V., *et al.* 1982, *Ap.J.* **289**, 81.
Rubin, V., Whitmore, B., and Ford, W.K. 1988, *Ap.J.*, in press.
Sandage, A. 1986, *Astr.Ap.* **161**, 89.
Schectman, S., and Dressler, A. 1988, *Ap.J.*, in press.
Silk, J., and Norman, C. 1981, *Ap.J.* **247**, 59.
Silk, J. 1985, *Ap.J.* **297**, 9.
Silk, J. 1987, in "Dark Matter in the Universe", J. Kormendy and G. Knapp (eds.), p. 335. Dordrecht: Reidel.
Silk, J. 1988, preprint.
Teerikorpi, P. 1987, *Astr.Ap.* **173**, 39.
Tinsley, B. 1981, *M.N.R.A.S.* **194**, 63.
Whitmore, B., Forbes, D., and Rubin, V. 1988, *Ap.J.*, in press.

DISCUSSION

J. WILLICK: To what extent might the variations in TF slope with cluster richness, velocity dispersion, etc., actually be due merely to curvature in the TF relation?

S. DJORGOVSKI: That would require a good correlation between the cluster properties and the redshift, and we doubt that all of the effects described above can be explained by the curvature alone.

M. CLUTTON-BROCK: The densities of loose and compact groups differ far more from each other than do poor and rich clusters. Would it be possible to use groups to check the variation of the TF and $D_n - \sigma$ relations with environment?

S. DJORGOVSKI: A very good idea. There is not enough homogeneous data at this moment for such a check, but it would be a wortwhile experiment to do in the future, when the data are available.

D. LYNDEN-BELL: (1) The TF Coma/Virgo distance ratio agrees with the $D_n - \sigma$ distance ratio very well, only 0.015^m different. This suggests that environmental differences in Coma and Virgo cannot be very large. (2) We got very similar streaming motions from 4 different distance indicators. It is unlikely that $D_n - \sigma$ and $Mg - L_B$ relations will be environmentally affected in such a way to give the same motions. (3) The slopes of the $D_n - \sigma$ relation are dependent on the measurement accuracy. The fainter members of the large clusters are not measured to the same accuracy as the main survey.

S. DJORGOVSKI: First, let me emphasize that some of the measured peculiar velocities must be real, due to the gravitational acceleration. Those components would bear an imprint of a large-scale coherence, suggesting that the entire observed (interpreted?) velocity field is real. Our claim here is that some part of the peculiar velocity amplitudes – perhaps as much as 50% on the average – may be spurious, riding atop of the real peculiar motions. But to answer your comments in order: (1) Both TF and $D_n - \sigma$ relations depend on a very similar (astro)physics, and may vary with the environment in a similar way. I am also not sure that everybody would agree that the differential distance moduli of Virgo and Coma are that well determined. (2) No other relation is as good as TF or $D_n - \sigma$, and many of them entail similar physics, and could be affected in similar ways. We would need a *truly independent, at least as good distance indicator* to check them, and there is none such at the moment. Perhaps with the HST in Virgo... (3) We just take the slopes of the $D_n - \sigma$ from you. If they are good enough to measure the peculiar velocities, they are good enough for our tests.

G. TAMMANN: Unfortunatelly, our ways to determine galaxy distances (or only relative distances) are still so crude that if one goes into finer details, problems turn up that we still do not understand. For instance, one expects relative distances to Coma and Virgo to be particularly well determined. Yet the $D_n - \sigma$ method yields $\Delta(m - M)_{Coma-Virgo} = 3.62$ (Lynden-Bell, priv.comm.), and the TF method requires a best value of 3.8 – 4.0 (Bottinelli *et al.* 1988, *Astr.Ap.* **196**, 17; Kraan-Korteweg *et al.* 1988, *Ap.J.*, in press). Moreover, Aaronson *et al.* (1986, *Ap.J.* **302**, 536) claimed from TF data to see our MWB motion reflected in the motion of distant clusters, while Kraan-Korteweg *et al.* (1988) did not see a trace of it from the same data. If you find here several subtle effects in the data on the 11 clusters, this does therefore not reflect any physics, but possibly only the enormous problems to define fair samples.

S. DJORGOVSKI: Certainly, the selection effects are at least as important as any physical variations in distance-indicator relations which may be present, and the trends which we observe are probably reflecting a combination of causes. Ideally, we would like large and homogeneous data samples with well-understood selection procedures, in order to test the environmental effects, but they are hard to get. We used the best data we could find, and our investigation is clearly very preliminary.

J. MOULD: I agree that in analysing data by the $D_n - \sigma$ and TF techniques, one has a choice of concluding that the galaxies are peculiar, or that the velocities are peculiar. However, there is one peculiar velocity that we can be fairly sure of, and that is the 600 km s^{-1} velocity of the Local Group relative to the MWB. In our analysis of the Arecibo clusters, we recovered that motion to ± 200 km s^{-1}. This confirmation, together with the economy of hypotheses suggests that this is the route to take.

S. DJORGOVSKI: It is my impression that explaining the MWB vector is a tricky problem, and a highly model-dependent one (e.g., the IRAS dipole controversy, etc.). Some of the Local Group motion is probably easily explained, but I doubt that the local acceleration field is so well known as to claim the complete success. The result you quote may be partly fortuitous. I don't think that the universality of the TF relation is an economical hypothesis, it is perhaps an insufficiently justified one.

R. NOLTHENIUS: I would worry that a significant part of the apparent flattening of the Tully-Fisher relation slope with spiral cluster richness is due to contamination. Spirals are notoriously difficult to group accurately, since they inhabit lower density regions, and too liberal a velocity linkage cutoff will produce severe contamination.

S. DJORGOVSKI: We've used Bothun's cluster identifications and cut out anything beyond 2σ.

DEVIATIONS FROM PURE HUBBLE FLOW: A REVIEW

JAMES E. GUNN
Princeton University, Princeton, NJ 08544

ABSTRACT. The possible origins of deviations from the Hubble flow in various scenarios is discussed, with emphasis on the gravitational perturbation picture of the growth of structure. The relevant observations are reviewed and the methodology considered in the light of possible effects of the initial conditions during galaxy formation. The agreement, which is marginal at best, between the velocity data and the mass distribution models inferred from the IRAS galaxy catalogues is considered, and finally the confrontation with theory briefly discussed.

1. INTRODUCTION

It has been known for a long time that the gross features of the large-scale motions in the universe are well-described by a pure expansion, the revered Hubble law. The distribution of luminous matter, however, is very far from uniform. If gravitational forces made it thus, or indeed if any forces produced it from a once more-uniform medium, the peculiar velocities which redistributed the matter should still manifest themselves. The nature, sizes, and spatial coherence of those velocities are, of course, the subject of a great deal of work at the present time, and this review is a feeble attempt to put these efforts in perspective. It is a field that I have not worked in directly for some years, and I apologize in advance for the omission of any important work which I have overlooked.

The organization of the review reflects an attempt at a methodical approach, and begins with a discussion of how peculiar velocities might arise in the universe, gravitationally and otherwise. We then proceed to discuss how one might measure them, and how in fact they have been measured. We then consider the results of those measurements, and the attempts which have been made to make sense of them in terms of various flow models. At this point we go back and review how one might attempt to predict what one *should* see from other observations. Then comes the confrontation of the results with the predictions, and the question of whether or not they agree at all satisfactorily, and a discussion of the inevitable question from the last comparison: *"Why not?"* We end with a brief discussion of comparison with theoretical predictions.

2. THE ORIGINS OF PECULIAR MOTIONS

I think it is fair to say that nearly everyone in the field agrees that galaxies and large-scale structure probably form by one of two mechanisms or various hybrids of the two. On the one hand is the picture of gravitational growth of small density perturbations which were either primordial or (as has become fashionable of late) grew from quantum/thermal fluctuations during inflation, or are the consequence of some kind of gravitating "seeds" produced in the early universe, such as cosmic strings.

A small but persuasive school, led by Ostriker (see, *e. g.* Weinberg *et. al.* 1988 and references therein) favor formation by explosive phenomena in the redshift range around 10 to 20, perhaps involving superconducting strings or perhaps nothing more exotic than supernovae; in the latter case large-scale structure and by inference large-scale velocity fields are probably *not* produced and one must fall back on gravity. If, however, superconducting strings exist with the correct parameters and if they exploit their ability to carry large currents, it seems possible that they could produce the large-scale structure seen today. The velocity data suggest, however, that peculiar velocities of the order of 500 $km\ s^{-1}$are present in the universe now, with coherence scales of some tens of Mpc, and it does not seem possible that any thermal process which rearranges the matter in the universe early in its history can produce velocities of this nature.

There are many variants on the scenario, and many variants on the arguments, but it would appear that one always violates some constraint set by the homogeneity or spectrum of the microwave background. In the simple case, for example, of an $\Omega = 1$ universe with $\Omega_{baryon} = 0.1$, or an $\Omega = \Omega_{baryon} = 0.1$, the universe is just optically thick to Thompson scattering at a redshift of 20; to produce the 500 $km\ s^{-1}$observed today, velocities of order 10^4 $km\ s^{-1}$were necessary in the explosion, or, equivalently, temperatures of order 10^{10} K. The associated angular scales are of the order of 0.3°, and the background fluctuations produced of the order of $\delta T/T \approx 10^{-3}$. Even if the density in baryons is high enough so that this epoch is hidden by Thompson scattering, those same electrons which hide the fluctuations produce by virtue of their high temperatures very large distortions of the spectrum, of the order of a few tenths in the Zeldovich-Sunyaev (1969) y parameter, for which there is a measured (value? upper limit?) of 0.028 (Hayakawa *et. al.* , 1987). These effects are ameliorated somewhat by gravitational growth subsequent to the seed explosion, lessening the necessary input pressures, but not by nearly enough to overcome the difficulty.

It thus seems likely that even if the large-scale structure *is* produced by explosions, they cannot produce the putative large-scale flows, so if the flows exist they provide an embarrasment to the explosion picture—as well as to many others, as we shall see.

Large-scale flows are, however, part and parcel of the gravitational scenario, the sticking point being that they are intimately tied to and generated by the present density distribution in the universe. Thus there must be consistency between the observed flows and such observations as there are of the density distribution, and unknown factors such as the value of Ω and biasing must conspire to give the correct amplitudes.

In linear perturbation theory, which should adequately describe the scales of interest, the equation which governs the growth of the density contrast $\delta = \delta\rho/\bar{\rho}$, written as a function of comoving coordinates $\mathbf{x}$ and time t, is

$$\frac{\partial^2 \delta}{\partial t^2} + 2H(t)\frac{\partial \delta}{\partial t} = 4\pi G\bar{\rho}\delta \tag{1}$$

(see, *e. g.* Peebles 1980). This equation contains no spatial derivatives and is homogeneous, of course, in δ. Thus the general solution is of the form

$$\delta(\mathbf{x}, \mathbf{t}) = \delta_0(\mathbf{x})\mathbf{f_g}(\mathbf{t}) + \delta_1(\mathbf{x})\mathbf{f_d}(\mathbf{t}), \tag{2}$$

where $f_g(t)$ and $f_d(t)$ are the two independent solutions of (1) regarded as an ordinary differential equation. For all cases of interest one of the solutions grows with time and the other decays, so we need only consider the growing solution $f_g(t)$. The relevant thing for the problem at hand is that the density grows *self-similarly.* Thus the perturbation *potential*, $\phi(\mathbf{x})$, the solution of the Poisson equation with the density perturbation as the source, $\nabla^2\phi = 4\pi G\bar{\rho}\delta$, is likewise self-similar, as is the peculiar gravity, its gradient. This means among other things that the *direction* of the acceleration does not change in the linear approximation, and has always been in the same direction one finds it now. Since the peculiar velocity is just the integral of the acceleration (taking the expansion properly into account) it points in the same direction as well and is proportional to the *present* gravitational acceleration. Note in passing that since the gravity is derived from a potential, both it and the velocity are curl-free, a condition which the velocity fields generated by explosions need not enjoy and which in some rosy future might provide a discriminant. Using an approximate but satisfactorily accurate solution of (1), one finds the standard result (see, *e. g.* Peebles 1980)

$$\mathbf{v}_p(\mathbf{x}) = \mathbf{g}_0(\mathbf{x}) \times \frac{2}{3H_0\Omega^{0.4}}, \tag{3}$$

where $\mathbf{g}_0$ is the present gravitational acceleration. If Ω is unity, the peculiar velocity is simply $\mathbf{g}_0 t_0$; for $\Omega = 0.1$, $\sim 1.8\mathbf{g}_0 t_0$.

For roughly *spherical* systems, one can do a little better, and introduce an approximate correction for the first-order deviations from linearity. It is also often helpful to write the peculiar velocity as if its source were density *contrast*, since that (or at least luminosity or galaxy number density contrast) is fairly easily measured. The expression for the peculiar velocity in terms of the unperturbed Hubble velocity at some distance from a spherical perturbation of mean density contrast $\bar{\delta}$ is

$$\frac{v_p}{v_H} \simeq \frac{-\bar{\delta}\Omega^{0.6}(1+\bar{\delta}/3)^{-1/2}}{3}. \tag{4}$$

A question which must concern us before applying these ideas is how well they work; this topic has been dealt with with some thoroughness in a recent paper by Villumsen and Davis (1986), who have compared linear theory and mildly nonlinear spherical theory to the flow into large peaks observed in n-body experiments with cold dark matter initial conditions. Their results can be summarized as follows: The mildly nonlinear expression for the magnitude of the velocity is an excellent approximation for mean density contrasts up to of order at least ten, which is far outside the range we need for the peculiar velocity problem. The linear theory gives an excellent approximation for direction of the flow up to density contrasts of a few, but the spherical approximation is pretty lousy. Both approximations are worse for simulations which resolve very small-scale structure, as might be expected; that is probably *not* of relevance, since we remove such effects at the outset in the observations (*i. e.* we discuss the motion of the local group, not the motion of the sun or the earth—it seems very unlikely that any serious nonlinearity is left in the motion of the center of mass of the local group.)

3. THE MEASUREMENT OF PECULIAR VELOCITIES

Since the peculiar velocity is defined as a deviation from the Hubble flow, the determination of distances, from which one can calculate the Hubble velocities, is part and parcel of any determination. Thus the basic ingredients of any determination are a set of radial velocities $v_{r,i}$ and a set of distances r_i, or, more precisely, a set of vector locations $\mathbf{r}_i$. The peculiar velocity is then

$$v_p = v_r - H_0 r, \tag{5}$$

of which we measure, of course, only the radial component. If we think we know our own peculiar velocity with respect to some absolute rest frame, for which the obvious choice for most would seem to be the microwave background, the radial component of that motion can be subtracted to obtain the peculiar radial velocity in that preferred frame.

It is important in any such endeavor to understand the errors in the distances, since almost any source of error leads to biased distance estimates. There are very many approaches to this problem, and the reader is referred to Lynden-Bell *et. al.* (1988, hereinafter 7S-V) and Faber and Burstein (1988, hereinafter FB) for particularly lucid discussions of the problem and possible approaches to circumvent it. The question is enormously complicated by the fact that the biases are influenced by gradients in the same galaxy distribution which is responsible for the peculiar forces, and exquisite care must be taken. We will not discuss this issue further, but it seems just possible (if unlikely) that lack of understanding of the details of the galaxy distribution might yet have serious effects on the conclusions about large-scale streaming. The importance of the statistics and bias therein is underscored by the required accuracy; to discuss 500 $km\ s^{-1}$effects at distances of 5000 $km\ s^{-1}$requires 10 per cent distances, a level of accuracy not obtainable for single galaxies by any extant technique.

The techniques used to obtain distances are many, but essentially all the recent work has focussed on two, often considered variants of one another (see the paper by Djorgovski in this volume) but which I think are fundamentally different: the Fisher-Tully (1977) technique in one of its many variants for spirals, and the D_n-σ relation (Dressler *et. al.* 1987a) for ellipticals. It is generally agreed that the most accurate form of the Fisher-Tully relation is the one used by Aaronson *et. al.* (1982) in which a suitably defined 1.6 μ luminosity is related to the 21-cm linewidth. FB have shown recently that when the data are restricted to those galaxies with well-determined inclinations and which have good H and B photometry the luminosity uncertainties using this technique are as small as 35 percent.

It has been pointed out by several authors including the present one (Gunn 1982, 1986, Faber 1982) that the relation follows closely that predicted by a density fluctuation spectrum with small negative power-law index, and follows even the curvature predicted by the cold dark matter spectrum. (Blumenthal et al 1984, Gunn 1986). It is difficult to understand how such a relation can come about without some such template in the initial conditions, though it is fair to say that even with such a template it is remarkable that the scatter in the relation is so small, and other effects not at present understood must intervene to make it so. Since the velocity width of spirals is at least approximately established by the form of the dark halo (for late type systems essentially exactly so) the relation almost certainly follows from some underlying relationship between the depth of the halo (dark matter) potential well and the mass of the baryonic (gaseous) matter which belongs to that well. It is on the face of it difficult to understand why this relationship is not strongly affected by environment; one would think that in regions of generally high density peaks of a given mass would be more tightly bound, but there seems to be no marked environmental effect. I think it is fair to say that we do not currently understand the origin of the Tully-Fisher relation in sufficient depth to have predicted (even after the fact) that it would be so useful as it is.

The D_n-σ relation of Dressler et al (1987a), hereinafter 7S-I; the group of Burstein, Davies, Dressler, Faber, Lynden-Bell, Terlevich, and Wegner, will be referred to as 7S, the Seven Samurai, as has become fashionable) is quite another kettle of fish, however. The original Faber-Jackson (1976) relation is the exact analogue of the Tully-Fisher relationship for ellipticals. It is not so accurate as T-F, probably at least in part because the velocity dispersion in the inner parts of ellipticals is not so slavishly dictated by the dynamics of the halo as is the rotation of the outer parts of spirals. Its origin, however, is probably similar to T-F. It can be shown quite simply, however, that the D_n-σ relation is a straightforward consequence of the observed facts that all (bright) elliptical galaxies are approximately deVaucouleurs laws when projected on the sky, that the mass-to-light ratios of their inner parts are very gentle functions of their luminosities ($M/L \propto L^{0.25}$) and the virial theorem, with no assumptions about the initial hierarchy. Armed with these observations it is simple to *predict* the form of the D_n-σ relation, and the only scatter allowed in it comes the deviations from deVaucouleurs-law form (small) and the scatter in the $M/L - L$ relation. The observed trend in the latter can be qualitatively explained by metallicity, though the residuals can evidently *not* be (7S-I), and must be ascribed either to variations in the IMF or to varying admixtures of dark matter in the inner parts of ellipticals, with the former being possibly the most palatable.

The argument, which is a rather backwards paraphrase of one presented in 7S-I, goes as follows: The diameter D_n is defined as the diameter within which the *mean* surface brightness has some fixed value μ_0 (20.75 $mag\ arcsec^{-2}$ in B in the 7S sample). Now it is very simple to show that if all ellipticals are deVaucouleurs laws (or any other universal form) the ratio of D_n to the scale radius r_e is a function only of the ratio of the scale surface brightness μ_e to the definining surface brightness μ_0, the latter being a constant for the sample, and to high accuracy one can write (note that μ_0 is very bright compared to μ_e)

$$D_n/r_e \propto \mu_e^{0.8}$$

over the range of interest. If we now assume that $M/L \propto L^\gamma$, the virial theorem becomes

$$\begin{aligned}\sigma^2 &= const \times \frac{M}{r_e} = const \times \frac{L(M/L)}{r_e} \\ &\propto r_e^{1+2\gamma}\mu_e^{1+\gamma} \\ &\propto D_n^{1+2\gamma}\mu_e^{(1+\gamma-0.8(1+2\gamma))}\end{aligned} \tag{6}$$

or, for $\gamma = 0.25$,

$$D_n \propto \sigma^{4/3}\mu_e^{-1/30}. \tag{7}$$

The trivial dependence on μ_e aside, this is the desired form, and is essentially the "true" form derived by 7S-I. Since the relation is not perfect and there are inevitable observational errors, the regression of D_n on σ, which is used as a distance estimator directly (one measures an *angular* D_n and a σ) has a somewhat shallower slope, and the relation actually used is $D_n \propto \sigma^{1.2}$, and the one-sigma error in $ln(D_n)$ is 0.21, of which about half is estimated to be observational error and half cosmic scatter.

4. THE 7S SAMPLE: RESULTS AND INTERPRETATION

There have been indications of the existence of large-scale flows since the important early work of Rubin and Ford and collaborators (Rubin et al 1976), and the importance of flows in general has been underscored by the now very accurate determination of the motion of the local group with respect to the microwave background (Fixsen, Cheng, and Wilkinson 1983). The most systematic study to date is that of the 7S with their elliptical galaxy sample using the D_n-σ relation, and it is the interpretation of those data which has been the subject of most of the recent activity in the field. FB have made an attempt to integrate other data sets into the 7S flow model, with conspicuous success in the case of the somewhat less deep but very high quality "Aaronson et al Good" IR Tully-Fisher data set alluded to above.

There have been essentially three steps along the way which have led to the most recent picture as presented in FB. It was first realized by performing what is essentially a classical solar motion solution in the frame defined by the microwave background that the motions to the limit of the survey, about 6000 $km\ s^{-1}$, were dominated by a very large-scale flow of some 600 ± 100 $km\ s^{-1}$toward $l = 312°$, $b = 6°$, with a probable error in the direction of about 10°. (Dressler *et. al.* 1987b, 7S-IV). Most of the weight of the determination came from within 5000 $km\ s^{-1}$, with some indication that the galaxies beyond might be essentially at rest with respect to the MWB. It appeared already at that time that there was some structure in the residuals, the dominant feature being a gradient in the direction of the flow, with velocities in the direction of the flow (in roughly the direction toward Hydra-Centaurus) larger than those opposite in the sky. A more complex model was tried with great success in 7S-V, in which all of the flow was attributed to the gravitational action of a "Great Attractor" (GA), an extended ($\rho \propto r^{-2}$) mass concentration of about 5×10^{16} $M_{\odot}$lying at a distance of about 4500 $km\ s^{-1}$in essentially the direction of the previously postulated flow, just beyond the Hydra-Centaurus complex, and responsible for the large positive residuals of the Centaurus galaxies. This model also accounted for a large-scale quadrupole in the velocity field which is present in the elliptical data and had been noted previously in T-F data at about the same level by Lilje, Yahil, and Jones (1986). The data are presented in Figure 1, from 7S-V; the distances in that figure and in all of this discussion are in velocity units essentially normalized to the Coma Cluster. This diagram plots all objects within 45° of the supergalactic plane projected on that plane. Note that there is some confusion about supergalactic coordinates, not only among different authors, but among the several 7S papers (!) themselves. This picture must be flipped right-to-left to compare with the IRAS density data to be discussed below or with the figures in FB; the similar pictures in 7S-IV must be *rotated* 180°. The prominent mass concentrations are identified: V=Virgo, C=Centaurus, H=Hydra, P=Perseus, Pi=Pisces, F=Fornax, and P-I=Pavo-Indus. The GA is just behind Centaurus at about 4500 $km\ s^{-1}$.

The next step made use of the higher spatial resolution afforded by the inclusion of the IR T-F data by FB, and included a "local anomaly" to account for the residual motion of the local group and, as it transpires, a volume of about 800 $km\ s^{-1}$radius surrounding it, comprising the "Coma Sculptor Cloud" of Tully and Fisher (1987). This motion is ascribed to the existence of a void called by Tully and Fisher (1987) the "Local Void" which has approximately the right direction and properties to account for the local anomaly. Also included is a Virgocentric flow model, which was also part of the model used in 7S-V, though used there with fixed input parameters. Remarkable in the fits to these models is the quietness of the flow within the local anomaly, the RMS one-dimensional dispersion amounting to only 80-90 $km\ s^{-1}$. The model finally used by FB includes the GA, the local anomaly, and a Virgocentric flow model, the GA model modified only to the extent of the inclusion of a core radius to flatten the r^{-2} density law used by 7S-V. They point out a result already seen but not discussed in 7S-V, that the result of the tide from the Great Attractor in the model is to lessen drastically the amplitude of the Virgo infall (Virgo being roughly perpendicular to the direction of the GA) to values in the neighborhood of 100 $km\ s^{-1}$. It would seem to me that some caution needs be exercised in the interpretation of this result, since the GA model is spherically symmetric and the distribution of mass in the local supercluster is very flattened; the quadrupole seen by Lilje, Yahil, and Jones is *not* axisymmetric about the flow direction, and clearly more work needs to be done. The lesson that larger-scale structure can significantly affect such "local" phenomena as Virgocentric flow must be heeded, however.

Before we proceed let us briefly summarize the 7S results. The flow is dominated by a mass concentration dubbed the "Great Attractor" by Dressler, which, in FB's last

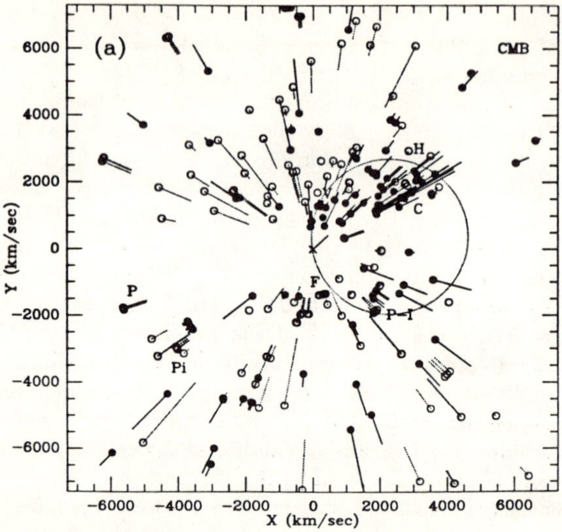

Figure 1. The 7S elliptical sample for all objects within 45° of the supergalactic plane. Open circles are galaxies with approaching peculiar motions, closed with receding. The lengths of the attached lines are the velocities. Velocities are in the CMB frame. From 7S-V.

model is at a distance of 4200 $km\ s^{-1}$at $l = 309°$, $b = 18°$. In that model the density fall-off is a power law with $\rho \propto r^{-2.7}$ and with a core radius about a third of its distance from us. The GA induces a peculiar velocity at the local group of 535 $km\ s^{-1}$. The parameters of the Virgocentric flow are highly dependent upon which data set is used and the method of fitting, but the best guess from these models of the amplitude of the infall to Virgo is in the range of 80-130 $km\ s^{-1}$, less than half the usual value. There is in addition a local superposed coherent (and very quiet) flow, the "Local Anomaly", of amplitude about 360 $km\ s^{-1}$toward $l = 199°$ and $b = 0$ and diameter about 1400 $km\ s^{-1}$. FB note that such regions seem common in the flow pattern indicated by the 7S data.

At still larger distances it would be reassuring if the mean motion with respect to the CMB were small, as it must be if the microwave dipole is a product of our peculiar motion. The data on this point are less than ultimately helpful. The distance accuracy required, of course, becomes very high as one moves to large distances, and as a result most of the work done on this point has revolved around clusters, in which one can use the distance data on many member galaxies to suppress the noise (so that the signal is entirely dominated by systematic errors one does not begin to understand??). The 7S data show some indication in their sample beyond about 3000 $km\ s^{-1}$that the velocities are small, but the errors are too large to say even that the 500 $km\ s^{-1}$bulk flow does not extend to the limits of the sample. Aaronson et al (1986), using the IR Tully-Fisher technique, have studied 10 clusters at a mean redshift near 5000 $km\ s^{-1}$, and find a result consistent with those clusters being at rest with respect to the CMB; the clusters, however, all lie in the Arecibo declination band, which is nearly perpendicular to the CMB dipole, and so the "leverage" on the interesting component is small. We must clearly wait for larger studies; since HI widths at the distances involved can really only be obtained with the Arecibo telescope, use will have to be made of the D_n-σ relation or TF via optical rotation curves—or, ideally, both in the same sets of aggregates just to lay to rest once and for all doubts that they are measuring the same thing.

5. CONFRONTATION WITH THE DENSITY DATA: WHAT DO WE EXPECT FROM OTHER OBSERVATIONS ?

If one believes the gravitational instability picture for the growth of perturbations, the velocity data must be consistent with the acceleration field generated by the density inhomogeneities in the universe, and indeed such consistency is a direct test of the gravitational origin of the velocities. There are, however, several fundmental problems with making the comparison between the density distribution and the velocity field.

First, and perhaps worst, is that we cannot observe the density field directly. We can count galaxies, but if galaxy formation is biassed toward regions of high density, which it must be if Ω is unity, the density of galaxies has higher contrast than the density of matter. In the linear regime, this is accounted for by a linear fudge factor, the biassing parameter b, such that $\delta N/N = b\delta\rho/\rho$ on some sufficiently large scale. If this were accurate, one would have a strength parameter relating the potential derived from the galaxy density contrast to the peculiar velocities of $\Omega^{0.6}/b$ (*c.f.* Eqn. 4), but hoping that this remains accurate up to density contrasts of a few, which is the relevant range, seems a bit optimistic. If we swallow this pill, however, the biassing (*IF it has the above simple form*) only introduces an overall scale factor which we have to determine in any case.

Second, even the distribution of galaxies is very problematical. The direction of the bulk flow and so the direction toward the Great Attractor is at very low galactic latitude, and though we can probably correct observations of individual galaxies for reddening to satisfactory accuracy, it is highly dubious whether we can correct catalogs like the UGC and the ESO for incompleteness caused by absorption. This problem has been largely overcome, at the expense of a few new ones, by a redshift survey of IRAS galaxies which we will discuss below.

Given the properly skeptical attitude which these remarks might usefully induce, we can ask two pertinent questions: First, is there any evidence that the putative Great Attractor in fact exists, and second, is the observed flow and structure therein in general consistent with the density field?

The answer to the first is clearly a qualified "yes", in that there is without doubt a large excess of galaxies in about the right redshift range in the right direction. The existence of an enhancement in the counts of galaxies in the direction of the GA has been claimed by Lynden-Bell and Lahav (1988), but the counts are subject to errors which this author feels are little understood. Much better evidence emerges from the spectroscopic studies of the IRAS sample alluded to above (Strauss and Davis 1988) and a survey directed to the purpose of studying the existence of the GA by Dressler (1988). The redshift distribution over about a steradian in the general direction of the GA shows two strong peaks, one at about 3000 $km\ s^{-1}$associated with the Centaurus complex and the other at about 4500, tentatively associated with the GA. The relative strength of the peaks is a matter of some dispute, and is not easily settled, since they overlap and are of poorly defined width. Dressler argues that the distant one contains about 4 times as many galaxies (and thus presumably mass) as the near one. He then uses arguments exactly analogous to the Virgocentric flow ones (and in fact simply scales to the Virgocentric flow, which would have seemed safe except for the unfortunate fact that the very existence of the GA might well completely destroy our simple understanding of the Virgocentric flow—see the remarks at the end of the last section) to show that such a concentration in fact yields flow velocities of the order of 500 $km\ s^{-1}$at the local group, *if* the Virgocentric flow is of the order of 250. The expected velocity scales with the Virgo infall, and if that velocity is as small as the 100 or so that the FB fits would lead us to, the concentration is totally inadequate to explain the observed flow— on the other hand, if the Virgo flow is as small as 100 $km\ s^{-1}$, either Ω is very small, smaller than can be justified by even the most straigtforward mass-to-light ratio arguments, or the biasing is huge, or the velocities have nothing to do with gravity. It would seem profitable to try to understand this issue.

It is worthwhile at this point to point out a difficulty with any endeavor which attempts to reconstruct the density distribution from spectroscopic surveys, which would seem to be the only method available. First, of course, one must take into account the very peculiar velocities one is trying to explain to assign distances to objects in such a survey in a self-consistent way. Such schemes are possible and have been implemented (see, *e. g.* , Yahil 1988, hereafter Y88). As long as there are no significant highly nonlinear structures in the sample, this would appear to be satisfactory. Individual dense clusters can be identified from their characteristic "finger of God" signatures in redshift space or simply from their enhancements on the sky, and assigned average positions and velocities, as has been done by 7S. What cannot be dealt with are slightly lesser nonlinearities which produce multiple-valued distance-redshift relations. This is familiar from the Virgocentric flow models (Aaronson et al 1982) but is much more difficult to deal with in general density distributions. It is pertinent here because the Centaurus region, upon which so much weight is placed in the GA models, shows this structure, with a foreground group (Cen45) at about 4500 $km\ s^{-1}$, the same redshift as attributed to the GA in the background, falling into the main Centaurus cluster (Cen30) at 3000 $km\ s^{-1}$(Lucey and Carter, 1988). Though this particular structure probably is too restricted in solid angle to perturb seriously the picture presented by 7S, it should perhaps be a cause of some concern. Very recent work by Mould *et. al.* (1988) on clusters in this region using 21-cm data from Parkes fails to reveal the very large velocities seen by 7S, though the data does show the approximately 500 $km\ s^{-1}$bulk flow. The velocity complexity of this region is also likely to confuse attempts to determine the density field; we return to this question below.

So what does the *general* density field look like over the volume sampled by the 7S velocity field? The best picture we have of it is probably that provided by the spectroscopic survey of IRAS galaxies (Yahil, 1988, Strauss and Davis, 1988a,b). This sample is defined by the set of IRAS point sources with $S_{60} < 2$ Jy and $S_{12} < S_{60}/3$ which can be identified with galaxies on the POSS, so that redshifts can be obtained. The sample discussed in the papers cited above are for the region $|b| > 10°$, but the extension to $|b| > 5°$ is almost complete. The density field is constructed from the counts by the assumptions of simple biasing as described above and the assignment to each galaxy counted a weight which is inversely proportional to the fraction of the cumulative luminosity function observable at the distance of that galaxy. This technique is optimal in the case that the luminosity function is constant in form everywhere, but is critically dependent on that assumption. Note that this technique is quite different from the naive one of assuming constant mass to light and simply counting light, which is what one does in the construction of flux dipoles, as first suggested some years ago by Gott and used for the IRAS sample and for optical samples by Lahav (see Kaiser and Lahav 1988, Harmon, Lahav and Meurs 1987). The latter technique would arguably be better for galaxies selected optically, but the enormous spread of IR luminosity and the uncertainties introduced by the flux limit make the counting technique rather more appealing for the IRAS sample.

The IRAS galaxies are associated with late-type systems and are almost certainly underrepresented in high-density regions. This is recognized (Strauss and Davis 1988a,b) but it is not clear what, if anything, should be done about it; it is quite possible that such systems trace the mass much better than galaxies in general and most biassing schemes would support that view. In any case the conclusions about biassing derived from the IRAS sample are not likely to apply to galaxies in general. This conclusion is supported by studies of the correlation function of the IRAS sample by Davis et al (1988), who find that the IRAS sample has weaker clustering than optically selected samples (but displays no luminosity effects within itself, which again argues that IRAS flux has little to do with mass, since the same study clearly shows the effect already recognized from many sources before that optically brighter and hence supposedly more massive galaxies are more strongly clustered than fainter ones.)

The distances to the galaxies in the sample are derived in a self- consistent application of linear perturbation theory, in which the amplitude of the velocity perturbations is set by the requirement that the *magnitude* of the flow at the local group is the observed motion with respect to the CMB; it is reassuring (or perhaps not, depending on your point of view) that the direction, over which one has no control, comes out *very* roughly right. In the density distribution derived in Y88 the counts in the unobserved region $|b| < 10°$ were derived by interpolating across the plane, and with that technique the misalignment of the predicted local group velocity with the observed CMB velocity is 35 degrees. While the *a priori* probability of getting the direction right to within 35 degrees purely by chance is small, about .09, it leaves an unexplained component of the motion more than half as large as the explained one. It is the latter view which I think is the relevant one. In the light of the discussion of the applicability of linear theory in section 2, such a discrepancy for the local group motion seems much too large, though it does not, apparently, unduly upset its author, who interprets the results of Villumsen and Davis a bit more liberally than I. We should remember that if the galaxy distribution is biassed the counts are probably related to the density in a complex nonlinear way and it is doubtful whether multivalued distance-redshift regions would have even appeared, much less been accurately represented, in the construction of the distances, so it is perhaps very gratifying that the results are as good as they are. In any case we are unlikely to do better by any technique now on the horizon and should be properly grateful for the enormous amount of work which went into constructing this sample instead of carping about its inadequacy. We must, however, remember that the directional disagreement probably means that there are vector errors of the order of 300 $km\ s^{-1}$or perhaps a bit larger which are to be expected when comparing the predicted IRAS velocity field with any peculiar velocity data.

The density distribution, smoothed in a manner variable with density in such a fashion to keep a roughly fixed number of observed galaxies per smoothing volume, is shown in Figure 2, reproduced from Y88, in a cut through the supergalactic plane; the primary mass concentrations are identified. The acceleration field derived from this (full 3-D) mass distribution is shown in Figure 3 from the same source, and the peculiar radial velocities inferred for the IRAS sample galaxies within 22° of the supergalactic plane in Figure 4. This is to be compared directly with Figure 1, the 7S peculiar velocities in the same plane.

Several things are to be noted from the map of the accelerations, Figure 3. Though there is a general convergence of the flow in the Hydra-Centaurus region (but at a distance considerably nearer than the putative Great Attractor) the influence of this mass concentration is very much less pervasive than in the GA model; indeed the flow

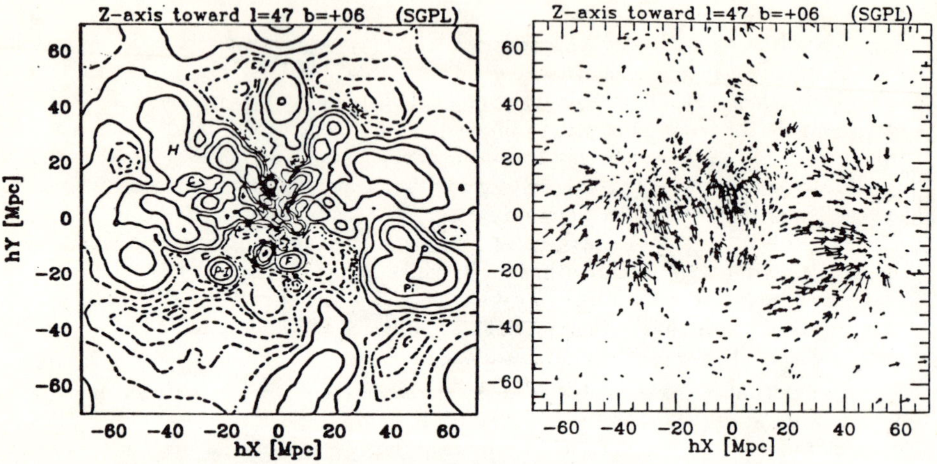

Figure 2(left), Figure 3(right) The smoothed densities and accelerations derived from the IRAS sample in the SG plane. The distances are for $H_0 = 100$, so one should multiply by 100 to get the velocity units of Figure 1. Remember that Figure 1 is reflected right-to-left. From Y88.

into the Perseus- Pisces complex in the opposite direction is predicted to be nearly as strong, with the bifurcation point between these "zones of influence" very near the local group. The strong (upward in the figure near the center) predicted flow locally into Virgo is seen, but the tide from the Hydra-Centaurus complex and the Perseus-Pisces complex, which should look at least qualitatively like the GA tide, is conspicuous by its absence; it is, indeed, cancelled by the rather weaker concentrations along the positive and negative Y axes. One (at least I) am impressed and dismayed by the general complexity of the picture, and to the extent that it represents even qualitatively the effects of the real density distribution it is amazing that as simple a model as that put forward by the 7S can cope with the velocity field, and that a simple model can adequately deal with such short-range features as tidal effects seems ludicrous. In view of these remarks it is not suprising that the observations and the predictions when viewed in the same kind of diagram, Figs 1, 4, do not show very satisfactory agreement. The vertical component of the tide in the predictions is missing, as is, indeed, the general bulk-flow character from right to left which is so striking in the 7S data. The situation is confused as it must be by the noise in the data, but it must be remembered that the GA model fits the data very well, *and that model does not really even qualitatively resemble the predictions.* The situation is well illustrated by the convergence of the IRAS predicted acceleration with the distance within which it is calculated; both Yahil(1988) and Strauss and Davis (1988b) discuss this point, and though they have treated the data somewhat differently the results are essentially identical: Most of the predicted acceleration of the local group, both component by component and the scalar amplitude, comes from mass within 3000 $km\ s^{-1}$. In the GA model very little of the acceleration comes from mass this nearby.

The source of the discrepancy is not clear. It *is* clear that much of the trouble is due to the fact that the IRAS data sees much less of the putative GA than the velocity data suggest it should. There is no question of the existence of the enhancement in the IRAS data, the spectroscopic sample agreeing beautifully, as mentioned above, with that from the optically selected survey of Dressler. Recall that the amplitude of the IRAS accelerations is set by making the magnitude of the local group motion agree with that of the CMB dipole; if that amplitude were cut by about a factor of two and the motion made up by adding more mass to the GA region, the agreement would be substantially improved. That statement must be taken with a grain of salt, however, since if the IRAS map contributed *nothing* and there were *only* a GA, the fit would be not bad, either. The behavior of the acceleration field does suggest that the normalization is too high, however, and that there is a component which does not appear in the IRAS maps. It seems just possible that the velocity field in the Hydra-Centaurus region is sufficiently complex that the distance-reconstruction algorithm used in the IRAS maps might fail. It is clear that the method does not "interpret" the velocity-count data in that direction as Dressler, using only the qualitative appearance of the redshift histogram, does, but it is not obvious whose interpretation is correct. To the extent that one or the other of the peaks is broadened by incipient virialization, I think it is impossible to disentangle them without independent distance data. It does seem clear that understanding this region is likely to be the key to undestanding the whole problem.

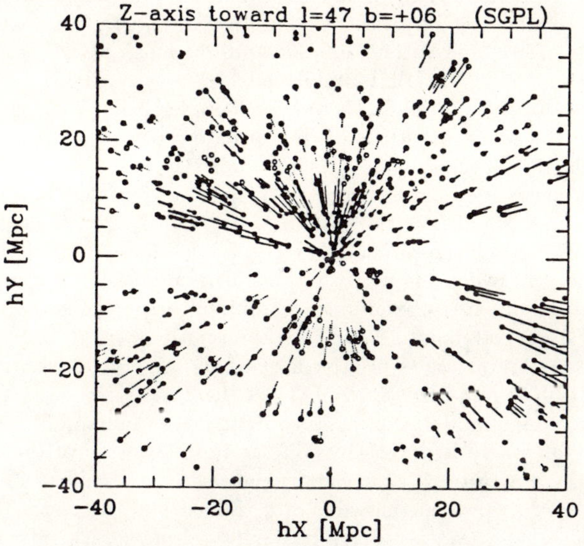

Figure 4. The peculiar velocity field in the SG plane predicted for the IRAS galaxies; the conventions are the same as for Figure 1 but the coordinate system is that of Figures 2, 3. Again recall the right-left reflection when comparing with Figure 1. From Y88.

If it turns out that the IRAS densities are roughly correct, then we must choose among a set of relatively unpalatable explanations for the phenomenon:

a). The 7S are not measuring peculiar velocities, but instead are measuring zero-point gradients in the D_n-σ relation. That such gradients might exist is perhaps not surprising, but I would find it so given the close connection of the method with the virial theorem (see the discussion in Section 3.) One must also consider the excellent agreement of distance ratios derived by this technique with those derived by IR T-F, which I have argued measures quite a different thing.

b). The density of galaxies is only very vaguely related to the density of matter; this is probably possible on the scales involved, and would essentially make it impossible to do dynamical measurements in cosmology.

c). Something is wrong in Centaurus with the 7S distances. This would remove the major signature of the GA model, and one would return to basically a bulk flow, for which one could invoke either a (much larger) mass concentration much more distant yet, or, heresy of heresies,

d). The MWB dipole is *not* a measure of our own peculiar velocity. We will return to this possibility below.

6. WHAT DO WE EXPECT FROM THEORY?

Rather than open and close this section with the glib response "Anything at all", let us focus on a question which has occupied a great deal of time and energy and is of very great interest to at least some of us: Are the 7S velocity observations consistent with the Cold Dark Matter (CDM) model for the formation of structure in the universe? There certainly are models for which it is not an embarassment, the isocurvature baryon models of Peebles (1987) and various *ad hoc* modifications of CDM (Bond, 1988) being the current favorites to display (Bertschinger and Juskiewicz, 1988). The only model we know which is definite enough to be able to make detailed calculations and which has met with conspicuous success in questions of the formation of galaxies and their small-scale distribution is CDM. Especially since at first glance it appears difficult to reconcile it with the velocity observations, it has been the focus of a great deal of attention. Close on the heels of the announcement of the first 7S results came a prescient paper by Vittorio et al (1986), pointing out the difficulty of obtaining coherent velocities on very large scales in *any* theory in which the perturbations arise in a scale-invariant, random-phase way during inflation, of which the premier candidate these days is CDM. Since then the arguments have degenerated into squabbles about just how big the velocities are and just how large the relevant scales are, both of which are crucial issues. The size of the MWB dipole is itself some worry. Since if $\Omega = 1$ there must be biassed galaxy formation in CDM in order to explain cluster mass-to-light ratios and small Ω values from such tests as Virgo Infall, there is uncertainty in the normalization of the amplitude of the perturbation spectrum through the biassing parameter b discussed in section 5 above. There seem to be two views on the value of b. Kaiser and associates (see, for example, Kaiser and Lahav, 1988) favor a value of b near 1.5, citing agreement with the frequency of rich clusters and a better fit to the velocities, the phenomenon under discussion. Davis, Efstathiou, Frenk, and White, (see, *e. g.* , White et al 1987) in a series of collaborations comparing detailed n-body models with observations, favor a larger value near 2.5, and thus a very much smaller amplitude of density fluctuations in the linear regime. The predicted velocities scale like $1/b$. Bertschinger and Juszkiewicz (1988) have recently attempted to assess the likelihood of a flow like that observed by 7S, using both the spherical GA model and a more restricted set of conditions (the "cylindrical flow model", FB) that attempt to fit only the region actually observed (there are no observations on the other side of the putative GA yet). There are uncertain phase space factors and probably a too-stringent imposition of conditions, but the b=2.5, $\Omega = 1$ CDM model appears to be ruled out. The existence of a mass concentration like the GA is about a 7.3 sigma event; there should not be even one such within the horizon. Even the more relaxed cylindrical flow model has a probability of less than one in 10^6. If $b = 1.5$ *and* the effective depth of the 7S data is as shallow as 1500 $km\ s^{-1}$(Kaiser 1988) the average flow over that region is a 2.4 sigma effect, and in fact only one Monte Carlo trial in 10^6

satisfied their constraint to "be like" the cylindrical flow model for $b = 1.5$. It is clear (and is recognized by the authors) that their techniques underestimate the probabilities; they argue that the factors involved are a few, and would have to be serious indeed to rescue $b = 2.5$. I think the situation for smaller b is much less clear, and Kaiser (1988) and Kaiser and Lahav (1988) in fact argue that $b = 1.5$ CDM is a quite acceptable fit to the "observations" of the IRAS velocity field predicted from the IRAS densities—but admit that the large coherence length arrived at from taking the 7S data at face value is a serious problem. It has seemed for some time that the velocities and associated very large-scale structure, perhaps exemplified by the cluster-cluster correlations (Bahcall 1988) pose the most serious problems for CDM, and that still seems to be the case. Its adherents (myself included, lest the reader think I am entirely unbiassed on the point) wait uncomfortably for some resolution of the observational problems.

Consider for a moment an extreme resolution. The striking feature of the 7S data is the bulk flow. The bulk flow is essentially absent in the IRAS accelerations. The level of coherence in the IRAS predictions are not embarrasing theoretically, but the large-scale flows of the 7S data are embarrasing to CDM and probably to any gaussian-inflation theory. Most of the problem, it seems to me, would disappear if the MWB did not, in fact, provide a rest frame. For it not to do so is, of course, not trivially easy to arrange. Arranging gradients across the horizon is easy; the first deviations from inflationary flatness would naturally be expected to be gradients, with curvature terms down in importance by of order the square of the gradient terms. A gradient in the metric perturbation, essentially a gradient in the curvature, would not do the trick, because matter geodesics and radiation "fall" equally in the presence of such a perturbation, and a piece of dipole only of the order of the quadrupole survives. But any gradient which manifests itself as a gradient in the equation of state, such as any gradient which modulates the entropy, or the baryogenesis, or the early annihilation of the dark matter, would produce a dipole. The probability is only one in ten that it would lie so close to the IRAS dipole as it does, but that does not seem so forbidding. As for the probability that we would now be just emerging from inflationary flatness at the 10^{-3} level, I think it is impossible to say, and depends upon the *a priori* distribution of the length of inflation. The idea can be laid to rest if it turns out that a shell at a few thousand $km\ s^{-1}$distance is in fact at rest with respect to the CMB, but it will become very much more attractive if the bulk flow persists to ever larger distances, and becomes compelling if those distances grow to the order of 10000 $km\ s^{-1}$, since then the potential fluctuations causing the flows could be ruled out by existing observations of the smoothness of the background.

7. SUMMARY

We have described how bulk peculiar motions might arise in the universe in a standard gravitational clustering picture and how one might measure them by measuring distances and redshifts to individual galaxies and clusters. The data of the "Seven Samurai" and the composite data set presented by a subset of them (Faber and Burstein 1988) represent the most homogeneous and complete picture to date of the observed motions. The agreement of those observations with the acceleration field produced by probably the best available density field, that produced from the spectroscopic IRAS survey, is less than satisfactory, and the possible reasons range from various technical difficulties with one or the other data sets, systematic variations of the zero-points of the distance relations used, to extreme possibilities like exceedingly poor correlations of galaxy density with mass density or the microwave background not being a satisfactory standard of rest. Finally, the possibility that the motions are not gravitationally induced must be considered, though simple explosive models cannot be made to work easily, and there are certainly some gravitational signatures such as local infall into galaxy concentrations which are clearly seen.

The subject is clearly very much in a state of flux at the moment, and catching it to write this review was not easy; the reader should be warned that the picture presented here is not necessarily a "synchronous" one, in that it probably does not represent the state of the subject (if indeed it does at all) at one particular time. I would like to thank several people for help: Amos Yahil and Sandra Faber for permission to reproduce diagrams from papers, some not yet published; Ed Bertschinger, Dave Burstein, George Djorgovski, Sandra Faber, John Huchra, John Lucey, Roman Juszkiewicz, Jeremy Mould, Jerry Ostriker, and David Weinberg for very illuminating discussions, and finally Sidney van den Bergh for inviting me to give the review.

REFERENCES

Aaronson, M., *et. al.* , 1982, *Ap. J. Suppl* **50**, 241.

Aaronson, M., Bothun, G. D., Mould, J. R., Huchra, J., Schommer, R., and Cornell, M., 1986, *Ap. J.* **302**, 536.

Bahcall, N. A., 1988, *Ann. Rev. Ast. and Ap.* **26**.

Bertschinger, E., and Juszkiewicz, R., 1988, preprint.

Blumenthal, G. R., Faber, S. M., Primack, J. R., and Rees, M. J., *Nature* **311**, 517.

Bond, J. R., in *Large Scale Motions in the Universe*, proceedings of the Vatican Study Week 27, Pontifical Academy of Sciences.

Davis, M., Meiksin, A., Strauss, M. A., and da Costa, L. N., 1988, *Ap. J. (Letters)*, to appear.

Dressler, A., 1988, *Ap. J.*, to appear.

Dressler, A., Lynden-Bell, D., Burstein, D., Davies, R. L., Faber, S. M., Terlevich, R., and Wegner, G., 1987a (7S-I), *Ap. J.* **313**, 42.

Dressler, A., Faber, S. M., Burstein, D., Davies, R. L., Lynden-Bell, D., Terlevich, R. J., and Wegner, G., 1987b (7S-IV), *Ap. J. (Letters)* **313**, 59.

Faber, S. M., 1982, in *Astrophysical Cosmology*, eds. H. A. Brück, G. V. Coyne, and M. S. Longair, proceedings of the Vatican Study Week 18, Pontifical Academy of Sciences, 191.

Faber, S. M., and Burstein, B., 1988, in *Large Scale Motions in the Universe*, proceedings of the Vatican Study Week 27, Pontifical Academy of Sciences.

Faber, S. M., and Jackson, R. E., 1976, *Ap. J.* **204**, 668.

Fixen, D. J., Cheng, E. S., and Wilkinson, D. T., 1983, *Phys Rev Letters* **50**, 620.

Gunn, J. E., 1982, in *Astrophysical Cosmology*, eds. H. A. Brück, G. V. Coyne, and M. S. Longair, proceedings of the Vatican Study Week 18, Pontifical Academy of Sciences, 233.

Gunn, J. E., 1986, in *The Galaxy*, eds. G. Gilmore and R. Carswell, Dordrecht:Reidel, 413.

Harmon, R. T., Lahav, O., and Meurs, E. J. A., 1987, *M. N. R. A. S.* **228**, 5p.

Hayakawa, S., Matsumoto, T., Matsuo, H., Murakami, H, and Sato, S., 1987, *Publ. Ast. Soc. Japan* **39**, 941.

Kaiser, N., 1988, *M. N. R. A. S.*, to appear.

Kaiser, N., and Lahav, O., 1988, in *Large Scale Motions in the Universe*, proceedings of the Vatican Study Week 27, Pontifical Academy of Sciences.

Lilje, P., Yahil, A., and Jones, B. J. T., 1986, *Ap. J.* **307**, 91.

Lucey, J. R., and Carter, D., 1988, preprint.

Lynden-Bell, D., Faber, S. M., Burstein, D., Davies, R. L., Dressler, A., Terlevich, R., and Wegner, G., 1988 (7S-V), *Ap. J.* **326**, 19.

Lynden-Bell, D., and Lahav, O., 1988, in *Large Scale Motions in the Universe*, proceedings of the Vatican Study Week 27, Pontifical Academy of Sciences.

Peebles, P. J. E., 1980, *The Large Scale Structure of the Universe*, Princeton: Princeton University Press.

Peebles, P. J. E., 1987, *Nature* **327**, 210.

Rubin, V. C., Thonnard, N., Ford, W. K., and Roberts, M. S., *A. J.* **81** 719.

Strauss, M. and Davis, M., 1988a, in *The Large Scale Structure of the Universe*, IAU Symposium 130, ed. J. Audouze.

Strauss, M. and Davis, M., 1988b, in *Large Scale Motions in the Universe*, proceedings of the Vatican Study Week 27, Pontifical Academy of Sciences.

Tully, R. B., and Fisher, J. R., *Astron. Astrophys.* **54**, 661.

Tully, R. B., and Fisher, J. R., 1987, *Nearby Galaxies Atlas*, Cambridge: Cambridge University Press.

Villumsen, J. V., and Davis, M., 1986 *Ap. J.* **308**, 499.

Vittorio, N., Juszkiewicz, R., and Davis, M., 1986, *Nature* **323**, 132.

Weinberg, D. H., Ostriker, J. P., and Dekel, A., 1988, *Ap. J.*, to appear.

White, S. D. M., Frenk, C. S., Davis, M., and Efstathiou, G., 1987, *Ap. J.* **313**, 505.

Yahil, A., 1988 (Y88), in *Large Scale Motions in the Universe*, proceedings of the Vatican Study Week 27, Pontifical Academy of Sciences.

Zel'dovich, Ya. B., and Sunyaev, R. A., 1969, *Astrophys. Space Sci.* **4**, 301.

DISCUSSION

B. LILJE: The low value of Virgocentric infall found by Faber and Burnstein (1988) is due to the large tidal velocity field component in the Local-Group-Virgo direction induced by the Great Attractor. In the work by Lilje, Yahil and Jones (1986) this component was found to be small. Could this discrepancy be due to an artificial axial symmetry in the quadropole of the Great Attractor model?

J. GUNN: I think so, and if the tide is, in fact, generated by the mass distribution, one would expect the result you found, since the distribution is so highly flattened.

S. DJORGOVSKI: I believe that your derivation of the $L \sim V^4$ relation from the Gott-Rees hierarchy (as did Faber in 1982 Vatican Conf.) is incorrect, and for two reasons: it is not $M \sim V^4$, but $L \sim V^{\sim 4}$; and that itself is a projection of a bivariate relation, involving also the surface brightness and (M/L). When all the terms are accounted for, the initial density spectrum index M drops out. These relations do not contain any memory of the initial density perturbation spectrum.

J. GUNN: The fact that the relation is with M and not L is certainly true. The rest I disagree with, but the whole story depends on how spirals and other types cut the hierachy. The dispersions of $F-J_{m\ell}T-F$ are both too small to be easily understood.

HI OBSERVATIONS IN THE HERCULES SUPERCLUSTER

WOLFRAM FREUDLING and MARTHA P. HAYNES
Cornell University/NAIC, Space Sciences Bldg.,
Ithaca, NY 14853

RICCARDO GIOVANELLI
Arecibo Observatory, P.O.B. 995, Arecibo, PR 00613

ABSTRACT HI spectral line observations have been obtained with the Arecibo 305m dish for more than 600 galaxies in the region of the Hercules Supercluster. For about 80 Sc's, we have obtained I band CCD images. The data will be used to search for a possible distance dependence of the Hubble ratio in that region.

INTRODUCTION

It has recently been argued that the observed Hubble ration v/d of galaxies (v=velocity in suitable restframe, d=distance of galaxy) is not a universal constant but rather increases with distance from the local group (Giraud, 1986; Tully, 1988, this volume). A different model of deviations from a smooth, linear Hubble flow is the "Great Attractor" (Lynden-Bell *et al.*, 1988) which predicts a different distance dependence of v/d for different regions in the sky. Theoretical models predict significant deviations from the Hubble flow around 'voids' in the distribution of galaxies (*e.g.* Davis *et al.*, 1982, Hoffman *et al.*, 1983). Observational confirmation of such streaming motions are difficult because of the uncertainty in the distance estimates obtained via the Tully-Fisher or Faber-Jackson relation. Surveys which are deep in redshift suffer from lack of quality magnitudes and heavy Malmquist bias. A plot of redshift versus deviation from a linear Hubble law for samples of galaxies strongly affected by Malmquist bias will show an apparent increase of the Hubble ratio with redshift (Bottinelli *et al.*, 1988). It might therefore disguise a local velocity anomaly which is mistaken for a Malmquist bias or vice versa.

THE HERCULES SUPERCLUSTER

The Hercules Supercluster is a grouping of rich clusters of galaxies at a redshift of $z \approx 0.034$ and extends at least across a region between $14h \leq \alpha \leq 17h$ and $0° \leq \delta \leq +40°$. The region of the Hercules Supercluster contains a large underdense region with a density of observed galaxies less than one half of the density in the surrounding volume. This void is confined by the Hercules Supercluster at a redshift of about 10000 km/s and by a foreground group at a redshift of about 5000 km/s. The Hercules regions is a good place to determine the Hubble constant and search for deviations from the Hubble flow because

- the Supercluster is at a redshift high enough so that any local velocity irregularities including our own velocity are only a negligible fraction of the total redshift. The Hercules Supercluster is at the highest redshift where systematic HI surveys are feasible with the Arecibo radiotelescope.
- there is significant structure on large scale in front of the supercluster so that a relatively good statistic of the galaxy distribution around a void can be obtained.
- the Hercules Supercluster has a large population of late type galaxies which makes it suitable for distance estimates via the Tully-Fisher relation.

STATUS OF OBSERVATIONS

We are currently obtaining HI spectra and I band CCD images of galaxies in the region of the Hercules Supercluster in order to trace the three dimensional structure in that region, investigate galaxy properties as a function of the environment and search for possible deviations from the Hubble flow. At the present time we have collected more than 600 HI spectra. For about 80 of the Sc galaxies we have also obtained I band CCD images, which we are currently reducing. In order to minimize the Malmquist bias in our sample, we select candidates for I band photometry by the estimate of the absolute magnitude based on Zwicky blue magnitudes. In this way, we hope to obtain a homogenous sample for a wide coverage of redshift. The Tully-Fisher relation will be calibrated on a sample of nearby Sc galaxies in and outside of the Hercules region, for which we will obtain I band CCD images and HI spectra as well.

Preliminary results and those of other workers indicate that a scatter in the Tully-Fisher relation of less than 0.3 magnitudes can be achieved. This would translate into a distance uncertainty of less than 15% for a *single* galaxy. A sufficient number of galaxies *e.g.* at the edge of the void will allow a search for smaller deviations from the Hubble flow.

REFERENCES

Bottinelli,L., Gouguenheim,L., Paturel,G., Teerikorpi,P. (1988), *Astrophys. J.*, **328**, 4.

Davis,M., Huchra,J., Latham,D.W., Tonry J. (1982), *Astrophys. J.*, **253**, 423.

Giraud,E. (1986), *Astrophys. J.*, **301**, 7.

Hoffman,G.L., Salpeter,E.E. (1982), *Astrophys. J.*, **263**, 485.

Lynden-Bell,D., Faber,S.M., Burstein,D., Davies,R.L., Dressler,A., Terlevich,R.J., Wegner,G. (1988), *Astrophys. J.*, **326**, 19.

PECULIAR MOTIONS OF THE LARGE SCALE STRUCTURES IN THE SOUTHERN HEMISPHERE

REINALDO R. DE CARVALHO*
California Institute of Technology
Pasadena, California 91125

LUIZ N. DA COSTA
Observatorio Nacional

ABSTRACT We have used independent photometric and spectroscopic data to derive the two-parameter distance indicator relation for elliptical galaxies. The distance indicator is a plane in the R - σ - μ space, similar to the "fundamental plane" solution by Faber, et al. (1987), or Djorgovski and Davis (1987). Our derived distances compare well to those measured by Lynden-Bell, et al. (1988), where our samples overlap. In the present work we have found some marginal evidence for systematic coherent motions in the surveyed volume, after subtracting out the motions induced by the Virgo cluster and the great attractor. This motion seems to be associated to the observed structures in the galaxy distribution, especially when we look at the Telescopium-Pavo-Indus region, in which the majority of our data concentrates. However, more reliable data in this system of galaxies is necessary to confirm this finding.

DISCUSSION

In this paper we have used independent photometric and spectroscopic data to derive the equation of the two-dimensional surface that describes the family of ellipticals. In particular, we have shown that although the coefficients are slightly different from those obtained in earlier works, the elliptical galaxies in our sample describe essentially the same surface. This has been demonstrated from the comparison between our estimated distances and those obtained by Lynden-Bell, et al. (1988). This result also indicates that despite the use of photographic material we can still determine good photometric

*On leave from Observatorio Nacional/CNPq - Brazil.

parameters that can be used to define a reliable distance indicator.

The previous conclusion is important because it justifies the use of the same techniques employed here to study the properties of the SO galaxies in the Southern Sky Redshift Survey sample. This work would complement that of Dressler (1987) for SO galaxies in the core of rich clusters. This extension would also be of interest in mapping of the peculiar velocity field since the SO galaxies delineate the structures more densely than ellipticals. The extension to the southern hemisphere is particularly important because in this region the influence of both Virgo and the great attractor are relatively small, thus allowing a better determination of the fundamental plane of SO galaxies in the general field. The more general question of uniform bulk motions, the existence of the great attractor and the influence of the environment on the fundamental plane should be pursued utilizing a larger number of groups and clusters of galaxies, for which the distance estimates are much more reliable, extending the observations to fainter members.

In the present work we have found some marginal evidence for systematic, coherent motions in the surveyed volume which may be associated to the observed structures in the galaxy distribution. Despite the large uncertainties, the peculiar motions detected exhibit a systematic trend and a certain degree of spatial coherence. Furthermore, the amplitudes of these residual velocities are also significantly larger than the expected velocity dispersion of field galaxies (Lynden-Bell et al. 1988). All of this may indicate that we have detected motions of, or within, the large-scale structures, relative to the microwave background reference frame, but the confirmation of this result will have to await until more data on groups, SO's and late-type galaxies become available. We are presently taking the first steps in those directions.

Finally, we emphasize that the detection of bulk motions of the large-scale structures, if confirmed, would be an important feature of the distribution of matter in the Universe that could help discriminating amongst competing models for formation and evolution of large-scale structures.

REFERENCES

Djorgovski, S., and Davis, M. 1987, Ap. J., 313, 59.

Dressler, A. 1987, Ap. J., 317, 1.

Faber, S. M., Burstein, D., Davis, R. L., Dressler, A., Lynden-Bell, D., Terlevich, R. J., and Wegner, G. 1987, in Proc. Workshop on Nearly Normal Galaxies, from the Planck Time to the Present, ed. S. M. Faber (New York: Springer-Verlag), p. 175.

Lynden-Bell, D. Faber, S. M., Burstein, D., Davis, R. L., Dressler, A., Terlevich, R., and Wegner, G. 1988, Ap. J., 326, 16.

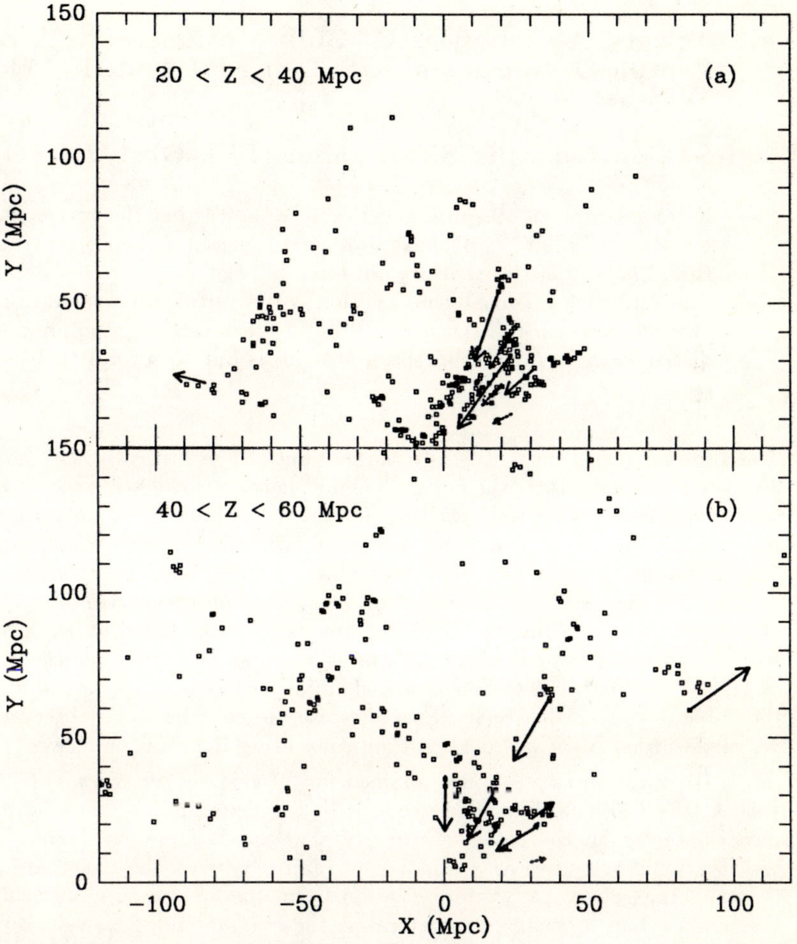

Fig. 1 Projection of the peculiar velocities onto the X-Y plane of the cartesian coordinate system defined with the X and Y axis in the equatorial plane and pointing on the directions $\alpha = 18^h$ and $\alpha = 10^h$, respectively, and Z axis pointing towards the south celestial pole. We plot the peculiar velocities in the MWB rest frame, after subtracting the flow generated by Virgo and great attractor. The residual velocities (solid arrows) are superposed to the distribution of galaxies (open squares) for slices 20 Mpc thick. We also plot in these figures the residual velocities obtained for the ellipticals, in the Lynden-Bell's sample (dotted arrows), overlapping the region covered in this study.

Toward An Analysis Of Bulk Motion From A Large Sample Of Spiral Galaxies Using HI And Hα Velocity Widths

S. Courteau and S. M. Faber (Lick Obs., UCSC)

From a sample of 19 spiral galaxies, it is shown that the correspondence between radio and optical ΔV's is good, differences being generally less than 7% for galaxies that are not directly edge-on ($55° \leq i \leq 80°$). The calibration of a Tully-Fisher relation with optical ΔV's should open up new windows on the sky and will be of great value in mapping out the distribution of galaxies in space around us out to several thousand km sec^{-1}.

The aim of this project is to study the motions of a statistically selected sample of ~ 300 field Sb-Sc galaxies in the northern hemisphere out to 7000 km sec^{-1} to test for large scale deviations from a uniform Hubble flow. We chose spiral galaxies because they populate the volume more densely than early-type galaxies, and because their distances can be measured more accurately. Together with the nearer Aaronson *et al.* spirals within 2500 km sec^{-1} , this catalog will provide a comprehensive map of the motions of northern spirals within the broad volume of space bounded by the Coma, Perseus-Pisces, and Local Superclusters. It will be the first uniform, statistically selected set of Tully-Fisher motions to be assembled. In order to derive distances, we make use of the R-band Tully-Fisher relation; to alleviate the need for radio observations of 21cm velocity widths, ΔV's are measured optically using Hα rotation curves.

Hα rotation curves were obtained for 19 galaxies on two nights in September 1988 at the 120-inch Shane Telescope with the Cassegrain CCD spectrograph. The physical center on the rotation curve of a galaxy is found by forcing symmetry in the rotational velocities as a function of radius between the oncoming and receding halves of the galaxy. A χ^2 minimization technique is used to numerically determine the center. Two procedures to determine the maximum optical velocity (V_{max}) have been adopted: (i) Add the peak velocity of both the oncoming ($|V^-|$) and the receding ($|V^+|$) parts of the galaxy and divide by two; (ii) Reflect the rotation curve through the origin to produce a one-sided unsmoothed curve and measure the peak velocity. An example for both methods is shown on the next page (also shown in the top figure is a comparison with Rubin *et al.*'s data). It is expected that the maximum optical velocity for galaxies with $i \geq 80°$ will be smaller than ΔV_{21} by many tens of km sec^{-1} because of optical depth effects. Therefore, we don't expect these galaxies to show a tight correspondence between 21cm and Hα ΔV's, as confirmed below.

On the next page, the bottom figure shows the correlation of V_{max} with V_{21} measured at 20% of peak intensity. The filled points are the ones used in the fitting procedure; the open points in the bottom quadrants correspond to galaxies with $i \geq 80°$. We find a least-square solution with the smallest scatter for galaxies with inclination $\leq 80°$ and velocities measured using the first method ($|V^+| + |V^-|$). The scatter is ~ 7% at 200 km sec^{-1}. With more data available in the future, a scatter of the order of a few percent should be attainable.

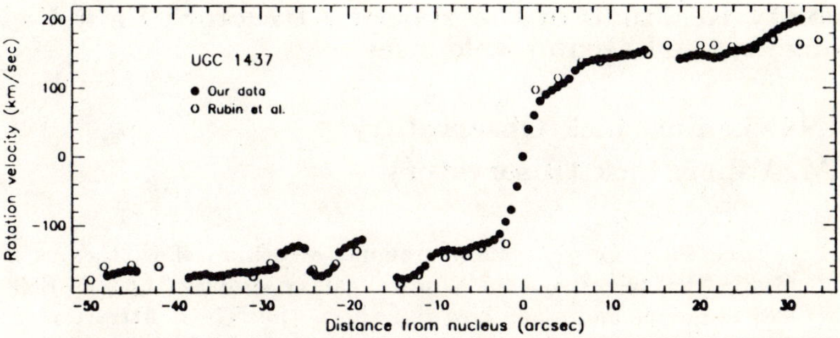
UGC 1437
Our data
Rubin et al.
Rotation velocity (km/sec)
Distance from nucleus (arcsec)

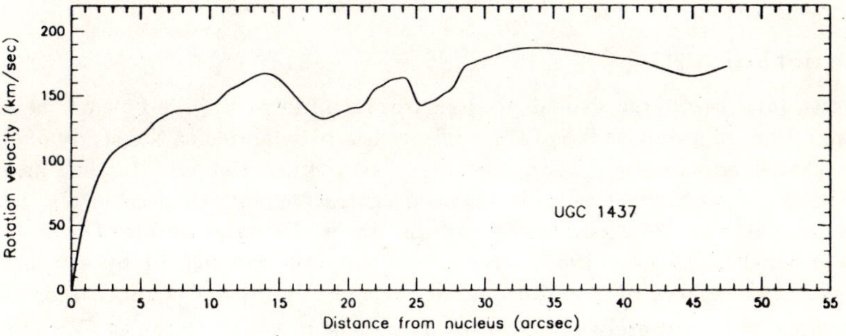
UGC 1437
Rotation velocity (km/sec)
Distance from nucleus (arcsec)

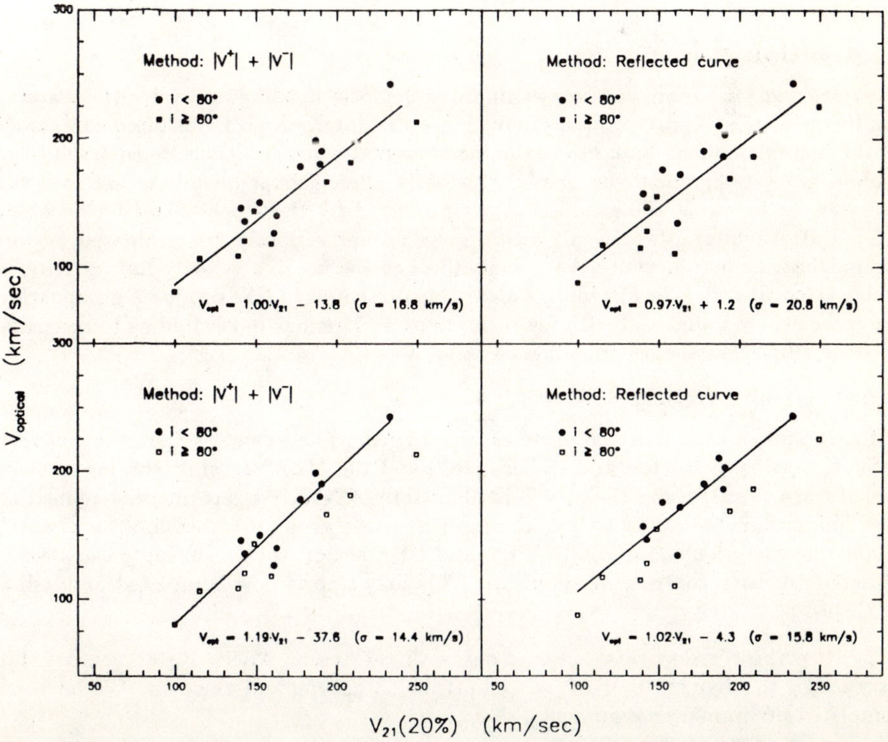
Method: |V⁺| + |V⁻|
i < 80°
i ≥ 80°
V_opt = 1.00·V_21 − 13.8 (σ = 16.8 km/s)
Method: Reflected curve
i < 80°
i ≥ 80°
V_opt = 0.97·V_21 + 1.2 (σ = 20.8 km/s)
Method: |V⁺| + |V⁻|
i < 80°
i ≥ 80°
V_opt = 1.19·V_21 − 37.6 (σ = 14.4 km/s)
Method: Reflected curve
i < 80°
i ≥ 80°
V_opt = 1.02·V_21 − 4.3 (σ = 15.8 km/s)
V_optical (km/sec)
V_21(20%) (km/sec)

Velocity Residuals in the "Great Attractor" Flow Field Using Revised Group Velocities

R. Nolthenius, Lick Observatory

S. M. Faber, Lick Observatory

We have used the improved galaxy grouping algorithm of Nolthenius and White (1987, NW) to reassign the elliptical galaxy sample of Lynden-Bell *et al.* (1988) to groups and study how this affects the "Great Attractor" flow model solution.

Introduction

Investigations of large scale deviations from a uniform Hubble flow are being studied by a number of groups. One of these efforts has been aimed at the study of elliptical galaxies (Faber and Burstein 1988, hereafter FB; Lynden-Bell *et al.* 1988 and earlier papers), using the observed relation between central velocity dispersion and isophotal diameter as a new, more accurate distance indicator. Using a sample of over 400 ellipticals, Lynden-Bell *et al.* (1988) have shown the data are well fit by a velocity flow model with two centers of gravitational attraction; Virgo, and a much more massive concentration approximately 4200 km sec^{-1} distant in the direction of Centaurus (the "Great Attractor").

Motivation

One can get a more accurate estimate of the local flow-induced velocity by averaging the velocities of galaxies in a group. Since this internal motion is often comparable to the highest induced bulk flows, the mass concentration solutions depend somewhat on how group assignments are made. Currently, these groups have been assigned subjectively, or by using the group catalogs produced by the Huchra and Geller (1982, HG). Unfortunately, the HG algorithm gives groups seriously contaminated by foreground/backround objects (NW), mainly through the use of a velocity linkage criterion which is too liberal. The alternative algorithm presented in NW produces groups which are more nearly round and with fewer interlopers. This has motivated us to reconsider how the elliptical galaxies are assigned to groups.

Goals of the Study

• **1.** Compile a new, more accurate catalog of groups and clusters from the "Nearby" redshift catalog of Huchra and Geller (1982) and the "CfA" redshift catalog of Geller and Huchra (1983) using the grouping algorithm of NW. We have further refined the NW link parameter values to give the most accurate group memberships for the well-studied nearby groups, especially Virgo and Ursa Major. The resulting groups have a minimum density contrast of about 20. This first step is now completed and will be published separately.

• **2.** Use these group catalogs together with the recent Abell cluster velocity data to reassign the Lynden-Bell *et al.* ellipticals to groups and clusters. We have just completed preliminary assignments.

• **3.** Use these new group assignments to do a new maximum liklihood analysis to determine the best Virgo + "Great Attractor" flow model parameters and the remaining dispersion σ_f of the ellipticals about this solution. This stage is not completed, but some tentative results are presented below.

Preliminary Results

The plot below shows the residuals from the new group data against those of the FB group data. If the group assignments had been identical between the two datasets, all points would fall on the 45° line. Most of the points with large residuals using the new groups represent galaxies mistakenly left out of distant rich clusters which are poorly sampled in the CfA survey. Any remaining larger scatter in the vertical direction may be due to a combination of using a flow model not yet optimized for the new dataset (we have used the FB model), and perhaps a few unspotted data errors.

However, it may be significant that the highest residuals seen in the old data remain high residuals in the new data. This suggests that the new group assignments are not likely to produce a significant drop in the derived field velocity dispersion, and that the old group assignments are not, on average, seriously in error.

References

Faber, S.M., and Burstein D., in "Large Scale Motions in the Universe", proceedings of the Pontifical Academy of Sciences Study Week no. 27, 1988 (FB)

Geller, M. and Huchra, J. 1983, Ap. J. **298**, 8

Huchra, J. and Geller, M. 1982, Ap.J. **257**, 423 (HG)

Lynden-Bell, D., Faber, S.M., Burstein, D., Davies, R.L., Dressler, A., Terlevich, R.J., and Wegner, G.W., 1988, Ap. J. in press

Nolthenius, R., and White, S.D.M., 1987, M.N.R.A.S. **225**, 505 (NW)

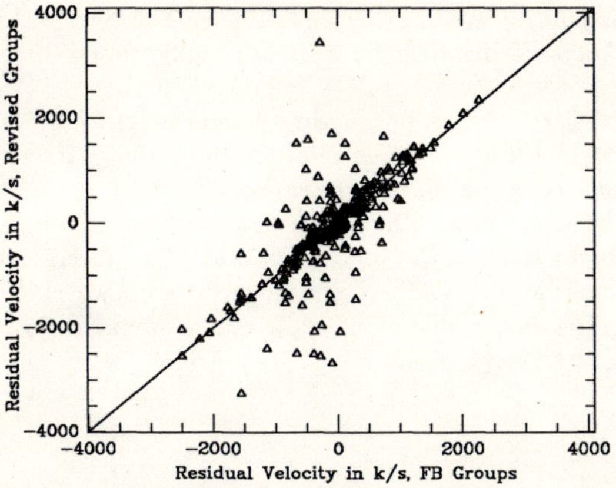

THE TULLY-FISHER RELATION, DARK MATTER AND H_0

Paolo Salucci and Carlos S. Frenk
Physics Department, University of Durham, UK

ABSTRACT. The aim of this paper is to investigate the effect of dark matter on the empirical Tully-Fisher (TF) relation for spiral galaxies and to propose a new relation between the luminosities of spirals and the dynamics of the *luminous* component. The classic TF relation suffers from three problems: *i)* It is affected by unexplained biases over and above the Malmquist bias. *ii)* It deviates strongly from linearity. *iii)* It has different effective slopes in different wavebands. All these three problems can be traced back to contamination by the non-uniform contribution of dark matter to the maximum velocity. We propose a method to eliminate this contamination. This leads to a modified Tully-Fisher relation which is free from these problems and has $0.2 - 0.3$ mag less scatter than the original relation. Applying this and a related relation to a set of 5 local calibrators (on Sandage's, 1988 local scale) and to our Galaxy we find $H_0 = 76 \pm 9$ km s^{-1} Mpc^{-1}

1. THE CLASSIC TULLY-FISHER RELATION

The use of the Tully-Fisher (hereafter TF) relation as a reliable distance indicator is compromised by the following drawbacks:

• *Biases* : For a given sample, the usual Malmquist bias can be estimated from the luminosity function and apparent magnitude distribution of the sample and from the scatter in the TF relation. However, a substantial bias remains even after the effect of Malmquist bias is subtracted (e.g. Giraud, 1986). The *observed* increase in H_0 as function of the kinematical distance $D_v = cz$ is usually about three times the *expected* Malmquist bias.

• *Non linearity*: The TF relation is not linear. Aaronson *et al* (1986) added a quadratic term $\propto (log V)^2$ which linearises the relation and reduces its scatter. This correction is empirical and has no physical justification.

• *Different slopes*: The slope of the TF relation is steeper, by almost a factor two, in the infrared waveband than in the blue waveband. This discrepancy reflects the different contributions of varying stellar populations to each waveband. To take this effect ito account requires knowledge of the disk mass-to-light ratio for galaxies of different types.

2. DARK MATTER AND THE TULLY-FISHER RELATION.

Only a fraction, V_*, of the total circular velocity, V_{tot}, is due to the luminous component. If V_*/V_{tot} varies from galaxy to galaxy, we expect the luminosity to show a significantly better correlation with V_* than with V_{tot}. If M_{tot} denotes the total mass inside the isophotal radius, R_{25}, we have, $M_{tot} \cong G^{-1}V_{25}^2 R_{25}$. Let us assume that the luminous matter is distributed in a exponential thin disk with a central surface density which is constant from galaxy to galaxy. Then, the disk luminosity, $L_B \propto R_{25}^2$. From the definition of V_* we have, $M_{disk} = \phi V_*^2 R_{25}$ where ϕ is a constant. We therefore find,

$$L_B \propto V_{25}^4 (M_{disk}/M_{tot})^2 \tag{1}$$

Scatter and systematic variations in M_{disk}/M_{tot} will be reflected as increased scatter and biases in the L_B - log V_{25} relation. Persic and Salucci (1988) studied a sample including all non-local galaxies with both good photometry and rotation curves and found that M_{disk}/M_{tot} does in fact vary from galaxy to galaxy; it is ~ 4 for bright galaxies ($M_B = -23$, $H_0 = 50$ km s^{-1} Mpc^{-1}) and decreases to $\sim 1/4$ for faint galaxies ($M_B = -19$). For V_* they obtained, $\log V_* = \log V_{25} - 0.04 - 0.58 <$ $d\log V_{tot}/d\log R >$, with the average calculated between $0.2R_{25}$ and R_{25}.

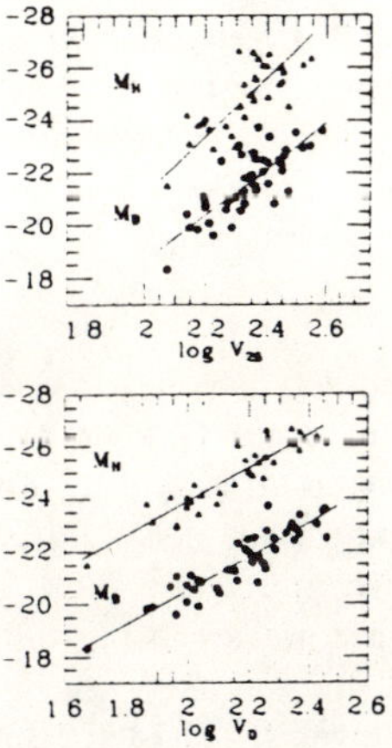

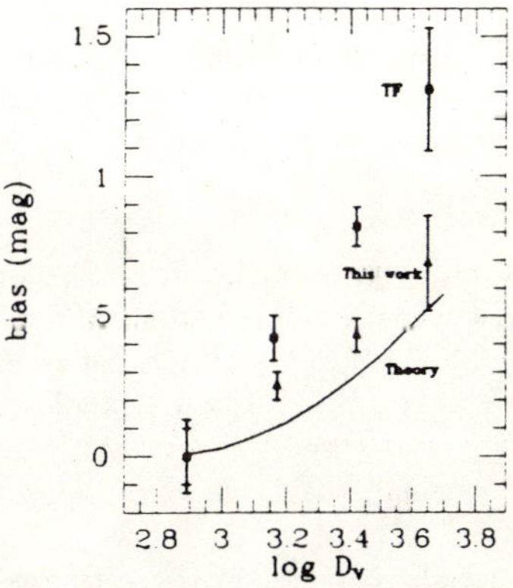

Figure 1 (*left*). The luminosity-kinematics relation for our sample, before (*top*) and after (*bottom*) removing the dark-matter contribution from the maximum velocity. Figure 2 (*right*). Bias as function of the kinematical distance D_v for our sample. The solid line is the theoretical expectation, squares and triangles refer to the bias observed by applying the TF and the corrected TF relation respectively.

If instead of V_{25} we plot V_* against luminosity (Figure 1) we find that: *a)* The scatter is reduced by 0.27 mag. *b)* The relation becomes linear. *c)* Most of the observed bias in the TF disappears and what remains agrees well with the theoretical expectation for the the Malmquist bias (Figure 2). *d)* Correcting for the different stellar populations using the derived values of M_{disk} and the Larson-Tinsley stellar evolution models (Tinsley, 1980), the slopes in the different wavebands are the same.

3. DETERMINATION OF H_0

In order to determine the value for the Hubble constant we use two methods.

1) Our corrected TF relation applied to the five (four) local calibrators in the B (IR) wavebands used by Sandage (1988) (Details will appear in a forthcoming paper, Salucci and Frenk, 1988). The relations we use are (see Figure 1):

$$M_H = -9.7 - 7\log V_* \tag{2}$$

$$M_B = 2.5\log(2/F(B-V)) - 6.3 - 7\log V_* \tag{3}$$

where we have assumed a reference value of 50 km s^{-1} Mpc^{-1} for H_0 and the function F is defined, according to Tinsley (1980), as $F(B-V) = 10^{-0.8+1.85(B-V)}$. The scatter is 0.51 mag in (2) and 0.54 mag in (3). We then measure H_0 from the residuals ΔM^i of the relations (2) and (3) when applied to our set of local calibrators with distances, extinctions and reddenings taken from Sandage (1988). With this procedure, $H_0 = 50 \times \sum_{i=1}^{N} 10^{0.2\Delta M^i}$kms^{-1}Mpc^{-1} with an rms uncertainty $\Delta H_0/H_0 = (10^{\sigma/N} - 1)/\sqrt{N}$, where σ is the intrinsic scatter in the relations and N is the number of local calibrators used. The residuals for each individual calibrator agree well with one another both in the IR-band and in the B-band: $< \Delta M_H^i >=$ 1.2±0.2, $< \Delta M_B^i >=$ 1.0±0.2 . The results, for the IR and B wavebands respectively are: $H_0 = 88 \pm 11$ km s^{-1} Mpc^{-1} and $H_0 = 79 \pm 10$ km s^{-1} Mpc^{-1}. The error ranges reflect the intrinsic scatter in the relation but neglect errors in the calibration. Averaging these two estimates and correcting to first order for Malmquist bias by applying a bulk correction $H_0^c = H_0 \times 10^{-0.28\sigma^2}$, we finally obtain, $H_0^c = 75 \pm 10$ km s^{-1} Mpc^{-1}

2) We obtain a relation between R_{25} and V_* (which, since R_{25} correlates very well with M_B, is closely related to the TF relation) and apply it to our Galaxy. We find, $\log R_{25} = 0.5\log(2/F(B-V)) + 1.6\log V_* - 2.1$ with a scatter of 0.12 dex. For our Galaxy we take $(B-V) = 0.8$, a disk scale lenght of 5 kpc (van der Kruit, 1986) so that $R_{25} = 15$ kpc, $V_{25} = 220, < dV/dr >= 0$ and we find $H_0 = 80 \pm 25$ km s^{-1} Mpc^{-1} where the large error range reflects the fact that we are using only one calibrator.

The weighted mean of the estimates from 1) and 2) gives $H_0 = 76 \pm 9$ km s^{-1} Mpc^{-1}

REFERENCES

Aaronson, M., Bothum, G., Mould, J., Huchra, J., Schommer, R.A. and Cornell, M.E. 1986, *Astrophys. J.*, **302**, 536.

Giraud, E. 1985 *Astr. and Astrophys.*, **153**, 125.

Persic, M. and Salucci, P. 1988. *Mon. Not. R. astr. Soc.*, in press.

Sandage, A. 1988, Space Science Telescope Institute preprint No. 208

Salucci, P. and Frenk, C.S. 1988, in preparation

Tinsley, B.M. 1981 *Mon. Not. R. astr. Soc.*, **194**, 63.

van der Kruit, K.C. 1987. *Astr. Astrophys.* , **157**, 230.

VII

SUMMARY

SUMMARY OF DISTANCE DETERMINATIONS

SIDNEY VAN DEN BERGH
Dominion Astrophysical Observatory, 5071 West Saanich Road, Victoria, B.C. Canada V8X 4M6

New results on the distances to Local Group galaxies, other relatively nearby galaxies, and the Virgo cluster are widely scattered throughout the proceedings of this symposium. For the convenience of the reader new determinations of distances to Local Group galaxies are summarized in Table I. Furthermore Table II lists new distances derived for individual galaxies situated outside the Local Group. A summary of distance determinations for the Virgo cluster is given in Table III. Finally Table IV lists new determinations of the Hubble parameter presented during this symposium. These values are seen to range from 51.8 km s^{-1} Mpc^{-1} (Tammann) to 91 km s^{-1} Mpc^{-1} (Willick). Determinations based on the Tully-Fisher technique are generally seen to fall in the upper end of this range.

TABLE I. Distances to Local Group members

Object	Author(s)	Method	True distance modulus
LMC	Feast	Cepheids, RR Lyrae, etc.	18.45±0.15
SMC	Feast	Cepheids, RR Lyrae, etc.	18.76±0.15
M33	Della Valle	Novae	24.5*±0.3
	Freedman	Cepheids	24.5 (preliminary)
	Pritchet	RR Lyrae	24.45:±0.2
NGC185	Saha	RR Lyrae	23.77
IC1613	Cook and Aaronson	Carbon stars	24.2
	Freedman	Cepheids	24.27
Phoenix	Caldwell and Schommer	B, V photometry	24.7

* A_B = 0.3 mag assumed

TABLE II. Distances of galaxies beyond the Local Group

Object	Author(s)	Method	True Distance Modulus
M81	Freedman and Madore	Cepheids	27.59±0.31
NGC1399	Geisler and Forte	Globulars	30.4±0.3
NGC2403	Freedman and Madore	Cepheids	27.51±0.24
NGC3109	Sandage and Carlson	Cepheids, bright stars	26.4*
NGC3377	Jacoby *et al.*	Planetaries	29.93
NGC3379	Jacoby *et al.*	Planetaries	29.80
NGC3384	Jacoby *et al.*	Planetaries	29.87
NGC5128	Jacoby *et al.*	Planetaries	27.90±0.09
IC5152	Caldwell and Schommer	B, V photometry	26.0
GR8	Ruotsalainen	C-M diagram	25.9±0.5
DDO53	Ruotsalainen	C-M diagram	27.2±0.4

* Apparent blue distance modulus

TABLE III. Distance to the Virgo cluster

Author(s)	Method	True Distance Modulus	
Harris	Globular clusters	31.70±0.22	21.9±2.2
Pierce	Tully-Fisher	30.96±0.20	15.6±1.5
Pritchet and van den Bergh	Novae	31.45±0.43	19.5±3.9
Tammann	Various	31.73±0.09	22.2±0.9

Inspection of Table III, and of Table I of Huchra's paper shows that all modern determinations of the distance to the Virgo cluster fall in the range $14 \lesssim D \lesssim 22$ Mpc. Conservatively adopting $D = 18 \pm 4$ Mpc and a cosmological Virgo redshift of 1250 km s^{-1}, with a generous uncertainty of ±150 km s^{-1}, yields $H_o = 69 \pm 18$ km s^{-1} Mpc^{-1} as the current "safe best buy" value for the Hubble parameters. Within the next few years new techniques, such as the use of planetary nebulae as standard candles, should greatly increase the accuracy with which the extra-Galactic distance scale can be determined.

TABLE IV. Determinations of the Hubble parameter

Author(s)	Method	H_o (km s^{-1} Mpc^{-1})
Branch	Supernovae Baade-Wesselink	57±10
Huchra	Cepheids + Virgo redshift	88
Melnick	HII regions	87±10
Pierce	Tully-Fisher	84±11
Pritchet and van den Bergh	Novae	69±14
Tammann	Various	51.8±5.1
Tendrup	Galaxy bulge size	55±10
Tully	Tully-Fisher	~90
Salucci and Frenk	Tully-Fisher	76±9
Willick	Tully-Fisher	91

OBJECT INDEX

STARS

STAR CLUSTERS AND ASSOCIATIONS

GALAXIES

CLUSTERS OF GALAXIES

AUTHOR INDEX